T0305502

Critical Thinking, Idea Innovation, and Creativity

Using topics of critical and creative thinking, *Critical Thinking, Idea Innovation, and Creativity* discusses methods of solving complex problems, demonstrates the benefits of using the methods of imaginative thinking, identifies ways to overcome problems and inhibitors such as a lack of confidence, provides guidelines for assessing creative experiences, and encourages the application of the methods to leadership, research, and decision-making. It allows readers to turn their unidimensional technical knowledge into a multi-dimensional knowledge framework that will provide a broader and more realistic framework for the solution of complex problems. Emphasis is placed on the fundamental concepts of critical and creative thinking and idea innovation, and each chapter presents numerous activities to accompany the knowledge-based educational material provided.

Features:

- Provides educational material on creativity in a format that stresses application. An array of creative thinking tools will enable the reader to develop imaginative ideas.
- Emphasizes ways that critical thinking, idea innovation, and creativity can enhance a reader's ability to solve problems related to leadership, the conduct of research, making decisions, and solving complex problems.
- Focuses on ways to improve the reader's thinking skills, which will enhance the likelihood of developing novel solutions to complex problems; this skill set includes skills like curiosity, questioning, and skepticism, which are central to efficiently solving complex problems and meeting the requirements of effective leadership.
- Includes numerous activities in each chapter that will enable readers to apply the methods and develop actual experience in critical and creative thinking; these activities are appropriate for use either by individuals or by small groups.

Critical Thinking, Idea Innovation, and Creativity

Richard H. McCuen

CRC Press
Taylor & Francis Group
Boca Raton London New York

CRC Press is an imprint of the
Taylor & Francis Group, an **Informa** business

Designed cover image: Shutterstock

First edition published 2023
by CRC Press
6000 Broken Sound Parkway NW, Suite 300, Boca Raton, FL 33487-2742

and by CRC Press
4 Park Square, Milton Park, Abingdon, Oxon, OX14 4RN

CRC Press is an imprint of Taylor & Francis Group, LLC

© 2023 Richard H. McCuen

ISBN: 978-1-032-46179-3 (hbk)
ISBN: 978-1-032-46180-9 (pbk)
ISBN: 978-1-003-38044-3 (ebk)

DOI: 10.1201/9781003380443

Typeset in Times
by codeMantra

Contents

Preface

I have been teaching and collecting material on creative thinking for decades, but I have never found a well-defined understanding of critical thinking. From the material that I have read, it seems that the definition of critical thinking is unique to each author, with each author having a narrow definition that focuses on the technical issue that was of interest to them. I use the word technical only in the sense that it applies to the central topic of interest to the user of critical thinking. Some problem solvers believed that critical thinking was little more than creative thinking. Before starting a seriously investigation of the topic of critical thinking, I was under the impression that critical thinking was much like logical thinking or creative thinking. Since then, I have arrived at the conclusion that critical thinking is much more than these unidimensional scopes of thinking; it is broader and deeper in perspective, more flexible in its application, and more effective and efficient for solving complex problems. Critical thinking is much more diverse than the simple application of logical thought. Critical thinking is a philosophy. In fact, I am not sure that I even like the term critical thinking—it does not do justice to the topic. At a minimum, I prefer to talk about critical problem-solving, but even that seems narrow and, therefore, deficient. If a person uses mental processes well beyond logical thinking, then he or she will be able to solve complex problems more effectively and efficiently. A critical thinker needs to rely on a wide array of attitudes and abilities, all supplemented with substantial experience as well as the ability to suppress the influence of thinking inhibitors. Experiences at creative and critical thinking are necessities if the ability to solve the complex problems of the future is to materialize.

What is critical thinking? More correctly, what is critical problem-solving? Critical thinking is just one aspect of critical problem-solving. The current uses of critical thinking seem to focus solely on the word thinking, even though solving problems is the ultimate objective of the critical thinker. Critical problem-solving is a process of solving complex problems that involve multiple dimensions with both significant conflicts within and between the dimensions and important constraints such as the lack of knowledge and uncertainties. Solving a unidimensional difficult technical problem is unlikely to be critical problem-solving. For a problem to be considered complex, the problem must be subject to multi-dimensional constraints that are inherent to the development of a solution. For example, a problem where multiple stakeholders are involved in a project and the stakeholders have very conflicting value needs would begin to suggest complexity. Similarly, a difficult technical problem that involves considerable levels of uncertainty in knowledge, data, and values may be a complex problem because of these other factors but not because of technical difficulties. Dimensions that are often involved in complex problems include technical details, economic constraints, risk, uncertainty, political restrictions, environmental concerns, and many others. As the number of dimensions relative to a problem and the restrictiveness of the constraints increase, the complexity increases and it is very doubtful that standard problem-solving approaches alone will yield a decision that satisfies all of the demands of the stakeholders. To be successful in solving such complex problems, a critical thinker must have a variety of experiences such that he or she can properly deal with each of the dimensions. While the technical problem may not be difficult to solve by itself, the multi-dimensionality is the fact that requires a critical problem-solving solution.

Unfortunately, the educational resources that discuss methods of critical thinking and the advancement of knowledge are minimal. Almost all books that are relevant to the topic concentrate on the unidimensional aspects of knowledge related to the underlying technical issue. The goal of this book is to provide introductory material on the multi-dimensionality of critical thinking, idea innovation, and creativity, as knowledge of these methods will be helpful in solving the incredibly complex problems that are currently confronting our global society and that will only increase in complexity in the coming decades. Students of any field of study will receive the technical knowledge related to their disciplines as part of their current curriculum; however, because of the multi-dimensionality of complex problems, the unidimensional approach to education will not be adequate to provide the educational foundation needed to prepare people for solving the complex problems that society will confront.

This book addresses some of the critical thinking dimensions that can supplement the technical knowledge of any discipline. These dimensions will be central to the solution of complex problems regardless of the discipline. To be an efficient and effective solver of complex problems requires knowledge that goes beyond technical foundations of a discipline, which is just one of the dimensions of complex problems that require critical thinking to solve. Problems are considered to be complex because issues related to other dimensions of problem-solving are important. For example, the balancing of human values is important in advancing knowledge in fields from medicine to the arts and to psychology, as well as others, not just in science and engineering. Complex problem-solving does not depend solely on the technical aspects of the issue. A failure to address the other dimensions will prevent the best solution for each problem from being developed. In addition to the technical dimension, knowledge of the following dimensions is usually necessary to solve complex problems in every discipline: (1) the mental mindset of the decision maker, (2) the critical thinker's ability to balance values that cause conflicts between the stakeholders who are involved in the problem, (3) a breadth of thinking types because it is important to match the types of thinking with the characteristics of the problem if the best solution is to be found, and (4) a sequential array of unique skills, such as curiosity and questioning, that a critical thinker must be able to apply in the proper way. This book provides detailed discussions and applications of these four dimensions of critical thinking.

How do students learn? The response to this question has been debated for centuries and will continue to be debated for centuries to come, especially because of the changes that we can expect in both knowledge and technology. The methods of Dewey differ significantly from the practices at the lyceums of Aristotle. Similarly, the methods of teaching in the twenty-first century are vastly different from those used in the classrooms of Dewey's time. Also, individuals learn in different ways, which is one of the reasons that we cannot achieve total agreement on an answer to the question about the way that students learn.

In this book, the fundamental concepts of critical and creative thinking, as well as innovation, are a primary emphasis. For each of these topics, the following elements are included: the basic concepts, relevant definitions, the benefits of being proficient in their application, practices that can increase the efficiency of their use, overcoming inhibitors to their use, their assessment from both personal and organizational perspectives, and their uses in solving complex problems. Counter-skills such as procrastination, pessimism, and a lack of self-confidence can limit the efficiency and effectiveness of the methods; guidelines are provided for overcoming these and other inhibitors. In addition to the four dimensions emphasized in this book, solving complex problems depends on knowledge of issues like inhibitors, assessment, uncertainties, and counter-skills, which is the reason that these topics are discussed at length.

Reading about critical thinking and creativity is a necessary but not sufficient condition for becoming an effective critical problem solver. Experience at these methods of problem-solving is a necessity. Existing books on the subject of creative thinking often use activities that intend to help the reader gain the necessary experience; however, the quality of an experience is very important. The type of the activity determines the quality of the problem-solving experience. Unfortunately, many books confine the activities to simple brain stimulators, or brainteasers, such as asking the reader to determine the number of squares shown in a figure or to determine which one of five symbols does not match the other four symbols. While these "warm up" activities can add a light moment to the reading, they do not provide the quality of experience that will lead to significant growth in critical thinking ability. This is the reason that I have included many minds-on activities with each chapter of this book. The activities stress the creation of new things—a metric, a process, or the advancement of an idea. The minds-on activities can be used either prior to or following any discussion of the topic. The activities that I use in my book attempt to challenge the reader to actually create something new, much as a person who works in a creative advertising agency, a think tank, or a research division of a company that develops new consumer products. People who hold positions such as these must be critical thinkers in order to achieve success. Activities that mimic their responsibilities will provide creative experiences and will teach the readers of this book to be critical thinkers. Minds-on activities are far better learning experiences than are brainteasers. On average, each chapter in my book presents eight activities to accompany the knowledge-based educational material provided in the chapter; thus, the book will provide both the fundamental knowledge and the experiences that are necessary to improve the reader's ability to create, innovate, and most importantly solve complex problems.

The goal of this book is to provide the basic concepts of critical and creative thinking for those who wish to be more creative and better at solving problems related to critical thinking, idea innovation, and creativity. This material should improve the efficiency of problem-solving and make any research experience more personally rewarding. The specific goals are:

1. To show that imaginative thinking has many benefits.
2. To understand commonalities and differences among problem-solving processes.

3. To encourage the belief that the generation of novel ideas can be learned, i.e., creativity is a mental activity that is not entirely innate.

4. To introduce synectics, which is an alternative idea generation method that improves the imaginative thinking process by making use of the participants' general knowledge.

5. To detail the Delphi procedure that extends the use of brainstorming to conditions where face-to-face sessions of group members would be inconvenient or conflicting.

6. To present checklisting as an applied tool for arousing both the breadth and depth of one's creative thinking ability.

7. To make users of imaginative thinking aware of the inhibitors that can reduce the efficiency of their applications of critical thinking methods.

8. To provide guidance on the assessment of critical and creative thinking activities.

9. To place critical thinking into a broader framework, one based on the diverse dimensions of values, thinking skills, mindset, and types of thinking.

10. To identify a process of mental skills that will improve the effectiveness and efficiency of problem-solving.

11. To show the importance of the value dimension in solving complex problems.

12. To show the ways that a person's mindset influences success in critical problem-solving, especially the benefits of learning to control one's mindset.

13. To show that successful problem-solving partly depends on selecting the thinking types that are most appropriate for the characteristics of the complex problem.

14. To identify ways that knowledge of critical problem-solving can advance the goals of an organization.

15. To identify the ways that critical thinking are important in every step of the research process.

16. To demonstrate that knowledge of critical problem-solving increases the effectiveness of decision-making.

17. To introduce a method of critiquing existing research that can produce ideas for novel research.

In meeting these goals, emphasis will focus on improving skills, such as critical analysis and self-confidence. Learning to be more imaginative is also vital to problem-solving. Since problems always arise when conducting state-of-knowledge problem-solving, these 17 goals focus on improving the efficiency of problem-solving. This document stresses the value of being creative and the ways that idea generation can lead to greater success in both problem-solving and research environments.

Acknowledgments

I would like to thank Ms. Madeline Jones for her very timely and competent administrative help in making this book possible.

I also very much appreciate the exceptional efforts of Mr. Joseph Clements of CRC Press for his continued support in editing this book and other books that I have authored.

About the Author

Dr. Richard H. McCuen is an Emeritus Professor of Civil and Environmental Engineering at the University of Maryland. He retired in 2020 as the Ben Dyer Professor of Civil & Environmental Engineering. He received a BSCE degree from Carnegie-Mellon University (1967) and the MSCE and PhD (1971) degrees from the Georgia Institute of Technology. He was a faculty member at the University of Maryland for 49 years and served as Director of the Engineering Honors Program for more than 35 years. Each year, Dr. McCuen taught an honors course that was centered about the topic of creative thinking.

1 Introduction

CHAPTER GOAL

To create an inquisitive feeling about the importance of imaginative thinking in complex problem-solving.

CHAPTER OBJECTIVES

1. To identify the ways that critical thinking, idea innovation, and creativity interact.
2. To show the benefits of creativity and critical thinking in the conduct of research and problem-solving.
3. To identify some of the factors that contribute to problem complexity.

1.1 INTRODUCTION

When we think of the greatest thinkers of all time, we would probably start with the three Greeks: Socrates, Plato, and Aristotle. Their influence on the state of knowledge extended even into the sixteenth century, maybe even today. We may want to include Archimedes and Ptolemy among the greatest thinkers. We might then move to the early seventeenth century and point to Francis Bacon, whom some credit with the development of the Scientific Method, but should we be passing over the likes of Copernicus, Galileo, and da Vinci? Their advances in knowledge clearly demonstrated thinking well beyond the norm. Is it right to ignore thinkers who were not scientists or mathematicians? What about the great thinkers in religion, politics, and the arts? Should we include the likes of Paul, Aquinas, and Luther? How about Rembrandt, Chaucer, and Beethoven? Are Constantine the Great and Charlemagne worthy of inclusion? How about William Harvey and Louis Pasteur who made significant advances in medicine? Gregory Mendel and Charles Darwin are giants in the broad field of genetics. The point is that periodically quantum leaps in knowledge and wisdom take place in every discipline. What enabled the great thinkers in each discipline to make these giant leaps for humankind? They obviously had a good understanding of the state of knowledge in their disciplines, but their thinking went well beyond their technical disciplines. This breadth of knowledge that extended beyond the technical issue probably contributed to their ability to recognize the real weakness that existed in knowledge and then to solve the problem and become one of the all-time greats. Did they have unique thinking skills that set them apart from their contemporaries? Did they have critical thinking skills? Did critical thinking even exist in those times of significant advances? Were the giants in all fields of study

DOI: 10.1201/9781003380443-1

just innovators who had the foresight to assemble bits of knowledge that enabled them to appear as great thinkers? It seems that they had unique ways of thinking that enabled them to become great thinkers. Maybe that is the nature of critical thinking! Can today's best thinkers capture the ways of the critical thinkers of the past?

Even in the world today, we are experiencing significant advances in almost all disciplines. As time progresses, both personal and societal problems are becoming more complex. As time progresses, the complexity of technology is increasing, and its effects are becoming more difficult to control. Advances in technology have had many positive effects. As time progresses, the quantity and quality of data are vastly increasing, which has both positives and negatives. We are living in the time of big data. We know that the climate is changing and the polar icecaps are shrinking. To solve these complex problems that are undergoing temporal changes, we will need greater knowledge of problem-solving methods not just greater technical knowledge in each discipline. If we are to successfully confront and solve the confounding problems that will materialize in the future, we will need critical thinkers to emerge from the masses and become all-time greats. Will we be able to solve society's complex problems if the critical thinkers do not emerge?

Achieving success in solving the complex problems of today and tomorrow will require advances in both knowledge and technology. Advances in methods of solving problems will also be necessary. Will mega-cognitive problem-solving methods be developed to enable problem solvers to advance beyond critical problem-solving? Society will need more critical thinkers and researchers who have unique thinking skills to investigate and solve the anticipated multitude of new problems. The problem solvers will need to know more than just the technical state of knowledge that underlies the issue, which is an emphasis of this focus on other dimensions of problem-solving. Breadth of both knowledge and thinking will likely be a requirement. The up-and-coming great thinkers will need to develop more accurate ways to balance the conflicting wants and needs of a diverse set of stakeholders. Innovative experimental designs and novel research methods will be necessary. Therefore, society must ensure that it is teaching the best methods of problem-solving. Planning and preparation will be keys to success! Will critical thinking be a contributor to major breakthroughs in diverse technical specialties or will the great thinkers of the future need to develop the mega-cognitive thinking methods?

1.2   WHAT IS CRITICAL THINKING?

Throughout this book, the underlying motive is to address the following question: "What is critical thinking?" Since the above question will be answered at length in later chapters, let it suffice to define critical thinking as a method of solving complex problems that are subject to a series of restrictions that act as restraints on a solution, but yield greater confidence in the solution because of the restrictions. These restrictions are referred to herein as dimensions because they provide specific constraints on any solution.

It is difficult to define critical thinking without first characterizing the word complexity, as critical thinking is used in the solution of complex problems. Complexity occurs when a problem involves many dimensions with some of the dimensions involving very conflicting issues. The most likely dimension is a technical issue associated with the problem. The word *technical* should be viewed broadly, as it is meant to refer to any discipline, not just STEM-related subjects. Environmental and economical restrictions could be other dimensions. For example, Copernicus had to contend with dimensions of knowledge limitations, religious-political constraints, and societal restrictions, with considerable conflicts inherent to each of these dimensions. In leading the building of the Panama Canal, George Goethals had to contend with dimensions of knowledge, public health, technological deficiencies and unknowns, risk, and political-military responsibilities, some of which involved significantly conflicting demands. In addition to the issues related to the technical discipline, complex problems are subject to the constraints of many other dimensions such as time-frame demands (i.e., short-term and long-term constraints), statistics, and uncertainties. It is not necessarily the total number of dimensions that determines complexity, but the extent and significance of the conflicts involved within each and between the dimensions.

1.3   THE INFLUENCE OF MYTHS

Creative thinking efforts are based partly on emotion, with emotional thoughts often being the basis for generating ideas. Creativity has also created some myths, which may have stymied the growth of its use. For example, many believe that creativity cannot be learned. The false belief goes something like: you are either born with creative ability or you are not. Think about children! Many of their day-to-day hours are spent in their fantasy world—they are the star quarterback or the teen idol. Yet as they mature, some allow their fantasy world to mature while others suppress their imagination because they believe that the world of emotional thinking is immature. This latter group actually allows their creative ability to deteriorate. But it can go the other way—a person's creative ability can actually be improved. So much for that myth.

Another myth is that good decisions only result from logical thinking, i.e., emotions only lead to illogical thinking and poor decisions. Unfortunately, for promoters of this myth, many good decisions have resulted from an emotional involvement in the person's thinking process. Imaginative thinking is just one alternative to logical thinking, and its use broadens a person's problem-solving ability. Fortunately, many types of thinking are available and the best solutions will materialize when the type of thinking is matched with the characteristics of the problem.

Should we believe the myth that innovations of past research that lead to new research constitute an act of plagiarism? As many people have stated: copying from one person is plagiarism, but copying the works of many people is research. Most published research that is recognized as being novel includes a long list of references, which clearly shows that the results of new novel research can be in some way similar to already published research. It is not plagiarism because some significant difference is quite evident. This mode of advancing knowledge is actually necessary. Thus, innovation does not automatically imply plagiarism. Where does one draw the line between valid innovation efforts and plagiarized work? The myth masks the answer to the question.

Is critical thinking nothing more than a new name for creative thinking? Some believe this myth. While cases of critical thinking often include some aspect of creative thinking, the former is much broader. Maybe it is the simplicity of the name critical thinking and the apparent similarity to the name creative thinking that has led some to believe this myth. While creativity can be a component of critical thinking, critical thinking is really a significant step above creative thinking.

The prevalence of myths like these and others serves as inhibitors to problem-solving efficiency. The myths can create a sense of fear that discourages people from using knowledge of problem-solving that can lead to advances in technical knowledge. Experience about the valuable talent of knowledge innovation is necessary. Attitudes that are essential to effective problem-solving need to be developed. Critical problem-solving skills can be learned. Negative attitudes such as pessimism and procrastination must be overcome in order to achieve such a goal. Academic knowledge and experiences can help overcome these anti-progress myths. One goal of this book is to provide knowledge of ways that deficiencies in problem-solving can be overcome.

1.4 CRITICAL THINKING IN PROBLEM-SOLVING

We are not all critical thinkers even though we continually confront issues that need decisions. Critical thinking is a process and should be an important part of our daily decision-making; therefore, it is the principal focus of this document. Recognizing the relevance of critical thinking to each step of the problem-solving process will show its broad application. Research is a special case of critical problem-solving. Critical thinking can be an important tool for producing novel research ideas and outcomes. Thus, those who are new to research should be introduced to critical thinking. It has a unique definition when it is applied to any part of the research process. Knowledge of critical thinking will enable knowledge of creativity to be put to better use. Critical thinking is a multi-dimensional process that incorporates values, mindset control, and special thinking skills in the development of solutions to problems. Thus, it provides the best means of developing solutions for complex problems.

As I view the topic of critical thinking or more correctly critical problem-solving, its primary distinguishing feature is its multi-dimensionality. Most day-to-day problem-solving is completed using logical thinking, without thought to the values and attitudes that must be considered in arriving at a decision. As will be discussed, critical problem-solving involves the active consideration of at least the following four dimensions: values, thinking types, skills, and mindsets. Of course, decision-making will also involve other dimensions based on technical and economic issues. However, since these two other dimensions are quite specific to each discipline, the focus will be on incorporating the other four dimensions, which are much less discipline specific.

Critical thinking is a term widely associated with the solution of complex problems; however, it is a somewhat restrictive term in that it refers to a small part of problem-solving. At a minimum, it is better to directly associate the word *critical* with the term *problem-solving*. Critical problem-solving is a more accurate descriptor than critical thinking. The term critical thinking is used throughout this book as a synonym for critical problem-solving. Even the term critical problem-solving does not do justice to the multi-dimensional approach to solving complex problems that is the primary purpose of this book.

1.5 SIMPLE AND COMPLEX PROBLEMS: DEFINITIONS

Solutions for simple problems may not require knowledge of critical thinking; however, solutions to the difficult problems in any discipline will require greater knowledge of problem-solving. Since the emphasis in this book is in developing solutions to complex problems, initiating the discussion of problem-solving needs an introduction to problem complexity.

1.5.1 BASIC DEFINITIONS

A few general definitions might be of value in providing a definition of problem complexity:

Simple: straightforward; unsophisticated.
Complex: compound; complicated; consisting of interconnected parts; a challenging thinking exercise.
Problem: an uncertain question or situation; difficult to deal with an activity that lacks an obvious solution.

These definitions can form the basis for two broad definitions, which can define the endpoints of a spectrum:

Simple problem: An activity that will only need a relatively straightforward solution.
Complex problem: A problem that needs a multi-dimensional solution procedure and has conflicting characteristics that must be balanced to find an acceptable solution.

Unfortunately, these definitions might suggest that the issue of problem complexity is dichotomous, with a problem being either simple or complex. This is not the case. Simplicity and complexity are the endpoints along a continuum without clear lines of demarcation between the endpoints. Complexity is a variable.

1.5.2 FACTORS THAT INFLUENCE PROBLEM COMPLEXITY

Problem complexity does much beyond just the complexity of the technical knowledge that is required to develop a solution. While technical knowledge contributes to complexity, it also depends on issues related to the problem solver and the method used in solving the problem. Problem complexity varies with a number of factors, each of which should be a factor in selecting the best person or team to assign the task of solving the problem. The problem solver should have experiences that are relevant to the factors that contribute to the complexity. The variations in the demands of the stakeholders who are involved in the issue will significantly influence the level of problem complexity. If stakeholders have quite varied and conflicting interests and decision criteria, then this discord will greatly influence the level of complexity, as problems that involve conflict are more troublesome to solve. When stakeholders promote irrational and biased viewpoints on the issues, problems can be more difficult to solve, i.e., the problem-solving is more complex. The level of problem complexity will influence the resources required to solve the problem, so resource requirements can be a factor in defining the level of problem complexity. Other factors include project risk levels and societal sensitivity. Cultural differences can be a factor. The problem-solving skills that will be needed to solve such problems will vary with such factors.

A number of other factors contribute to both the assessment of complexity and the difficulty in solving a problem. The number of stakeholders and the diversity in their demands are often a primary determinant of problem complexity. The number of objectives associated with the problem statement and the uncertainty of each will influence the level of complexity. For example, profits and losses are much easier to assess and quantify than would be the effects of a project on public safety or environmental health. The latter two may have a more significant influence on the assessment of complexity than would the profits and losses because of their effects on the people, not their pocketbooks. The completeness of current knowledge would also be a factor; if the knowledge base required to solve a problem is largely lacking, then the complexity of finding a solution will be greater, as the development of new knowledge will be needed to provide an acceptable solution. The same is true for the database or resources needed to formulate and verify a solution; an insufficient database will require greater effort to collect additional data and thus make a problem more difficult to solve. The complexity of the experimental design can be a contributing factor. The quality of a solution may be better for a high-quality experimental design, but only when the problem solver has the ability to conduct the experimental analysis; this is not a trivial point. If the elements involved in the measurement of the experimental analyses are imprecise, then decision-making is more difficult, i.e., more complex. Project uncertainties, such as the measurement accuracy of input data or the quality of the theory used in model development, can be a significant indicator of complexity. It is evident that many factors contribute to problem complexity.

1.5.3 PROBLEM-SOLVING PERSPECTIVE ON COMPLEXITY

The best perspective on the issue of problem complexity is through a comparison of simplicity and complexity at each step of the problem-solving process (see Chapter 2 for details of the process). Table 1.1 outlines the distinguishing aspects of simple and complex problems for each of the six steps. The most obvious distinguishing characteristic is the dimensionality of the two problem types. Complex problem-solving must deal with multiple dimensions that are typically uncertain in their ability to be classified. Simple problems are characterized by fewer dimensions, and are often defined by only one dominant dimension. It is important to note that the table specifies the two ends of the complex-simple continuum. Points along the continuum would be characterized by levels between the extremes.

TABLE 1.1

Criteria to Distinguish Problem Complexity and the Steps of the Problem-Solving Process

Step	Simple Problems	Complex Problems
Problem statement	Straightforward; easy to compose the problem statement; generally, one-dimension dominates the problem; very little stakeholder conflict.	Problem is multi-dimensional, with several diverse objectives; multiple stakeholders contribute conflicting assessments of the needed problem statement.
Knowledge and data	Data are readily available; minimal theory required; minimal, independent decision criterion.	Extensive theory involved, but multiple elements of important knowledge are missing. Sparse data with considerable uncertainty in values. Stakeholders want to adopt conflicting decision criteria.
Experimental design	Only commonly used methods of analysis are needed; uncomplicated hypotheses.	Multiple dimensionality of hypotheses; interconnectedness of model components; conflicts about methods of analysis.
Analysis	Generally based on a small set of unconnected advancements of knowledge; narrowly focused use of logical thinking.	Must formulate sophisticated decision components for interconnected dimensions; broad array of thinking types required; imaginative thinking is used.
Synthesis	Narrow set of decision criterion and little value conflict; alternatives easily evaluated because of simplicity of the decision criterion.	Controversial evaluation of alternatives, as the decision criterion reflects numerous dimensions; difficulty in balancing value issues.
Decision	Straightforward solution based on minimal decision criterion and commonality of stakeholder needs; little uncertainty in the analysis.	Uncertainty in decision due to (1) the complicated decision criterion, (2) difficulty in balancing stakeholder demands, and (3) decision criterion with similar assessment rankings.

Complex problems require the use of multiple dimensions about which knowledge must be collected and evaluated. The most obvious dimension would center about the primary technical issue. Economic issues could form another dimension, which in many cases is important. While the technical and economic dimensions are usually very important, they may not be a prime determinant of problem complexity because they are often more directly included into the decision process. Discussions of the technical and economic dimensions are largely ignored herein, as this document is intended to be broadly applicable across a wide array of disciplines. Instead, the focus will be on four dimensions that are important across multiple disciplines: values, skills, thinking type, and mindset. These four dimensions are generally subject to considerably more uncertainty, which can complicate decision-making when decision makers do not understand the effects of these dimensions.

1.6 CONCLUDING COMMENTS

Critical problem-solving is the most effective and the most efficient way of solving complex problems. As the complexity of problems increases, the methods of solving such problems will need to increase simultaneously. Problem-solving needs to be efficient, so problem-solving requires more than just having the skills or knowing the values. Critical problem-solving implicitly implies that the skills are applied in the proper sequence and the value issues are properly balanced. A critical problem solver also needs to have knowledge of multiple thinking types and the understanding that the thinking type must be matched to the characteristics of the problem that needs a solution. Short-term critical problem-solving requires the problem solvers to have control over their mindsets, i.e., the individuals must knowingly control their moods, dispositions, and attitudes that direct both the thinking and decision-making throughout the problem-solving process. When confronted by a problem, the dimensions that should be addressed will depend on the nature and the breadth of the problem statement. Any societal concern could be presented as a dimension. In some cases, the environment might be part of the problem statement. Crime and criminality could be a dimension. Mental health, wildlife conservation, or the preservation of forests can be significantly relevant to a problem and thus warrant being dimensions. The dimensions that involve the greatest conflicts will usually have the greatest influence on the decision. Thus, critical problem-solving is multi-dimensional, and the problem solvers will need experience in all relevant dimensions, not just the technical aspects of the issue.

To become a critical thinker requires advances in knowledge of many dimensions. A primary focus of this book is on ways of improving the skills and attitudes needed to solve complex problems. Ways to overcome weaknesses such as procrastination are discussed. Effective problem-solving often requires the problem solver to be self-confident; ways of improving self-confidence are discussed. The advancement of knowledge requires a critical attitude. Criticism should seek out both the strengths and weaknesses of the current knowledge base. Moods and attitudes are very influential in making decisions. Pessimism can be very constraining to a problem solver. Methods of transforming a pessimistic attitude to one of optimism are presented. Ways of improving numerous skills and attitudes are discussed. These self-help discussions will assist in transforming the reader to one who has the attitudes and abilities of a critical problem solver.

1.7 EXERCISES

1.1.   Provide brief definitions of the following terms: research, thinking, innate, knowledge, innovation, and creativity. Discuss the ways that they are similar and the ways that they differ.

1.2.   What evidence would be necessary to show that creative ability is not entirely innate, i.e., it can be learned?

1.3.   What factors inhibit problem-solving? This could involve solving personal problems, school assignments, or work-related problems.

1.4.   When you solve a problem, whether it is a personal problem, work-related, or a math problem, what are the common characteristics used to arrive at a solution to the problem?

1.5.   Find definitions of the words *research* and *innovation*. Discuss the ways that they are similar and the ways that they differ.

1.6.   How are research and problem-solving related?

1.7.   In constructing a bridge to span a river, what decision criteria are used in making a decision about its construction and design? Rate the complexity of estimation.

1.8.   Why is a piece of artwork considered creative but a modern bridge is not viewed as being creative?

1.9.   Why do many companies have divisions of research?

1.10. Define the following words: attitude, value, mood, skill, talent, and ability. Briefly indicate the way that each word relates to problem-solving.

1.8 ACTIVITIES

1.8.1 ACTIVITY 1A: AMATEUR ART CRITIC

Go online and obtain a copy of Picasso's "Weeping Woman." Respond to the following questions:

1. What idea or thought is Picasso trying to convey?
2. What aspects of the painting suggest Picasso's creative ability?
3. What creative change could be made to the painting that would enhance Picasso's idea?
4. What human values does the art work suggest?
5. What is the imaginative aspect of this painting?

This activity is designed to emphasize critiquing and innovation.

1.8.2 ACTIVITY 1B: INNOVATIVE SONG WRITING

SELECT A SONG THAT YOU ARE NOT FOND OF BECAUSE OF SOME OF THE WORDING. CRITIQUE THE SONG. THINK ABOUT WHY YOU DISLIKE THE SPECIFIC PARTS OF THE WORDING. THEN CHANGE THESE WORDS TO SOMETHING THAT YOU BELIEVE WILL IMPROVE THE SONG. THIS ACTIVITY WILL INTRODUCE YOU TO THE SKILL OF CRITIQUING AND CREATIVE IDEA DEVELOPMENT. 1.8.3 ACTIVITY 1C: FOLLOWING IN FLEMING'S FOOTSTEPS

Using the internet, find details about Alexander Fleming's discovery of penicillin. What was he thinking when he recognized the change in the mold? Is this an example of a curious mind? Why did other people not recognize the same change? This activity should introduce you to important skills such as curiosity, critiquing, skepticism, and questioning. Discuss the way that Fleming used these four skills.

1.8.4   ACTIVITY 1D: POKER MARRIES HOPSCOTCH

Find the directions for both the card game Poker and a school kids' activity called hopscotch. Take any of the rules of poker and use them to modify the rules of hopscotch to make it more fun or more challenging. Then take any rule of hopscotch and use it to modify the rules of poker. This activity should introduce the innovation process, i.e., the idea of taking ideas from one source and using them to make improvements to another activity.

1.8.5   ACTIVITY 1E: $, $, AND $

Determine the way that benefits and costs are used in making decisions about public work projects. How should values such as aesthetics and public safety be incorporated into the decision process? Discuss the use of qualitative and quantitative criteria in public-project decision-making. This activity illustrates the problem of weighting and balancing quantitative and qualitative values.

2 Processes for Problem-Solving

CHAPTER GOAL

To understand commonalities and differences among problem-solving processes.

CHAPTER OBJECTIVES

1. To draw parallels between the concept of machine efficiency and the efficiencies of different problem-solving processes.
2. To summarize the research, design, decision-making, innovation, and creative problem-solving processes.
3. To present de Bono's thinking process for comparison to decision-making processes.
4. To introduce a new perspective on the critical thinking process by including four generalized restrictions to ensure quality decisions.

2.1 INTRODUCTION

We are personally concerned with the energy efficiency of our homes, and when purchasing a new car, gas mileage is an important consideration. Even a hungry lion is concerned about efficiency; how much effort will the lion have to expend to bring down that gnu and how satisfying will the lion find the gnu as the dinner? Any engineer will tell you that efficiency is a dimensionless quantity, but miles per gallon, therms per BTU, and gnus per day are not dimensionless but show some measure of efficiency. How can they represent efficiency?

We have to view efficiency in broader terms than as a dimensionless work-out/work-in ratio that is used in recording the efficiency of machines. When the input and output are not in the same units, it may be necessary to use a percentage change to reflect efficiency. For example, if car model X provides 35 miles per gallon and model Y yields 32 miles per gallon, then if we purchase brand X, we will get 9% better mileage efficiency. The comparison approach is a different way of measuring efficiency.

We also have to recognize that the plans for solving a problem influence both the system output and input, thus the efficiency. A lion thinks of the best plan needed to take down the gnu. The lion assesses factors such as the height of the grass in which he is hiding, the footing provided by the soil, the way that the soil will influence his and the gnu's running, and the way that the health of the gnu will enable it to run. Lions that

DOI: 10.1201/9781003380443-2

lay out the best plans will have the best dinners for the least effort. A lion without a plan will likely not catch as many gnus as lions who have plans. Society faces many problems and the efficiency with which society fulfills their goals depends on the quality of their plans. The lion's plan for killing the gnu has many differences from the car owner's plan to maximize fuel efficiency, but their plans do have some important similarities.

The Scientific Method, the problem-solving process, and the experimental design procedure are all well-known and widely used. While words like method, process, and procedure are used in naming them, the three methods have many similarities. Let us assume that the word *process* can be used for any of them and reflects their similarities. The following is a reasonable attempt to define the word *process*, as we will apply it to our discussion of critical thinking:

A logically ordered, systematically carried out set of interdependent operations.

The ordering of the steps is usually necessary because the successful completion of one operation depends on the successful completion of previous operations; this illustrates the interdependency of the operations. The operations are conducted systematically because that is expected to ensure the best efficiency and greatest chance for success; however, this does not preclude the use of stochastic operations as part of a problem-solving process.

Francis Bacon was a strong supporter of the Scientific Method, which greatly contributed to the expansion of scientific knowledge during the sixteenth and seventeenth centuries. Business leaders believe that establishing systematic procedures for activities, such as planning, organization, and decision-making, increases both the efficiency and effectiveness of company operations. Engineers use a process called the design process for solving problems in their fields. Many of these processes adhere to the same philosophy, even to the point of having steps that differ in name only. The processes may differ in the amount of detail per step, rather than in the intent behind the processes.

Process is a commonly used term. Many of the processes, such as the Scientific Method, follow operations that directly correspond to the operations in other processes; however, some tasks have unique purposes. Therefore, it is worthwhile reviewing some of the more widely used processes, with special note of their commonalities and their differences. A critical thinker will take advantage of the differences and by doing so will improve the efficiency of the problem-solving effort. Critical thinking, innovation, and creative thinking are processes that can increase problem-solving efficiency, improve research results, and simplify decision-making.

2.2 DEFINITION OF PROBLEM-SOLVING EFFICIENCY

Effective problem-solving assumes the achievement of the best possible efficiency. In many cases, the best efficiency is achieved when a systematic process is followed. Consider the example of machine efficiency. If the work input (W_i) and the work output (W_o) of a simple machine can be measured, the machine efficiency (ε) can be computed by

$$\varepsilon = W_o/W_i \qquad (2.1)$$

Work output cannot exceed work input, and assuming that work output cannot be negative, the efficiency is constrained to the range 0–1, or 0–100%. While some might argue that a machine cannot be 100% efficient, one objective in any machine use is to take the actions that will make the efficiency as large as possible. This objective would mean that the loss of resources must be held to a minimum. We can also discuss the inefficiency of a process. Inefficiency is equal to 1.0 minus the efficiency. One goal is often to maximize the efficiency of a system. The efficiency is often used as a performance criterion. We desire to maximize the efficiency of any system, unless other criteria such as the benefit-cost criterion would show that the efficiency criterion does not lead to the best solution based on other decision criteria.

The efficiency of any operation depends on a number of factors. For example, the efficiency of a machine shop depends on factors such as the diligence and intelligence of the personnel and the quality of the machinery. The efficiency of the service at a fast food restaurant depends on the attitudes of the serving personnel, but the organization of the operations and the serviceability of any equipment are additional factors. Similarly, the efficiency of a creative problem-solving activity depends on the ability and attitude of the facilitator, i.e., the person who conducts the session. In terms of the generalized efficiency model of Eq. 2.1, these factors would influence the value of the output. The problem statement and required resources would be reflected in the denominator. While an exact numerical value of efficiency may not be possible when solving many problems, users of critical and creative problem-solving often believe that these practices provide the most efficient solution to a problem.

2.3 THE SCIENTIFIC METHOD

The process known as the Scientific Method has been credited with the rapid expansion of scientific knowledge that occurred during the sixteenth and seventeenth centuries. At that time, a goal of scientific inquiry was to advance the state of knowledge for the field of science through the development of new knowledge. Francis is often credited with its wide-scale implementation. Several versions of the Scientific Method have been proposed. Table 2.1 shows a four-step process. Other versions with a greater number of steps have been proposed primarily to show an expansion of the objectives of each of the four steps of Table 2.1. For example, the four-step process does not specifically indicate the collection of experimental data, the communication of results, or the opportunity for feedback. Yet, they are inherently part of the four-step model. The four-step process does not show that a problem definition follows the observation step, but it does assume that it is a part of Step 1. To clarify the needed tasks, it might be better to replace the first step with two steps: observation and problem definition. Also, the version of Table 2.1 does not suggest the need for unbiasedness, and it does not specify the methods of analysis. Decision criteria that would be used to make judgments at each step are not explicitly noted, but obviously, any application of the Scientific Method must detail the criteria that are applied in the analyses and in selecting the best alternative. While these additional actions are inherent to the four-step process in Table 2.1, the method would probably be a better educational tool if more steps were included.

TABLE 2.1
The Scientific Method

1. Observation.
2. Hypothesis formation.
3. Experimental analysis.
4. Induction.

The second step of the Scientific Method involves formulating a hypothesis. Francis Bacon wanted to stress the use of experimental analyses. A particular problem may actually involve more than one need, with each need leading to a specific objective in Step 1 and each objective should be transformed into a research hypothesis. A hypothesis can be defined as:

- An assumption that needs to be tested.
- A statement that reflects the known knowledge and will give direction to a decision that reflects the corresponding objective.
- A linkage between the problem and the decision.

The idea is that if you can justify the analysis of Step 3, then the hypothesis of Step 2 is adequately supported, and it is safe to make a generalization in Step 4 for each hypothesis. Step 4 is the induction step, which allows the investigator to reason from the particular result of Step 3 to a broad, general conclusion. Based on the generalization, the result can be applied in other times and other places where the same conditions used in the analysis exist. The validity of such application depends on the quality of the hypothesis and the experimental analyses. For example, if the data used in Step 3 were not broadly applicable, then the broad use of the generalized result could be quite erroneous.

2.4   A MODEL OF THE RESEARCH PROCESS

To what do we associate the word research? Medical research would be one common answer. Some people may use the word research to describe the effort that they put forth to prepare for placing bets on weekend football games. Obviously, research on medical issues and research on football games are quite different examples of research. Yet surprisingly, these two examples of the concept of research share many of the same tasks. The same steps that are used in medical research can be applied to research being conducted on improving the safety of infrastructure and solving environmental problems such as climate change and improving the quality of water in small ponds. The diversity of these topics for research suggests that it may be worthwhile having a general model of the research process; however, the model should be flexible rather than a model with an inflexible set of steps.

Research can be defined as (1) a *scholarly or scientific investigation,* (2) *an inquiry, or* (3) *a study thoroughly.* The need for researchers has grown considerably as the problems of society have become more complex, but also because research leads to advances that can have considerable positive economic and societal implications. Research is not limited to the sciences. It is important in a variety of fields from criminal justice to economics and from music to education. Certainly, the benefits of medical research are evident to all of society. Most graduate degrees have a research requirement. Those involved in research try to complete the research as efficiently as possible. Research problems are less difficult to solve when a systematic solution procedure is followed, as this suppresses disorganization and maximizes efficiency.

The quality of research can vary considerably, with the worth of the outcomes depending on the excellence of both the ideas and the effort. While not ignoring the influence of uncertainties, the quality of the research will be influenced by the thoroughness of the investigation and the conscientiousness of the researcher. Both the foresight of the investigator and his or her knowledge will influence the quality of the result. When the researcher uses good experimental practices, the quality of the results is enhanced and the overall efficiency is maximized.

Many versions of the research process have been proposed. A six-step model of the research process is given in Table 2.2. In some ways, Step 1 is the most important, as a poor problem statement cannot lead to a worthwhile outcome. Step 2 involves the establishment of the current knowledge on the subject, including a review of the literature that summarizes all past research on the subject; resources, both knowledge and date, are important inputs to successful research efforts. The quantity and quality of the data are resources that influence the quality of the result. In Step 3, an experimental plan is developed for conducting research. The plan should identify the methods of analysis and the decision criteria that will be used in decision-making. In Step 4, the experimental data are analyzed for the purpose of advancing knowledge that is related to the problem. Step 4 efforts lead to a set of results from the experimental analyses. Following the analysis phase, Step 5 provides the synthesis of potential outcomes, i.e., conclusions, from the results of Step 4. Step 6 involves the application of decision criteria, which are used to select the most feasible outcome. Any feedback loop will provide the option to return to a previous step and modify the actions taken. When a decision does not seem feasible or optimum, a feedback loop should be used to consider alternative actions. For most research efforts, feedback can be used between any of the steps.

The same concept of efficiency that was applied to machines can be applied to creative endeavors, such as research. In general, research efficiency (ε_R) or

---

TABLE 2.2

Steps of the Research Process

1. Problem identification.
2. Resource collection (knowledge and data).
3. Research plan development (experimental plan).
4. Analysis of results.
5. Synthesis of outcomes.
6. Make a decision and report.

---

problem-solving efficiency can be assessed using the effort expended (E_R) to reach a solution and the quality of the solution (Q_R), i.e., the output:

$$\varepsilon_R = Q_R / E_R \qquad (2.2)$$

The input and output of Eq. 2.2 are generally more difficult to quantify than the work values of Eq. 2.1 for a simple machine. It may be necessary to account for various factors such as a person's mindset at the time, with a poor attitude causing Q_R to decrease. Also, the person's research experiences, technical knowledge, and creative ability will influence efficiency. Quantitatively measuring the numerator and denominator of Eq. 2.2 would be subject to considerable uncertainty; however, the concept of efficiency is very relevant to research. Steps can be taken to improve problem-solving efficiency, including:

• Learn and believe in methods of creative thinking.
• Gain experience in creative practices prior to the time when they are needed.
• Overcome dominant creativity inhibitors of oneself, of colleagues, and of the organization.
• Develop good mindset characteristics.
• Identify relevant values and incorporate these into the decision criteria.
• Develop appropriate assessment criteria.
• Develop skills such as curiosity, critiquing, and questioning.

Productivity in a work plant depends on the efficiency of the machines, the workers, and the management, both the planning and the staff. Whether the issue is a machine shop or a research activity, all inputs are factors that can be changed to improve the efficiency. The same is true for innovation activities. The efficiency of research and innovation depends on the attitudes of the people involved, as well as the resources provided. The level of creativity promoted by the research laboratory or group will also influence the efficiency and effectiveness of the research.

Giving thought to the efficiency of a plan for problem-solving can be beneficial. Planning is especially important for problem-solving in a research environment. As an example, if the outcome of a research effort has implications to public safety, then the most beneficial outcome, i.e., Q_R of Eq. 2.2, and the maximum efficiency ε_R can be extremely important. Thus, the measures of research efficiency for alternative solutions are important and can be used to evaluate the outputs and the personnel involved in the activity.

2.5 THE DECISION PROCESS

While the design process is often associated with engineering, the decision process is more closely associated with the business world. Organizations invoke their version of the decision process when a company recognizes a performance or opportunity gap that could act as a business stimulator. The company then sets specific goals and formulates decision criteria that can be used to set targets for reducing the gap. A set of alternative decisions are identified, data are collected to use in evaluating the different alternatives, and the expected performance of each alternative is evaluated using the decision criteria. The alternative that shows the best performance value is chosen as the alternative to be implemented. These steps are summarized in Table 2.3.

Performance criteria are metrics that indicate some measure of performance quality; a correlation coefficient and the relative bias are examples. Decision criteria are conditions used to decide on the best alternative; a benefit-cost ratio is an example of a decision criterion. Step 6 indicates that a feedback loop would be part of the process. If after implementation, the alternative that yielded the best values of the decision criteria was less effective than expected, then changes would be necessary and a new alternative implemented. While the process is presented here as consisting of six steps, other models of the decision process can legitimately be presented.

2.6   THE DESIGN PROCESS

The design process is commonly used in engineering. It is used in a wide variety of activities, from the design of space shuttles to the design of bridges, from the design of electronic equipment to the design of power plants, and from the design of chemical processing plants to the design of fire suppression equipment. Design is a process and can be represented by a series of steps, with some representations having as few as four steps and others as many as 20 steps. Obviously, as the number of steps is increased, the increase reflects the amount of detail provided. A six-step version of the design process is given in Table 2.4. Note the similarity to the steps of the research process. To provide more detail, steps such as the following could have

---

TABLE 2.3

Steps of the Decision Process

1. Identify the stimulus.
2. Identify the goals and performance criteria.
3. Formulate the decision criteria.
4. Identify alternative solutions.
5. Collect relative data to use in evaluating the alternatives.
6. Make a decision and assess the results after implementation.

TABLE 2.4

Steps of the Design Process

1. Identify the need or the problem.
2. Identify relevant knowledge.
3. Generate alternative designs.
4. Completely analyze each design.
5. Use criteria to evaluate the alternatives and select the best design.
6. Communicate the results.

been included: (a) assemble the necessary data for evaluation; (b) identify the design variables to be included in the analysis; (c) provide a statistical analysis of the data; (d) develop decision criteria; and (e) verify the adequacy of the results. Increasing the number of steps provides for more systematic application, but simultaneously could act as a constraint if the application of the process is binding and not applied using a flexible philosophy.

2.7 THE MODELING PROCESS

Models are commonly used to make important decisions. Weather bureaus use models to project the pathways of hurricanes and tornados. Businesses use models to improve efficiency and maximize profits. Criminal justice professionals use models to assess plans for distributing the spatial positioning of both manpower and dollars. A model is a properly developed representation of an actual object or system, such that it provides a rational explanation of the system and can accurately mimic the functioning of the system. Models are usually simplifications of the real system, but they hopefully are close enough in performance so that they can provide reasonably accurate estimates of the outcomes of the real system.

The modeling process is a series of steps that is used in the development and application of models. Different modelers develop different models of the same system. Of course, no one model is universally accepted as the perfect representation of the real system, so the different models may provide quite different estimates of the output variable even when the same input is used for the different models. Models are used to forecast the future responses of the corresponding system, predict average responses of the system, study effects of changes to the real system, and explain past events that are not fully understood.

One possible depiction of the modeling process includes the following four phases: model conceptualization, model formulation, model calibration, and model verification. While different forms of the modeling process are available, they usually have the same intent even when the number of phases differs. Different types of models are available, including field models, physical models, and computer models, with the emphasis here centered on mathematical models. The objective of Phase I of the modeling process is to identify the problem and the processes and variables that should be included in the model for it to provide an accurate representation of the system. This phase would include the following steps: identify the problem, identify the needed model outputs, identify the important system processes, identify the inputs that will be needed and the inputs for which data will be available, and identify any relevant theory.

The second phase of the modeling process, model formulation, uses the information from Phase I to formulate a representation of the system. The objective of Phase II is to develop mathematical representations for each of the model components or processes identified in Phase I and assemble them into a representation of the system. The steps to accomplish this task include:

1. *Collect the resources*: Compile the necessary data, while maximizing the sample size and the range of each of the variables.

2. *Analyze the data*: Check for outliers that could distort the model, assess inter-correlations between the variables, evaluate the sampling variation of each variable, assess the adequacy of the range of data for each variable, and identify uncertainties with the database.
3. *Establish functional forms*: Identify alternative mathematical functions to represent each of the system components; when mathematical functions are used, the functions should be evaluated as to the way that they will cause the model to perform outside the range of the measured data.

Theory is rarely adequate to provide accurate estimators of all of the coefficients of an empirical model. Thus, Phase III of the modeling process involves fitting the model coefficients using measured data. Values of the model coefficients are generally fitted using measured data. Analytical, numerical, and subjective methods of fitting are available to analyze data in a way that provides estimated values of the coefficients. The values are considered best according to some assumed decision criterion. Regression analysis is a commonly used analytical fitting method, with the best fit coefficients being the ones that minimized the sums of the squares of errors, where an error is the difference between an estimated value of the output variable and the corresponding measured value, which is included in the database. For more complex functions, nonlinear least squares or the method of steepest descent can be used to provide estimates of the coefficients. For very complex models, it may be necessary to use a subjective method of fitting the coefficients.

Once the model is fitted with the measured data, it should be verified, which is Phase IV. Numerous methods have been used for the verification phase, including statistical methods such as split-sample testing and the more accurate jackknifing method. Some studies use an independently collected set of data for verification. For Phase IV, it is also important to perform a sensitivity analysis to ensure that values predicted with the model will be rational at the design conditions, which are often at or beyond the extreme limits of the data used in Phase III.

The modeling process is widely used, but simultaneously widely criticized. For many complex processes, the models provide fairly accurate reproductions of the measured data, but applying the fitted model to data for other locations often leads to inaccurate reproductions, which is a result that suggests concern for the accuracy of the model. In spite of this problem, fitted models are widely used to make important forecasts.

2.8   INNOVATION: DEFINITION AND PROCESS

A problem can be defined as *a condition or state that presents a difficulty or inconvenience*. Obviously, some problems are more complex than other problems. Complex problems generally take more time and resources to resolve than less burdensome problems. Solutions to the more difficult problems will require greater effort, knowledge, and resources. Complex problems are those that involve many conflicting dimensions and will generally require a more involved process to solve. The solution process for the less complex problems will require less effort and can be represented by a correspondingly less complex problem-solving model. That is, the model will need fewer components because the variables and interdependencies are fewer. Such problems generally need a less complex solution rather than the application of a complex problem-solving process.

Some research investigations want to apply a model that already exists, but is deemed deficient in some minor way. The existing model may not provide specific outputs or does not account for a process that is relevant to a specific problem. Thus, a new model is not needed. A modification of the existing model could solve the problem. This type of problem could be referred as model innovation.

Innovation can be viewed as a problem type that involves a less detailed solution procedure than that needed to solve the more complex problems. Innovation can be defined as *the introduction of something new*. New is interpreted as meaning either *used for the first time* or *unfamiliar*. The definition of innovation is imprecise because it is difficult to decide where *new* begins or ends. Unfortunately, applying the word *new* to something is not indisputable. For example, innovations to cell phones are not new. An added option may not have been available on the most recent version, and in that sense, it can be considered as being new; however, since adding it does not reflect complexity, it is only considered as an innovation. Note the difficulty in applying the definition of innovation.

The problem of precisely defining innovation can be circumvented by considering innovation to be a one-dimensional issue. To illustrate the idea, some innovations would be quite simple in that the problem and possible solution are easily recognized. In such cases, the first step of the research process, i.e., the identification of the problem, is obvious. Even the second step, resource identification, does not need a detailed description because the problem is so obvious. The sixth step of the research process is also trivial for simple modifications. For example, if a new screen lighting option is added to a cell phone, the user would recognize the change as an improvement, so the step "make a decision" is not overly relevant. It seems that some modifications suggest a level of newness for which a detailed research effort is not required. Then, innovation should be seen as a process where the number of levels needed for problem-solving can vary, as some steps of the research process are not really necessary to develop a solution.

Which of the steps of the research process will be necessary to formulate a model of the innovation process? Since innovations can vary in complexity, it may not be possible to identify just one process that applies to every case of innovation. Let's assume that the problem is sufficiently simple that it needs nothing more than a simple statement; thus, Step 1 of the research process is obvious and is not really necessary to define the innovation process. Similarly, the decision step of the research process may not be relevant to innovation because the decision is a direct result of the synthesis step. Such a situation would only require a four-step process, with the following being considered as a reasonable model of the innovation process:

1. *Resource collection*: Obtain facts and data that address the topic of interest.
2. *Weakness identification*: Peruse the resources of Step 1 and identify either the weaknesses of the problem statement or restrictive assumptions.
3. *Innovation analysis*: Establish an experimental plan that is based on the problem statement of Step 2 and conduct the analysis.

4. *Innovation synthesis*: Communicate the results of the innovation analysis
   and identify the broad implications of the results.

For very simple problems, even the collect resource step may not be necessary. Such
problems would only require a three-step model. When confronted by a problem, it is
best to begin the solution process with the six-step model of Table 2.4 and eliminate
the steps only when the elimination can be justified.

   In summary, the innovation process can be presented as a four-step process
because the problem statement (Step 1) and decision step (Step 6) are usually obvious. Keep in mind that problems cannot be classified discretely as simple versus
complex. Complex-simple is a continuum. Thus, it would be wrong to imply that all
problems of innovation only require the four-step process while all complex problems should use the six-step problem-solving process. Flexibility is needed, so unless
the problem is universally agreed to be simple, it may be best to approach a solution
using the process with all six steps that are associated with complex problem-solving.

2.9   THE CREATIVE THINKING PROCESS

The term *creative thinking* has many interpretations. Some interpret it as nothing
more than the imaginative generation of ideas using brainstorming. Others define it
as a method of thinking and often assume that it is nothing more than an alternative to logical thinking. Creativity should be viewed as a process that involves some
form of imaginative thinking. The so-called creative process is nothing more than the
problem-solving process that integrates imaginativeness with logical thinking, which
makes the process more flexible than logical problem-solving. For someone to be considered as a creative thinker, they should routinely incorporate imaginative problem-
solving methods into the decision-making. Just because someone incorporates a
technique, such as brainstorming, into their decision-making does not qualify him or
her to be labeled as a creative thinker. Only the continual, routine use of imaginative
thinking skills and attitudes qualifies a person to be labeled as a creative thinker.

   As I view the nature of creative thinking, I associate it with the solution of narrowly focused problems, which are simple enough to be solved using a method based
on idea generation. The problems to be solved are not complex; therefore, they do
not need to be restricted in any way. A list of ideas is generated and then evaluated
using some pre-selected decision criterion. The creative problem-solving procedure
is summarized in Table 2.5. Steps 1–4 can be considered as Phase I, with Phase II
being Steps 5 and 6.

   The efficiency of the various factors involved in creative analyses can be assessed
individually. For example, the facilitator of a creative activity is very important. The
efficiency of the facilitator could be expressed by

$$\varepsilon_f = P_f / R_o \qquad (2.3)$$

where P_f is the facilitator performance and R_o is the organizational resources provided to the facilitator for the creative activity, which could include the use of facilities, release time for participants of the activity, and personnel support. Similarly,

---

TABLE 2.5

The Creative Thinking Process

1. Identify the problem.
2. Select the decision criterion.
3. Select the idea generation method.
4. Generate a list of possible solutions.
5. Evaluate each item on the list using the decision criterion.
6. Based on the evaluation identify the best solution.

---

the efficiency (ε_g) of a group that is involved in creative thinking activities could be assessed using the following metric:

$$\varepsilon_g = P_g / R_o \tag{2.4}$$

where R_o would be a measure of the motivation and effective use of imaginative idea generation methods used in their attempt to solve the problem and P_g is some measure of the extent to which the activity fulfills some criterion. The performance of the group would depend on the values, attitudes, and experiences of the individuals who comprise the group.

2.10   THE CRITICAL PROBLEM-SOLVING PROCESS

It is not uncommon to believe that critical thinking is nothing more than another name for creative thinking. In reality, the two methods of problem-solving are quite different. Creative thinking is used for simple, narrowly focused problems, i.e., little more than a method for generating ideas. Conversely, critical thinking is used for solving complex problems. Thus, critical thinking is highly constrained to ensure that the best solution is identified; however, the restrictions serve very beneficial purposes.

From my perspective, critical thinking is, like creative thinking, not a unique process that is applied in the exact same way each time that it is used to solve a complex problem. Instead, critical thinking is a problem-solving philosophy that uses a wide array of thinking practices during the solution of a problem. Critical problem-solving is most effective for solving complex problems. Critical thinking is more appropriate than creative thinking when the problems are complex and have more significant, broader implications than problems of minimal significance.

I will present critical problem-solving as a method for addressing important, complex problems. It goes beyond the creative problem-solving approach as other factors are considered. From my perspective, critical problem-solving has the following four dimensions that go beyond the dimensions that are specific to the problem; the technical dimension would be the most obvious dimension that is added to the following four dimensions:

1. *Thinking set*: Uses multiple types of thinking, not just logical thinking, and matches the thinking type to the characteristics of the problem.
2. *Skill set*: Applies a sequential array of skills acquired through experiences to understand the current deficiency in the state of knowledge and ultimately solve the problem.
3. *Value set*: Incorporates an array of values relevant to conflicting stakeholders with knowledge of the proper way to balance values when some of the values are in conflict.
4. *Mind set*: Mental processes can encourage or constrict problem-solving as they influence decisions and actions. Short-term states of mind, such as dispositions and moods, are controlled in critical problem-solving.

While these four added dimensions act as restrictions, they ensure decisions are not haphazardly made; however, the added dimensions will add confidence to decisions as uncertainties due to mental state variations are under greater control. Critical problem-solving is influenced by one's mindset. A person's mood or disposition may cause a distortion of the person's thinking based on his or her beliefs. Solving critical problems is best achieved when a person uses all four dimensions as well as the other dimensions that are relevant to the problem.

These four dimensions do not include any reference to other dimensions of decision-making such as technical competency or economics. The other issues require knowledge specific to the technical specialty and the problem. Technical knowledge would be quite different for those in the medical profession versus those in engineering or one of the sciences. Technical knowledge might be considered as a fifth dimension. Economics, environmental concerns, politics, and risk are other possible dimensions, but like the technical dimensions, they are unique to the problem and the stakeholders. Since this book is broadly applicable, the dimensions beyond the four listed above are not discussed herein.

Unlike the Scientific Method, the research process, or the decision-making process, the critical problem-solving process does not have a series of steps like these previously mentioned processes. These four dimensions can be used with any process. For example, a complex research problem could apply the four dimensions and the steps in Table 2.2, while a business-oriented problem could apply the four dimensions with the decision process in Table 2.3. The critical problems solving process is not a unique set of steps, but a process that requires consideration of the additional dimensions.

2.11 DE BONO'S MODEL OF THINKING

Thinking is actually a multi-step process, and appreciating the individual steps will lead to more productive idea-generating activities. Edward de Bono created a six-step process to reflect the individual elements that take place during a decision-making activity. His conceptual model of thinking related each step to a different colored hat. While a number of different interpretations of the expected outcome that each hat represents have been proposed, the following series is one possible interpretation of the colors and hats that are relevant to the objectives of this document:

1A. *Blue hat*: Facilitator provides big picture; sets agenda; states problem; controls the sequence.
 2. *White hat*: Collects data; establishes facts and knowledge; makes logical analyses.
 3. *Red hat*: Introduces affective feelings, opinions, and subjective ideas.
 4. *Black hat*: Acts negatively; identifies weaknesses and fatal flaws; makes critical analyses; skeptic.
 5. *Green hat*: Creative; enters new concepts; free-wheeling; identifies out-of-the-box facts.
 6. *Yellow hat*: Provides a positive view; optimistic; speculates on outcomes.
1B. *Blue hat*: Summarizes; provides conclusions; makes decision.

Note that the blue hat appears twice because the wearer of the blue hat acts as the facilitator. One responsibility of the facilitator is to complete the planning activities and monitor the activities of those responsible for the tasks of the other hat wearers. When the process has sequenced through the different activities associated with the other hats, the wearer of the blue hat has the responsibility to assess the process used and to ensure that the final solution is feasible. The facilitator summarizes the important aspects of each hat in a final report.

In any one execution of de Bono's method, all of the hats may not be needed. For example, a specific problem, such as that associated with logical thinking, may only require the blue, white, red, and black hats. Additionally, the facilitator may decide that a different sequence may be more productive. For example, the facilitator may decide to have the green hat participate before the red hat. The facilitator has the responsibility to make the final decision on which steps are used and the order of the activities to use in developing a solution to the problem. Since the order of the actions will influence the overall efficiency and the quality of the results, the facilitator must have the experience needed to ensure the best results.

Should de Bono's six steps be viewed as a process for thinking or just a process for problem-solving? Is de Bono's sequence just appropriate for brainstorming (see Chapter 3) or can it be used with other idea-generating methods, such as synectics (see Chapter 4) and the Delphi method (see Chapter 5)? If the white-hat step is included as the second step of the process, then it could be used for brainstorming, but not for a synectics activity. The white-hat step suggests that facts and knowledge of the problem are collected and known to the group. In a synectics activity, the real problem should not be known, so the group would not have the specific data and knowledge. It may be possible to distribute knowledge and data that apply to the synectics problem rather than the actual problem. However, if the white-hat step were moved such that it followed the red-hat step, then de Bono's method could be used for a synectics activity. The point is: de Bono's approach should be viewed as a flexible process that can be modified to meet the needs of the problem statement and characteristics of the problem-solving group.

Thinking is a process, design is a process, and decision-making is a process. Therefore, each of them should be presented as a series of steps. Before making a decision about a design problem, various types of thinking should take place. Therefore, it is not surprising that the steps of these three activities are often quite

---

TABLE 2.6
de Bono's Six-Hat Thinking Process

1A. Identify the problem (blue).
  2. Gather resources: data, facts, and knowledge (white).
  3. Incorporate an emotional involvement (red).
  4. Apply counterfactual/negative thinking (black).
  5. Identify new concepts using a creative attitude (green).
  6. Place ideas in an optimistic perspective (yellow).
1B. Solution adoption (blue).

---

similar but may differ depending on the problem that needs a solution. A process may differ slightly in the number of steps that are included in the model of the process, but the specific steps of any activity should be developed with a certain type of problem as the motivation. Problem complexity may influence the number of steps. Table 2.6 places de Bono's method in terms of the fundamental tasks needed in problem-solving. Note the similarity of de Bono's steps to the steps of the research process (see Table 2.2).

2.12   THE INTELLECTUAL DEVELOPMENT MODEL

The Scientific Method and all of its problem-solving offspring are processes that describe a sequence of steps for solving problems. Unless a feedback activity is necessary, the steps are completed in order from identifying the problem (Step 1) until the selection of the alternative that best meets the adopted decision criteria (Step 6). When using any of these processes to solve a problem, it is assumed that the user has the necessary intellectual, problem-solving capacity to execute each step; however, the decision identified as being the best will depend on the intellectual abilities and attitudes of the problem solver. A person who has a relatively low level of intellectual capacity and little practical experience for problem-solving may conclude that a truly non-optimum solution is actually the best option. In order to ensure that the best decision is reached, it is important for the problem solver to have the highest level of beliefs and skills. Scholarship and experience at developing knowledge can be measured using a method like Bloom's taxonomy.

Bloom's taxonomy is a widely used framework for classifying educational goals. Bloom's taxonomy includes the following six steps: knowledge, comprehension, application, analysis, synthesis, and evaluation. The six-step procedure has several variations, as it has undergone some revisions since it was introduced in 1956. Bloom's taxonomy is not a problem-solving process in the same way as the research process. It is more of a pathway to gaining knowledge of a topic. In fact, Bloom's first step is not to identify a problem. The method assumes that the topic is pre-specified, essentially Step 0. Bloom's first three steps seemed to have been oriented toward identifying the depth of current knowledge. Starting with Step 4, Bloom was integrating more breadth of knowledge about the topic—depth before breadth.

The intent herein is to use some of Bloom's ideas to begin establishing criteria that reflect the ability for a person to become a critical problem solver. Bloom used a six-step format. The proposed model uses five steps. Some variations in the names are adopted herein to more closely associate with critical problem-solving. Distinctive characteristics of a five-step format are as follows:

- *Comprehend*: To identify relevant knowledge, such as definitions, basic theory, or previously completed work relevant to the issue under investigation, and to establish the state of knowledge by assembling the elements of related knowledge. To understand the ways that the elements of the knowledge fit together.
- *Experience*: To use the assembled understanding in a basic application of the fundamentals; the objective of the application is for the user to gain experience at using the knowledge for problem-solving.
- *Analyze*: To critique and question the existing state of the art in order to identify its important parts, its weaknesses and its deficiencies; this knowledge will serve to direct future activities needed to advance the state of knowledge.
- *Synthesize*: To combine existing knowledge with the new knowledge to form a distinctive set of conclusions.
- *Create*: To use the synthesized knowledge to develop a novel or unique solution to a complex problem.

Note the progressive problem-solving: gain knowledge, and place the knowledge in a format for solving problems, gain experience via applying the knowledge, identify weaknesses in the existing knowledge, develop distinctive results that advance the state of knowledge, and then use the new knowledge to provide a better indication of the expected responses of the system in the future. The first three steps characterize the existing state of knowledge. Steps 4–6 seem to identify a weakness in the topic and aim to advance the state of knowledge. This characterize-and-advance process would be a valid approach to making advances to any subject; therefore, it should be a valid approach to presenting the growth of knowledge of critical problem-solving.

Both Steps 2 and 3 are meant to provide experience in the use of the topic; however, Step 2 provides depth while Step 3 provides breadth. The applications of Step 2 focus primarily on the topic of interest. The applications of Step 3 bring a broader array of decision criteria into the learning. The breadth of Step 3 is intended to teach the novice that an issue may need to be viewed in a broader context.

Steps 3–5, while very different, represent a form of assembling, with the difference being the level of newness. Truly novel solutions are the domain of Step 5. In Step 2, the synthesis is nothing more than an application based on the existing state of knowledge, while the products of Steps 3–5 reflect advanced study of the problem, such that a weakness of or constraint on the state of knowledge was detected and corrected via Steps 3 and 4. Step 4 reflects a distinctive application based on new knowledge or data, and Step 5 involves a novel or unique solution to the original problem.

2.13   CONCLUDING COMMENTS

Solving complex problems will most likely be more effective and completed more efficiently if a systematic procedure is used to identify the best solution. This approach
does not require the same steps to be used to solve every problem, but if the steps of a
systematic process are considered in the planning of the solution, then the solution to
the problem will likely better meet the needs of the stakeholders. Additionally, some
problems will require very novel solutions and the outcomes might be identified more
efficiently if imaginative solutions are considered. For complex problems, it may be
best to assume one of the processes, e.g., the research or decision process, and expand
the number of steps to include steps that reflect the uniqueness of the problem. When
solving any problem, it may be best to initially assume that the problem is complex. It is
better to discover that the problem is less complex than expected, than to learn that it is
actually more complex than initially assumed. All of the dimensions of critical problem-
solving should be considered when trying to find solutions to complex problems.

If the different steps of the processes discussed in this chapter were placed side
by side, a pattern of similarities would be obvious. For processes that are defined by
only four steps, a user could reasonably assume that some of the steps assume multiple aspects of problem-solving and that the seemingly missing steps are inherent to
the steps shown. For example, if the first step of a process is to identify the problem,
it should be assumed that this implies stating the problem in terms of the knowledge
that is unknown and sought, as well as stating the research goal and research objectives that will direct the investigation. For educational purposes, it may be better to
present a process using a large number of steps so that a novice recognizes all aspects
of problem-solving. Then the first step might be to identify the problem while the
second step would be to specify the research goal and objectives.

Similarly, Step 2 of a six-step process might be to obtain the necessary inputs.
This step could be expanded into multiple, more specific steps, such as the following,
which would be one possible set: (1) identify the existing theory related to the problem; (2) identify the relevant data that currently exist; and (3) identify all variables
and factors that are relevant to the problem. The advantage of using a larger number
of steps is that a novice to problem-solving would have a clearer understanding of the
actions that need to be taken. Regardless of the number of steps that are needed to
identify the actions, it is important for the novice problem solver to understand that
following the sequence is the most efficient way of solving the problem.

The third step of the six-step process for problem-solving is usually something
like *design the experiment*. A novice might not recognize that this step inherently
involves the following four sub-steps: (1) for each specific research objective of Step
1, a research hypothesis needs to be formulated; (2) then the specific theory that is
relevant to the hypothesis must be identified and placed in a form that is appropriate
for the problem; (3) then the data that would be needed to test the hypothesis need to
be compiled; and (4) the performance criteria and the decision criterion that will be
used in any data analyses must be identified.

The bottom line is that regardless of the number of steps used to represent a
problem-solving process, each step usually has sub-tasks inherent to it. A novice to
problem-solving needs to recognize the importance of the sub-tasks and, at the same

time, recognize the value of the concise statements that make up any six-step model of the problem-solving process.

2.14   EXERCISES

2.1.   Find definitions for the words process, method, procedure, and system. Use these definitions to improve the definition of a process given in Section 2.1.

2.2.   Present the Scientific Method as an eight-step process.

2.3.   Find an alternative description of the Scientific Method that is based on more than four steps. Discuss the implications of the disparity in number of steps.

2.4.   Find definitions for induction and deduction. How do they apply to the Scientific Method and the research process?

2.5.   Find the definition of the efficiency of a lever as a machine. Adapt this definition to the measurement of the efficiency of problem-solving.

2.6.   What skills would influence the efficiency of the problem-solving process? Define each skill.

2.7.   What values would influence the problem-solving process? Define each value.

2.8.   Develop a ten-step model of the research process by expanding the six-step model presented in Section 2.4.

2.9.   Discuss the meanings of the words *analysis* and *synthesis* in reference to the research process.

2.10. Using the six-step model of the research process (see Section 2.4), outline a research plan to improve the accuracy of Manning's roughness coefficient for flow in rivers.

2.11. Under what circumstances would a person's research efficiency change?

2.12. Discuss ways that a person's negative feelings on the value of creative thinking could influence the numerator and denominator of Eq. 2.2.

2.13. What factors would need to be considered to quantify the creativity efficiency of Eq. 2.2?

2.14. Develop a metric for measuring research efficiency that is based on research quantity. Develop a second metric that is based on research quality. Propose a novel way of weighting the two metrics that would provide a single measure of research efficiency.

2.15. Step 2 of the research process indicates the importance of knowledge. Identify both general and specific sources of knowledge that can assist in research.

2.16. PhDs who apply for faculty positions are often rated on the quality and quantity of the journal papers that they co-authored. Will research efficiency influence the quantity and quality of a person's research? How?

2.17. How are the efficiencies of Eqs 2.1 and 2.2 similar? How do they differ?

2.18. What is the intent of a research plan? What components would you include in a research plan?

2.19. Define the word feedback. Why might it be necessary when conducting research?

2.20. Identify a problem that relates to one of the concepts that you have learned in one of your engineering classes. Use de Bono's six-hat model to illustrate each of the underlying idea generation steps for solving the problem.
2.21. Because of climate change, the extinction of polar bears is possible. Use de Bono's method to analyze the issue.
2.22. Models are used as research tools and their development is part of Steps 4 and 5 of the research process. Discuss the use of the steps of the modeling process as parts of the research process.
2.23. Models need to be verified. Where would model verification fit into the steps of the research process?
2.24. Expand the design process of Table 2.4 to 12 steps. Discuss the benefits and drawbacks of using more steps.
2.25. Where in the design process should risk to society be considered? Explain.
2.26. Discuss the meaning and importance of feedback in the decision process.
2.27. Without losing any of the intent of the decision process, create a four-step version. What is lost from its clarity by this reduction?
2.28. Identify ways that a person through their own efforts can gain experience in innovating ideas.
2.29. Recall one specific example of cell phone innovation. Outline the steps that you believe were used in this innovation.
2.30. Why do we infer that creative thinking must involve imaginative thinking?
2.31. Some people view creative thinking that involves imaginative idea generation to be a waste of time. Is this belief true or false? Explain.
2.32. What skills are relevant to the *Experience* step of the process of Section 2.12? Define each skill and indicate its application.
2.33. Discuss the association between the Induction step of the Scientific Method and the *Create* step of the model of Section 2.12.

2.15 ACTIVITIES

2.15.1 ACTIVITY 2A: THE PROCESS OF ART

The objective of this activity is to develop a general, multi-step process for creating a work of art, which could be a painting, a piece of sculpture, or a musical composition. Identify the objective of each step. Describe the thinking that would be involved and ways that the use of creative methods can be used. Compare your proposed process with the six-step process of problem-solving.

2.15.2 ACTIVITY 2B: PROCESSING CHEESE

Find a *prose* description of a food processing operation, such as for the processing of cheese or wine or strawberry jam. Convert the *prose* description into a concise, numbered set of interdependent operations. Compare the steps of the proposed set of operations to the steps of the Scientific Method. Discuss the similarities and differences.

2.15.3   Activity 2C: The Yellowish-Green Hat

Modify de Bono's problem-solving model to incorporate decision criteria such as the benefit-cost ratio. This step will be referred to as the yellowish-green hat. Indicate where it will be placed and the benefit of having such a task involved in the process.

2.15.4   Activity 2D: Where Are the Doctors?

The state has one major medical school; however, too many of the graduates leave the state for better paying jobs in other states, which is producing a brain drain on medical health suppliers within the state. As governor of the state, what can you do to reverse the migration of doctors from the state? Represent the problem and the analysis using the steps of the problem-solving process.

2.15.5   Activity 2E: Robots or Humans?

A company realizes that to remain competitive they will need to modernize their production processes, which may invoke the replacement of some workers with robots. However, the workers have a long contract and a strong union. If production is not modernized, the profits will decline and the company may even need to close, which would mean that all jobs would be lost. You are the owner and CEO of the company, what will you do? Represent the problem and the analysis using the steps of the problem-solving process.

2.15.6   Activity 2F: HA! HA! HA!

One of the toughest jobs is to be a stand-up comic. Getting a sober audience to laugh is not easy; in fact, it is easier to rob a bank! The better comics know that joke making follows a process, with the intended end result being an audience that is rolling in the aisles with laughter. The funniest jokes generally have an unpredictable punch line but one that makes sense. Very often, the best jokes make the people laugh at themselves as they were not expecting the outcome. The objective for this activity is to develop a multi-step process that a comic would follow to develop humorous material for his or her show.

2.15.7   Activity 2G: Mini-You

The objective of this activity is to show the solution of a problem in terms of the six steps of the decision process. You are a scientist involved in solving problems of the human circulatory system. Using a new opportunity that allows you to miniaturize yourself, you are now able to enter the circulatory system of another person who has a critical problem that needs to be corrected. You are planning to enter a patient's system to correct a life-threatening problem. Identify the planning process that you would follow prior to entering the person's system to correct the problem. Compare your steps to those of the decision process.

2.15.8   ACTIVITY 2H: IS MERCURY A MONKEY

As a youth, science teachers provided a method for remembering the order of the planets, such as *Men Very Early Make Jars Stand Up Nearly Perfect*. Since the P for Pluto can now be dropped, create a sequence using only animal names, but with a way that a young person could remember the order of the animal names and, therefore, the names of the planets. The objective of this activity is to create a new method. Be creative!

2.15.9   ACTIVITY 2I: A DEAR JOHN LETTER

Part I: Mary has been going steady with John for three months, but as of late, their relationship has been rocky. Mary decides to write a letter to John telling him the reasons that she is not going to continue dating him. Identify the steps of the process that Mary should use to develop the *Dear John* letter.

Part II: Compare the steps that you recommend for writing a *Dear John* letter with the steps of the decision process. Note the reasons for any differences.

2.15.10   ACTIVITY 2J: ANYONE FOR WINDOWS AND SIDING?

Part I: You are considering the start of a home-improvement company that would specialize in the redesign of closets in homes and offices. You believe that a systematic process of the design and construction of closets is needed and should be followed in order to maximize profits. Develop the steps that you would have employees of your new company use when they are hired to build some one's closet.

Part II: How do your steps of the process compare to the steps of the decision process?

2.15.11   ACTIVITY 2K: GET INTO THE TIME TUNNEL

Propose a method of revising Eq. 2.2 so that research efficiency is an explicit function of time.

3 Fundamentals of Imaginative Idea Generation

CHAPTER GOAL

To stress that the generation of novel ideas can be learned, i.e., creativity is not something that is entirely innate.

CHAPTER OBJECTIVES

1. To introduce the rules for directing brainstorming sessions.
2. To discuss the responsibilities of the facilitator.
3. To present ways of stimulating imaginative ideas.
4. To introduce the method of brainwriting and its benefits for generating ideas.
5. To show ways that brainwriting can be used in research.
6. To discuss various creative visualization methods.

3.1 INTRODUCTION

In my classes on creativity, I often conducted a brainstorming session on ways to peel a banana. I encouraged wild-and-crazy ideas. While many of the responses were traditional, some were wild and crazy. One student recommended playing sexy music to the banana such that the music would encourage the banana to do a strip much like the burlesque dancer does to her music. Of course, the response created some snickering that sounded like criticism, which is a no-no during the first phase of a brainstorming session. When we initiated Phase II, which is the idea evaluation phase, one student said that maybe sound waves, i.e., the sexy music played to a banana, may be a way of slicing the outer skin of the banana along the seams. A more common idea was to have a chimp strip the peel while having duct tape placed over the chimp's mouth to prevent it from eating the fruit. The sound waves (i.e., sexy music) idea seemed to be a more feasible option, and it is also much more imaginative.

Brainstorming is the most commonly used method of imaginative idea generation. Other methods of idea generation, such as synectics and the Delphi method, are variations on the traditional method of brainstorming. Why the name *brainstorming*? Guidelines for conducting brainstorming sessions emphasize

DOI: 10.1201/9781003380443-3

imagination, unpredictability, free-flowing idea generation, and uninhibited action. These are also characteristics of cyclonic storms that have major, often unusual, societal consequences. Users who brainstorm for the purpose of improving their problem-solving hope that the activity leads to unique results in many ways similar to the unique results of a cyclonic storm. Thus, the name brainstorming reflects the association of uninhibited mental processes and uninhibited unpredictable storm action.

3.2 BRAINSTORMING

When a person has a problem, he or she will mentally make a list of potential solutions. While the list may not be in written form, the practice of generating potential solutions to a problem represents an informal instance of brainstorming. This format of brainstorming will likely be most effective in finding a solution if the person allows his or her emotions to contribute to the idea generation. In formal brainstorming sessions that are successful can encourage a person to adopt formal brainstorming as an effective mental tool for problem-solving.

3.2.1 DEFINITIONS

Brainstorming is a group activity in which ideas that are broadly related to a problem are generated, recorded, and evaluated. A brainstorming session consists of two phases: Phase I is the idea generation phase, while Phase II is the idea evaluation phase, where the ideas from Phase I are evaluated to find the best solution. A brainstorming session involves a facilitator and a group of participants. During the first phase, the individuals in the group work together to generate an extensive list of ideas, while not criticizing or evaluating any of the proposed ideas. Ideas proposed by members of the group can be used to identify additional, similar ideas, which are referred to as *piggybacked* or *hitchhiked* ideas. The leader of a brainstorming session is referred to as the *facilitator*.

An efficiently conducted brainstorming session is the ideal. From Eq. 2.2, efficiency would be the ratio of the worth of the output of the session to the extent of the resources needed to produce the resulting decision. Both the facilitator and the participants influence the efficiency of a brainstorming session. The facilitator must prod the group to produce unique ideas, while the participants must enthusiastically endorse the effort by generating unique ideas. While commonplace ideas can be of value, unusual, emotionally encouraged, ideas are the ones that usually lead to novel solutions. Imaginative ideas that are generated in Phase I may not lead to novel solutions to a problem unless the evaluation phase is conducted properly.

The participants in the group can positively or adversely influence both the outcome and the efficiency of a brainstorming session. They need to take the session seriously, avoid critical analysis of ideas during Phase I, uniformly participate in idea generation, and act to minimize downtime, which is a time during the session when ideas are not being actively generated. Creativity inhibitors are *factors that limit creative efficiency and productivity*. Too often, a few individuals in the group do not actively participate in the idea generation; they constitute the primary inhibitors;

however, peers and even the facilities where the session is held can act as inhibitors. Both the facilitator and the participants need to be aware of all possible inhibitors, especially those that are self-imposed by either the individuals or the group collectively.

3.2.2 RULES FOR BRAINSTORMING

Brainstorming is intended to be an unconstrained mental activity, so the rules are minimal and are only intended to encourage the free flow of ideas. The basic rules for conducting a brainstorming session are:

1. To encourage the generation of a large quantity of ideas.
2. To promote imaginative (i.e., wild-and-crazy) ideas.
3. To support the combination and improvement of ideas using ideas that have already been generated.
4. To discourage criticism of ideas during the idea generation phase.

Rules 1–3 reflect activities that generally lead to a high creative efficiency and useful output. Rule 1 is intended to inspire the group to actively generate as many ideas as possible in the allotted time. Rule 2 stresses the importance of emotion and imagination. Rule 3 is intended to encourage fluency and to minimize downtime. Generally the best solution to the problem will be found if the group generates a long list of diverse ideas. A wild-and-crazy idea may, at first, seem just that, but in the second phase of a brainstorming session, specifically in the evaluation phase, a wild-and-crazy idea may lead to a very novel solution, one that is far superior to the solutions that would result from in-the-box thinking. Encouraging the combination of ideas, i.e., piggybacking, is especially helpful as it often encourages the more timid participants to contribute ideas. Piggybacking is also an effective practice to minimize downtime.

Rule 4, which concerns the criticism of ideas during Phase I, is intended to ensure an environment where the participants feel free to contribute ideas, especially imaginative ideas. Criticism is discouraged during Phase I because many participants do not like criticism of their ideas and would stop contributing if their ideas received negative comments. Criticism takes time, which could be better used on the generation of new ideas; thus, untimely criticism during Phase I can reduce the overall efficiency and effectiveness of a brainstorming session. Self-criticism by a group member, i.e., the reluctance to contribute ideas because of a pessimistic mental assessment of ideas, can be a major inhibitor to the generation of a large number of ideas. Inhibiting factors are discussed in Chapter 7.

A fifth rule could be added to the four rules cited. Specifically, the efficiency of a brainstorming session will improve when the responses are kept short. Temporally long statements by a participant will detract from the flow of the idea generation process. That is, longwinded statements by one person can be counterproductive, as the delay can distract other participants from engaging their imagination and contributing ideas. It is the facilitator's responsibility to prevent long-winded statements and condense ideas into short phrases. The facilitator needs to keep the group members who tend to give longwinded statements from dominating the session. Detailed responses from anyone can appear to be idea evaluation, which will divert the attention of other participants who will then likely prematurely lose their enthusiasm for generating ideas.

Critiquing is an important skill for a problem solver. Constructive criticism is only allowed, even encouraged, in the second phase of brainstorming, i.e., idea evaluation. Every idea on the Phase I list needs to be evaluated from both positive and negative perspectives for its potential to contribute to a solution to the problem. Phase II can be time consuming, but it must be conducted with the thought that every generated idea may be the one that will lead to the most novel and best solution.

3.2.3 Facilitating a Brainstorming Session

Each brainstorming session is led by a facilitator, which is the person who has the responsibility to plan the event, select the group of participants, direct the session, and submit the decision to the stakeholders who generated the problem statement and initiated the task. Idea stimulation is the most obvious responsibility of the facilitator. The ability of the facilitator to prod the participants into generating a large quantity of high-quality ideas will ultimately determine both the efficiency and the success of the session. Prior to the group convening for the session, the facilitator should do everything possible to improve the facilities.

A major responsibility of a facilitator is to ensure that criticism of ideas is suppressed during Phase I, so that ideas will flow freely from all participants. Participants who are reluctant to contribute ideas may act as inhibitors in that they appear to believe the practice of brainstorming is not a worthwhile activity; this apparent attitude can reduce the efficiency of the group effort. Therefore, the facilitator may need to specifically call on the timid individuals. It may be helpful to compliment a timid individual's idea. The facilitator must use the position and his or her own talents to encourage as many ideas as possible, both unique and commonplace ideas. The facilitator must adhere to the principles of idea generation, i.e., Phase I, and idea evaluation, i.e., Phase II. The following are a few ways that a facilitator can stimulate the participants to generate ideas:

- If the initial statement of the brainstorming problem seems to be too narrow or ordinary, the facilitator can adjust the problem statement to broaden it.
- The facilitator could select one of the best ideas to date and ask the group to generate piggybacked ideas on that item.
- The facilitator can ask for very, very wild ideas; this may put pressure on the group to be more affective in their thinking.
- The facilitator can provoke thinking by asking a novel question, such as "What would Superman do?" or "What would Cleopatra have done?"
- When someone generates a silly or obvious idea, some of the participants may moan, which is a form of criticism. A facilitator should discourage all criticism during Phase 1. The facilitator should show ways that the silly idea can lead to a good add-on idea; hopefully, this will suppress future criticism of ideas and encourage piggybacking.

- The facilitator can suggest assessing the problem from the standpoint of a classic movie (e.g., *Gone with the Wind*) or a current movie (e.g., one of the *Star Wars*, *Spider-Man*, or James Bond series).
- Efficiency can be enhanced by having the participants provide concise responses, as this practice will suppress counterproductive evaluation of ideas.

3.2.4   MANAGING QUIET PERIODS

Quiet periods, i.e., times when the participants are not contributing, will almost always occur. The facilitator has the responsibility to reignite the group's active participation. This can be done by:

- Adding her or his own ideas, especially those that will encourage piggybacking.
- Pointing to one of the ideas on the list and ask for piggybacked ideas.
- Asking specific individuals for ideas.
- Identifying a very general topic that is similar to the original brainstorming idea and encouraging idea generation on the analogous topic.
- Temporarily suspending the group brainstorming activity and asking each person to continue generating ideas via brainwriting for three minutes; then returning to the group brainstorming mode and asking for some of the items that are on their brainwritten lists.
- Using a stochastic thinking method such as selecting a random word from a dictionary as a crutch to a new idea.
- Challenging the two sides of the audience, with the winning side being the side that generates three new ideas the fastest in the shortest amount of time.
- Having everyone stand and face the back of the room. A person can only sit back down when he or she has contributed an idea.
- Having the participants form a line and each person must add another idea to the list on either the board or flip chart.
- Playing a song and having the group develop ideas based on the words of the song.
- Forming pairs of participants and challenging each pair to identify an idea.

Effective facilitating will greatly influence the efficiency and effectiveness of the overall effort.

3.2.5   WARM-UP ACTIVITIES

A group that initially shows enthusiasm is more likely to generate a set of high-quality ideas. A facilitator cannot expect all participants to arrive at a brainstorming session prepared to enthusiastically generate imaginative ideas. A facilitator can encourage enthusiasm for the session by starting the session with a warm-up activity. The nature of the activity should be designed to mimic the intent of the session. If the session will need to generate many wild-and-crazy ideas, then the group could be confronted by a solution that would place them in that mood. For example, they could be asked to describe the appearance of the man from Mars. If the session is intended to be on problem-solving, then the facilitator could show a series of drawings and ask for interpretations. For example, one figure could show the letters WAVE sketched vertically and the group would be expected to indicate that it stands for a standing wave. Another example could be to draw a square and write the word AIR along the bottom but in a way that the letters are considerably wider than taller; this would indicate compressed air. Numerous warm-up activities are available, and they can be used to inspire the group to have an affective mindset for the idea generation activity that is to take place at the end of the warm-up period.

3.3 BRAINWRITING: INDIVIDUAL BRAINSTORMING

Brainwriting is a single-person method for generating ideas and can be very useful for problem-solving; it is a pencil-and-paper version of brainstorming.

3.3.1 BRAINWRITING BASICS

Brainwriting is an important tool for those involved in research, especially when the day-to-day research problems that need solutions are the primary responsibility of one person. When performing novel research, problems continually arise in the pursuit of new knowledge, and most often these problems do not lend themselves to quick solutions. Many alternative solutions to a problem need to be identified and evaluated before the best alternative can be selected. Solutions that have not been proposed can obviously not be identified as the best solution, so generating all potential solutions must be an initial objective.

Brainwriting is essentially brainstorming by an individual, just a paper-and-pencil version rather than a group effort. It follows the same procedure and uses the same rules as brainstorming; it has the same objectives. In brainwriting, a person generates ideas and records them. Thus, the person acts as both the facilitator and the group. Brainwriting can be used either to develop new ideas or to solve specific problems that arise during a problem-solving session. Brainwriting generally increases one's research efficiency, and more importantly, it is likely to increase the number and breadth of alternative solutions identified.

3.3.2 PROCESS OF BRAINWRITING

One approach to brainwriting is to start with a blank sheet of paper. Place a concise statement of the problem at the top of the page. The problem may even be stated as a question that describes the specific concern. Then without any intermediate assessment or criticism of ideas, the brainwriter should fill up the page or multiple pages with simple statements related to the concern; even one or two words can adequately summarize an idea. Using complete sentences for entries to your brainwriting list is discouraged, as this requires excessive time and distracts from the free flow of thoughts. The ideas do not need to be organized while they are being generated. Also, the rules of grammar can be ignored. In the idea generation phase of brainwriting, piggybacking of ideas is strongly encouraged, as this helps to maintain the flow of ideas and may provide multiple perspectives on the problem. The idea generation session can be considered as Phase I of the brainwriting method. Phase II will be the evaluation phase.

A tape recorder is another way of keeping track of generated ideas. Instead of writing the ideas on paper, the item could be spoken into the recorder with the list printed out at the end of Phase I. The advantage of this method would be that piggybacked ideas would less likely be lost due to the time required to write out the original idea. The disadvantages of this approach are that the taping equipment must be available and the brainwriter does not have the written list in front of him or her while doing the brainwriting.

It is sometimes helpful to present the problem as a question. Questions appear to be a challenge in that an answer is expected; most people enjoy a challenge. The brainwriting list becomes a list of possible answers to the question. If this question approach to brainwriting is used, the question should have the following characteristics:

1. The question should be a summary of the real problem.
2. The question should be short in length; this forces the question to be worded in a way that focuses on the central concern.
3. Answering the question will provide a possible solution to the problem that originally motivated the brainwriting activity.

When the question is drafted, it may spawn additional related but different questions. The additional questions can be written onto different sheets of paper with solutions identified later. When responding to a question, it is important to focus on the question at the top of the page and not allow other issues to act as mental distractions.

The purpose of Phase II of a brainwriting session is to assess the ideas generated during Phase I; the ultimate purpose is to identify the best solution to the problem. The assessment of the generated ideas should be confined to the idea evaluation phase, i.e., Phase II. Assessment should not be conducted during Phase I—the idea generation phase. Phase II can be started by organizing the ideas into groups that show some measure of commonality; some of the ideas may appear in multiple groups, which is acceptable. Once the ideas are grouped, other ideas may be generated to fill in gaps. Once groups of ideas have been formed, positive and negative characteristics of each idea should be identified. Finally, the most reasonable solution should be selected, which is the outcome of Phase IV.

3.3.3 BRAINWRITING: CRITICAL ASSESSMENT

A brainwriting session should be conducted using the same two phases as discussed for brainstorming: Phase I, idea generation, and Phase II, idea evaluation. During Phase I of a brainwriting session, the person must act in the same manner as the facilitator of a brainstorming session, but at the same time, the person is also a participant. Initially, a time frame for conducting Phase I should be stated. When a person is generating ideas, any urge to criticize or evaluate an idea should be repressed. A brainwriter should *not* stop brainwriting if a particular idea seems to be especially promising, as the hiatus used to evaluate the idea will stymie the continued flow of ideas. Instead, the initially apparent good idea should lead to several piggybacked ideas, which will strengthen the case or the idea. Even a potentially good idea can be improved, so stopping Phase I before the specified time may prevent the expansion of promising ideas.

Premature idea assessment is counterproductive. During Phase I, criticism that seems positive (e.g., "Aha! I have found the solution!") can be just as detrimental as negative criticism (e.g., "This brainwriting is not getting me any closer to a workable solution!"). Both of these types of criticism are emotional responses to the immediate state of idea generation. Affective involvement in the idea generation phase should not be repressed, as some emotional thinking can be a positive attribute to idea generation. Critical feelings, either positive or negative, during the idea generation phase of brainwriting can reduce the overall efficiency of the research effort. A brainwriter can suppress on-the-spot self-criticism by pre-specifying a time duration during which only ideas will be generated. While the time frame for idea generation can be extended, it should not be shortened during the activity. Critical assessment of ideas is the second phase of a brainwriting activity, just as it is with brainstorming.

3.4 PICTOSTORMING (A.K.A., SKETCHSTORMING)

Is it true that "A picture is worth 1000 words?" If the adage is true, then pictostorming may be a useful method of imaginative idea generation. Pictostorming is essentially *brainwriting that focuses on the use of artwork rather than words*. So instead of brainstorming a list of words, the outcome of a pictostorming session is a set of sketches, each of which suggests a solution to the problem or a principal component of the problem. Each image will lead to a potential solution to the problem that the pictostormer believes could assist in solving the problem.

Leonardo da Vinci (1456–1519) is well known for his imaginative artwork. Some of his creations suggested solutions to problems that would not become reality for more than hundred years after his lifetime. He was an exceptional pictostormer. His pictographs are not known for their clarity, but they are widely respected for their imaginative origin and usefulness. His creative genius was evident from his efforts at pictostorming.

A pictostorming session can be conducted either by an individual or as a group effort. Of course, in group sessions, all artwork must focus on solving the problem at hand. The facilitator of the session states the problem and each participant then makes one or more sketches that reflect the problem. The individual entries will be completely unique as the individuals will have different life experiences and levels of imaginativeness. They may also have unique perspectives on the problem that are based on their life experiences. Diversity of thought is always valuable in creative problem-solving. The format for a group session has not been standardized, so it is the facilitator's decision on the best way to conduct the session.

During a group pictostorming session, each participant should develop a sketch that represents a solution to the problem. Then the facilitator instructs the group on the procedure to follow. One alternative would be to have the individual participants generate ideas independently of each other, and then coordinate the alternatives into a set that will lead to a solution to the problem. An alternative format would be to have each individual draw a pictograph and then have the other participants individually critique a partner's pictograph. An alternative would be a multi-pane pictographs that could then be assessed and revised as a group effort. Ideas from two or more submissions could be combined to create a more descriptive sketch. This combining of panes would be similar to piggybacking in a brainstorming session. Another alternative would be for the entire group to develop one pictograph, with proposals for a rough idea as a starter and then followed by the group making changes. A facilitator might decide to begin a session by asking for sketches that are wild and crazy rather than as logically descriptive as might normally be done. Ultimately, the group would need to select the pictograph that offers the best chance of providing a unique solution to the problem. Just as all ideas from a brainstorming session should be evaluated in Phase II, all pictographs should be fully evaluated following the generation phase.

One class was given the following problem to solve using a pictostorming exercise: a picnic is scheduled for a group of about 30 people, but the weather report for the day of the picnic calls for rain. Using a sketch without a written or spoken explanation, provide ideas for keeping the group dry. The ideas included sketches of a large umbrella (three entries), a circus tent (two entries), a large fan that blows the raindrops to the side, an extremely large Canadian goose with its wing spread over the picnic area to cover the group, a shower cap, and the option of letting the group get wet but with a prize for the person who was most affected by the rain. The amount of creativity obviously varied. What could the group do with the goose idea?

The two greatest inhibitors to pictostorming are the attitudes of the participants toward the use of artwork to solve a problem and the time required to solve a problem. A brainwriting or brainstorming list would likely be completed in much less time than the completion of a set of pictographs. But a picture might be worth a 1,000 ideas on a brainstorming list. Obviously, those who doubt the merits of pictostorming will act as creativity inhibitors. A lack of enthusiasm by the facilitator could also be an inhibitor.

The assessment criteria for a pictostorming session are much the same as for any brainstorming session. Fluency (i.e., the quantity of ideas) would be an important criterion, as it increases the chance of useful pictographs being generated. The clarity of the artwork is generally not an important assessment criterion; however, the clarity of intent, but not the clarity of the artwork, should be used in assessing the idea and for transforming a pictograph into a solution of the problem. For group sessions, the diversity of imagination among the individual ideas and the seriousness of the individual participants should be used as assessment criteria. The facilitator should be assessed using the same criteria as used for conducting a group brainstorming session. Some of the other criteria in Table 8.1 may be used in making a comprehensive assessment.

A variant of pictostorming can be useful for generating divergent ideas. Problem-solving begins with a recognition that a problem exists. The stakeholder provides a concise but definitive statement of the problem. The facilitator assembles the group. Instead of immediately brainstorming on the problem statement, the facilitator asks the group to develop an image or caricature of the problem statement. The pictorial representation of the problem statement is then used as a basis for brainstorming or brainwriting. The idea behind this approach is that the image may provide characteristics of the problem statement that would not be recognized from the statement of words. This approach could also be used by a facilitator as a tool to overcome quiet periods during a brainstorming or synectics activity.

3.5   BRAIN DIAGRAMMING

Some people are visual learners. They may feel more creative dealing with a problem when a visual portrayal is used rather than being constrained by the use of words. Brain diagramming is a method that combines logical thinking and imaginative thinking. The purpose is to solve a problem by augmenting a free-form diagram with brainstorming. The diagram takes the form of an octopus, with its head in the middle of the diagram and multiple legs emanating from the head. The main issue can be enclosed in a geometric figure, such as a circle or rectangle. Each of the legs, or rays, represents a principal factor related to the issue. Shorter rays, which can be referred to as secondary rays, can emanate from each of the principal rays that extend out from the geometric figure, with the smaller rays being used to show specific aspects of the principal ray to which they are attached.

Consider the case of a brain diagram for the redesign of an office desk. The main issue is shown as a sketch of the actual object, i.e., a desk. Rays are drawn from each part of a desk for analysis, with each ray indicating a part of the desk that could be improved (e.g., drawers, top, legs). Other principal rays could be added such as one for the unused areas under the bottom drawers; even the area immediately above the desk user could be studied. The diagram would use secondary rays to allow for the inclusion of accessories, e.g., if a primary ray is used to represent the top drawer, then a paper clip bin in the drawer could be a secondary ray. If the image of the desk had not been shown, this idea may not have been considered.

Brain diagramming has several positive merits. The primary benefit of the diagram is that the individual parts of the object become individual focal points rather than having just one focal point, such as if the problem was presented as a verbal statement, i.e., redesign a desk. Brain diagramming encourages a systematic analysis, with imaginative thinking systematically applied to the individual parts. The visual image may help a person recall past frustrations with some aspect of any of the rays. Personal experience can be a good teacher!

3.6   CONCLUDING COMMENTS

A diverse set of topics have been addressed in this chapter, including brainstorming, brainwriting, pictostorming, and brain diagramming.

3.6.1   FAILURE OF BRAINSTORMING SESSIONS

When a brainstorming session has been successful, the facilitator is usually given primary credit; however, the facilitator gets the blame if the session does not succeed. When a group is blamed for a poor performance, it may be the result of self and/or group inhibition. Negative attitudes, conflicts between participants, and an organizational culture that fails to promote and reward imaginative thinking can be responsible for failure. Thus, a facilitator needs to be fully aware of the wide array of potentially inhibiting factors and be prepared to overcome them while the session is in progress.

In Chapter 13, several types of thinking will be discussed. Logical thinking is considered the standard type that most people try to use. When we think of brainstorming and brainwriting, we assume that logical thinking will be sidelined, but not eliminated, and replaced with some of the imagination-oriented thinking types. A brainstorming session will be most effective if all types of thinking are applied. Each type of thinking will place the problem in a different perspective, thus providing a more diverse list. A facilitator will be most successful when he or she prods the group for ideas related to the different types of thinking. This can be an effective strategy for periods when the group is struggling for generating ideas. The facilitator can ask for ideas using a specific thinking type, such as counterfactual or stochastic thinking.

It is difficult to pre-set the length of time of a group brainstorming session. Yet, a pre-set length of time is important because the failure to set a duration can lead to an early discontinuance of Phase I, which can lead to an inadequate set of ideas. Setting a time limit to brainwrite is also more effective because the writer will likely want to use all of the specified time, thus using the time to generate novel ideas for the list. The facilitator can discontinue Phase I or extend its length if the actions of the group suggest that a change is warranted.

3.6.2 BRAINWRITING OUTLINES

Most people gravitate into positions where some writing is an important part of the job description. The responsibilities could include reports, job summaries, completing forms, job performance reports, and many others. Brainwriting can be very useful to a leader and, therefore, it is likely a part of all aspects of leadership. Good writers begin with a list of issues to include and then turn the list into an outline. For any particular session, a person may go through several outlines, with each outline being progressively more detailed than the previous outline. Then the actual writing can begin. The end product will be better if the assignment starts with a brainwriting exercise to ensure that all of the important points are addressed.

What is the primary inhibitor to learning to use brainwriting? Some might believe that brainwriting wastes time. If the responsibility is to write an essay, they believe that it is easier to just start writing the first paragraph and continue on with the essay. Actually, that approach is inefficient, as it leads to a disorganized first draft and more revisions are needed to achieve a final product. Initially, relying on the brainwriting of an outline will lead to greater efficiency and a better end product.

3.6.3 IMAGINATIVE IDEA VISUALIZATION

A group of methods that are intended to inspire creative thought are referred to as methods of visualization. The basis for visualization is that it is supposed to improve memory, simplify complex information, and add a dimension of visual effects. It is supposedly best for visual learners. Two visualization methods were discussed, specifically pictostorming and brain diagramming. It should be evident that these two methods would appear to be applicable to different types of problems.

The pictostorming is most appropriate where the artwork would help inspire thoughts that would be more creative than a simple brainstorming session. The brain diagramming method would be more appropriate where the problem can be associated with a number of distinct attributes. Brain diagramming improves organization by focusing on the individual components of the problem. Other visualization methods are available, including morphological analysis, mind mapping, visual thinking, and brain graphing. Each of the methods is most appropriate for a specific type of problem. A mini-brain visualization activity could be included as part of a brainstorming session as a means of diverting the attention of participants during quiet periods.

3.7   EXERCISES

3.1. Develop two piggybacked ideas for each of the following: (a) when brainstorming on ways to peel a banana, one idea is to play sexy music to it and make the banana strip. (b) When brainstorming on the ways to cross a river, one idea is to have a Pogo stick that freezes the water just before the Pogo stick hits a portion of the river. Try to generate imaginative ideas.

3.2. Brainstorming on ways of crossing a river.

3.3. Brainstorm on the creation of a new religion.

3.4. Brainstorm on the use of plastic bags.

3.5. Develop a more specific set of rules for brainstorming than the list provided in Section 3.2.2.

3.6. Discuss efficiency (Eq. 2.2) relative to the second phase of brainstorming (i.e., idea evaluation). What factors would influence the efficiency?

3.7. Why is criticism of ideas during Phase I of a brainstorming session discouraged?

3.8. Why are wild-and-crazy ideas promoted in a brainstorming session?

3.9. Brainstorming based on imaginative thinking seems counterproductive to advancing knowledge. Discuss this seemingly wrong thinking.

3.10. What factors should be used in assembling a brainstorming group?

3.11. What factors contribute to a facilitator doing a poor job of conducting a brainstorming session?

3.12. Can a facilitator's body language influence the success or failure of a brainstorming session? Explain.

3.13. Using all of the rules of a brainstorming session, develop a three-tiered model of the rules for facilitating brainstorming sessions.

3.14. Identify ways that a facilitator can use to discourage criticism of ideas during a group brainstorming session?

3.15. Identify and provide a brief summary of three new methods for generating ideas.

3.16. What factors limit the ability of a person to be creative?

3.17. Humor is one of the most creative of the arts. Draw a picture of Goldilocks and the three bears. (a) Brainstorm on what Papa Bear is saying to Goldilocks. (b) Brainstorm captions of Goldilocks comment to Baby Bear. For both cases, the responses can be of any subject, such as politics, getting vaccinated, and religion.

3.18. Humor is one of the most creative of the arts. Identify your least favorite politician. Create a caricature of the politician. Brainstorm on captions that indicate the politician's most recent campaign issue.

3.19. Create a funny story in 50 words or less.

3.20. Bill meets Monica in a fast food restaurant. What would he say to her? Develop a list of ideas.

3.21. What are the advantages and disadvantages of brainwriting?

3.22. What factors would be appropriate for setting the time duration of a brainwriting session?

3.23. What factors should be considered in selecting the best location to hold an individual brainwriting session?

3.24. Brainwrite a list of creative ideas on ways to cut toenails.

3.25. Brainwrite on ways that the Earth could be shielded from the rays of the Sun to combat the rise of the Earth's temperature. Avoid ideas that are commonly cited by the news media.

3.26. Discuss inhibitors to the successful use of brainwriting.

3.27. What values are relevant to brainwriting?

3.28. Brainwrite on functional forms that could be used in modeling increases in world temperature with time. For example, a linear function is possible.

3.29. Brainwrite on creative ways of coring an apple.

3.30. What assessment criteria would be most useful for an individual assessing his or her own brainwriting session?

3.31. Brainwrite on the content and layout of the perfect dorm room.

3.32. How can the facilitator of a pictostorming session assist in the clarity of the pictorgraphs?

3.33. If pictostorming takes longer per idea than the idea generation in brainstorming, why could pictostorming be beneficial?

3.34. Identify and discuss inhibitors to pictostorming.

3.35. Define the word imagination. How can a person improve his or her imagination.

3.8 ACTIVITIES

3.8.1 ACTIVITY 3A: BRAINTSUNAMI

Major storms are characterized by randomness, uncertainty, and uncommon damage. Other types of natural hazards occur and have different types of damages. The purpose of brainstorming is to use mental powers to generate ideas in a manner that is similar to major storms. Using this idea, create an idea generation method based on the conditions of a tsunami.

3.8.2 ACTIVITY 3B: HERE COMES THE JUDGE!

Brainstorming sessions vary in effectiveness, so they should receive a post-session assessment using criteria that can judge the effectiveness. Develop a program for the assessment of a brainstorming activity. Discuss how each factor is rated and the way that an overall assessment will be made.

3.8.3   ACTIVITY 3C: WHO'S THE BOSS!

Identify all of the factors that you would use in the selection of a facilitator for a brainstorming session. Provide a quantitative measure of importance of each factor and a decision criterion for selecting the best candidate.

3.8.4   ACTIVITY 3D: A FIVE-STAR RATING

Identify and rank all factors that prevent brainstorming sessions from being successful and ways that each factor can be prevented. Discuss the ways that each of the problems can be overcome.

3.8.9   ACTIVITY 3E: MUD

Microplastics are polluting our water bodies. A proposal is made by your company president to develop a swimming drone that could roam through a water body from small ponds to the ocean and intercept microplastic objects. A detector would determine if an intercepted object qualifies to be retained. If the drone senses that it is made of a material that should be captured, the microplastic underwater drone (MUD) inhales the microplastic and decomposes it so that it will not pollute the water body. To make the project cost-effective, the president indicates that your brainstorming committee should find additional underwater jobs for the drone.

Brainstorm on the issue and identify new activities for the drone.

3.8.10   ACTIVITY 3F: MAKING THE IMPOSSIBLE, POSSIBLE

Select a technical topic that interests you, but one about which you are not familiar with the many commonly recommended solutions. Topics such as the following could be used: damaging mud slides, control of large area forest fires, space debris, the plight of polar bears, or excessive deaths of bees. Allow a 20-minute brainwriting session to generate solutions to the problem.

3.8.11   ACTIVITY 3G: PUT YOUR HEADS TOGETHER

Develop a set of rules for holding a group brainwriting session. Identify the ways that a facilitator would lead the session.

3.8.12   ACTIVITY 3H: SAVE THE BIRDS

Over the next few decades, we can expect significant changes to biodiversity even if humans do not change their current lifestyles and practices. Human populations are expected to increase, with CO_2 amounts also increasing. Forested areas will decrease in size along with non-human populations of the forests. Some species will become extinct. The health of marine ecosystems will decline, especially that of large predators, sharks, and whales. Regions will experience quite different changes.

First, brainwrite a list of three research topics that you believe could lead to invest-ments in the state of knowledge of biodiversity. Identify your research objectives for each topic and briefly describe the types of analyses that you would conduct.

3.8.13 ACTIVITY 3I: I CAN SEE ALL OF THE WAY TO THE BOTTOM

Go online and find basic information about the Secchi disk, which is a disk that is used for measuring water transparency or turbidity. After reading the material, brain-write a list of ideas on other ways of measuring water transparency. Wild-and-crazy ideas are encouraged.

3.8.14 ACTIVITY 3J: DE BONO'S APPLE

Identify a way that de Bono's method (see Section 2.5) can be used for brainwriting. Discuss the role of each hat. For each hat, provide two or more possible answers related to the issue of peeling an apple.

3.8.15 ACTIVITY 3K: SUMMER JOB

Use brain diagramming to develop a brain diagram for use in providing guidance to the problem of preparing to find a summer job. This might include principal rays such as prepare the resume and identify firms.

3.8.16 ACTIVITY 3L: GIVE ME SOME MONEY

Use brain diagramming to develop a brain diagram for use in providing guidance to the problem of fund raising for a community center rehabilitation. This might include principal rays such as prepare announcements and compose a list of potential donors.

3.8.17 ACTIVITY 3M: MERRY-GO-ROUND

This is a small group brainwriting activity that shows problem-solving, team work, and planning. There must be absolutely no talking between the group members before or during play. Each person in the group is given a small sheet of paper, maybe one-sixth of a regular sheet. The objective is to create the longest word on each sheet of paper. Each person in the group writes a letter on the top left-hand side of their piece of paper. When each person in the group has written one letter on their sheet of paper, the papers are passed counterclockwise to the next team member; that person writes the second letter and then when all group members have finished their letter, the sheets are passed counterclockwise. This continues until adding a letter will cre-ate a string of letters that is not a word. The longest word is the winner.

3.8.18 ACTIVITY 3N: THE INVISIBLE FLY

Part I: Brainstorm on naming a new superhero. Record the length of time spent on creating the list.

Part II: Review the list of Part I and place a check mark next to all of those that relate to the name of a known superhero. Compute the number of piggybacked ideas as a fraction of the ideas.

Part III: For the same length of time devoted to Part I, develop a new list that avoids piggybacking items. Compare the number in the list to the number of *un*checked items in the Part I list. Discuss the meaning of the results.

4 Synectics
Analogical Brainstorming

CHAPTER GOAL

To introduce synectics, which is an imaginative idea generation method that improves the thinking process by making use of the participants' general knowledge.

CHAPTER OBJECTIVES

1. To present the procedure and benefits of using the synectics method.
2. To discuss ways of conducting either group or individual synectics activities.
3. To compare synectics and brainstorming.
4. To present a way of developing a synectics topic for a given problem.

4.1 INTRODUCTION

Synectics is essentially analogical brainstorming. It is generally a group activity that is operated like a brainstorming session, but where the group does not know the real problem. Only the facilitator knows the real problem. Instead, the synectics group is given a different topic, which is referred to as the *synectics problem* or *synectics topic*, which is similar in some ways to the real problem but different in other ways. The synectics topic often centers on a more common event, one with which the group would be more familiar than they would be with the real problem. The greater familiarity is a ploy to increase the breadth and depth of the ideas generated. For example, the real problem might be to hold a professional meeting with potential clients; you want your group to identify ways of making the meeting more effective. One possible synectics topic would be to brainstorm on holding a kid's birthday party. The kid's birthday party is similar to the professional meeting in that both are meetings of peers, both have objectives, and both will be held for approximately the same length of time. Yes, differences between the two "meetings" are obvious. Since a kid's birthday party is less formal than a meeting of adult professionals, the participants of the synectics session might be more willing to generate wild-and-crazy ideas, with the thought that better ideas may be generated when using imaginative thinking. Additionally, the participants of the synectics session are more likely familiar with children's parties, so novel ideas will more readily come to mind. Note that the two

DOI: 10.1201/9781003380443-4

parties have elements in common, but differ in one or more elements, specifically the nature of the audience and the specific objective.

Users of synectics usually generate a long list of ideas, many of which are imaginative ideas. To achieve this, the following questions need to be addressed:

- How are synectics sessions conducted?
- How is a synectics topic selected?
- How can a facilitator be most effective?
- What are the benefits of using synectics?

Answers to these questions should enable a person to effectively plan and conduct a synectics session.

4.2   THE SYNECTICS PROCEDURE

Synectics is an idea generation method that replaces the underlying problem with a more common, more understandable setting, which enables the problem to be internalized through an analogy. A synectics session is conducted in two phases. Phase I is the brainstorming session using the synectics topic. Phase II is the idea evaluation activity in which the items on the Phase I list are evaluated, which is much like Phase II of a brainstorming session; however, in a synectics session, Phase II requires both the transition from the synectics topic to the real problem and the transformation of the ideas into a solution. When solving a problem with brainstorming, the synectics part of transitioning is not needed as the idea generation is done directly on the real problem.

Synectics is an idea-generating method in which the problem is stated in analogous terms, such that the participants in the group will not recognize the real problem that needs to be solved. For example, assume that the real problem is to decide on the complexity of a proposed climate change model. For the Phase I activity, a brainstorming session might be held on the needed complexity of a new board game. Only in Phase II would the group learn the nature of the real problem. The ideas developed based on the proposed board game would then be evaluated in the second phase relevant to the climate change model.

4.3   DEVELOPMENT OF A SYNECTICS TOPIC

Synectics and brainstorming use the same two phases: idea generation and idea evaluation. The development of the synectics topic is an important step in the process. A poorly stated topic can mislead the participants, especially if the synectics topic is too dissimilar from the real problem in the supposedly similar concepts. Then the group might struggle to connect the ideas generated in Phase I and the original problem discussed in Phase II. The facilitator must ensure that the synectics topic will properly direct the generation of ideas toward the real problem.

In an idea generation session, the breadth of generated ideas is important. Encouraging wild-and-crazy ideas in Phase I will usually enable the participants to justify thinking beyond practical solutions, which is an important reason that a synectics topic is used to generate the list of ideas rather than brainstorming on the real problem. If the participants knew the real problem, then some of them might believe that out-of-the-box ideas are a waste of time, so they would only state ideas that have a direct connection to the real problem. This constrained type of thinking is counter to the philosophy of imaginative problem-solving and can stymie the identification of novel solutions to a problem. Ultimately, a synectics-based effort may yield more novel ideas than ideas generated through brainstorming on the real problem. Table 4.1 gives several other examples of a specific problem and the corresponding synectics topic.

A real-world example might assist in understanding the benefits of using the synectics topic. Assume that the problem is to clean up debris in space, much of which are the pieces of old satellites that have disintegrated while in orbit. This problem is a real-life concern as the debris is traveling at high speeds and could cause damage to functioning satellites. The facilitator of a synectics session might use a synectics topic such as the gathering up of leaves that are scattered about the lawn or cleaning up trash on a public beach. Note that both of the two synectics topics address the general problem of assembling items spread about an area, but they differ in the seriousness of the issue and the level of technical knowledge that the participants would need to know about the material, i.e., space debris versus leaves or trash. Participants would certainly be more familiar with the trash and leaves problems, so they would likely generate more ideas since they can visualize the trash and the leaves. For the space debris problem, the participants might need to know some technical details such as the sizes of the pieces, the velocities at which the pieces were traveling, and the altitudes of the satellites and the debris. Use of the synectics topic eliminates the need for such technical knowledge but maintains the collecting of materials aspect of the real problem.

Another example can illustrate the idea of using an analogy or metaphor as a synectics topic. For example, a leader who has an organizational planning problem may want to have his or her staff plan the baptism of a pet rather than focus on the specific planning problem of the organization. A baptism even of a pet may inspire a greater number of ideas because baptisms are more familiar situations than is the

---

TABLE 4 1

Actual Problems (Phase II) and the Corresponding Synectics Topics (Phase I)

Problem Statement	Synectics Topic
Keep trash from entering a reservoir	Prevent leaves from getting under the azaleas
Create a new bottle cap	Design a new roof
Design a new food blender	Cutting bushes
Improve toll booth collection efficiency	Efficiently accumulate objects
Improve urban flood protection	Stop an overflowing toilet
Reduce air pollution	Prevent tear gas discharges
Reduce air pollution	Reduce the number of mosquitoes

actual planning problem of an organizational meeting; also, a humorous element is involved, as pets are not normally baptized. Since the object of the baptism is a pet rather than a person, the obviously irrational situation would encourage a greater emotional involvement than would a brainstorming session that is centered on the actual organizational planning problem. The two planning problems have similarities, which is a positive. Specifically, both cases involve the need to plan and both involve the assembling of groups. While the organization meeting is for humans, the synectics group topic involves a pet. Suggesting that a band made up of dogs might be a synectics idea that would lead to the thought of peaceful music to temper contentious parts of the planning meeting. The familiarity of the baptism of humans would enable the group to appreciate activities that are associated with baptisms. A greater number of ideas would likely be generated for the synectics topic than there would be for the real problem. In Phase II, the synectics ideas for a pet's baptism can be transferred to the process of solving the organizational planning problem.

To develop a synectics topic, the facilitator should identify a common event, thing, or idea that has characteristics that are similar to those of the real problem. Preferably, it should be a topic with which the participants have some familiarity. The group will be able to generate more ideas if the topic relates to something about which they have experience. Ideally, the synectics topic that the facilitator uses is somewhat lighthearted, so that the group will not generate ideas that are based only on logical thinking. Instead, the participants will use their emotions and be predisposed to generating novel ideas.

4.4   ROLE OF THE FACILITATOR

Like a brainstorming session, a synectics session is led by a facilitator who is usually selected by the stakeholders. The selection of the facilitator is a very important decision, as the quality of the output partly depends on the facilitator's skills and leadership. Many of the roles of the facilitator of a brainstorming session are relevant to the facilitation of synectics activities. The facilitator should have some broad technical knowledge of the problem being investigated, but detailed knowledge should be the responsibility of the group of participants. It is important for the facilitator to have had past experiences as a facilitator. The person selected to be the facilitator should be free of biases toward any of the participants, any group of participants, or the subject of the problem. For example, if the participants are drawn from two competing divisions of a company, the facilitator should not be aligned with one of the two divisions.

The facilitator is possibly the most important person who is involved in a synectics activity. He or she must ensure that the session is properly conducted. The facilitator's responsibilities include:

1. To meet with the stakeholders to ensure that the problem statement is an accurate description of the real problem.
2. To select the synectics topic.
3. To develop the performance criteria that will be used to rate the quality of each generated idea.

4. To develop the decision criterion that will be used to rank the ideas, such that the best alternative can be identified.
5. To decide on the time durations allotted to each phase.
6. To select the participants who will be generating the ideas and assisting in evaluating each idea.
7. To serve as leader of the synectics session.
8. To write the report that identifies the best alternative.

The session could fail if any of these responsibilities are not properly addressed.

Initially, the facilitator knows the problem for which a solution is needed. The facilitator then selects the synectics topic on which the Phase I brainstorming will be done. The facilitator does not make the group aware of the real problem. In Phase I, the group takes all of the allotted time to brainstorm on the synectics topic and generate a list of ideas. Once the facilitator is satisfied with the outcome of Phase I and a good list has been generated, the facilitator begins Phase II by telling the group the actual problem. During Phase II each of the Phase I ideas is evaluated using the performance criteria and transformed into a possible solution of the actual problem. Of course, additional ideas can be added to the list during Phase II. Each idea is evaluated on the basis of pre-selected performance criteria. Using some pre-specified decision criterion, the best solution should be selected from the Phase II evaluation results.

4.5   ADVANTAGES OF SYNECTICS

Synectics is an idea generation method that has the following advantages:

1. The synectics topic is more familiar to the participants than the real problem; the connection between the real problem and the synectics topic is not immediately obvious, but it will be in Phase II. Therefore, all (or most) of the participants can relate to the synectics topic.
2. The synectics topic should be less sophisticated and more recognizable to the group; therefore, it is acceptable for the group to generate silly ideas.
3. A synectics topic should be more entertaining than a brainstorming topic that describes the actual technical problem.
4. The difference between the problem and the synectics topic might be based on age (e.g., youth vs. adult); personal vs. professional or job related; time (e.g., ancient or current or future); economics (e.g., rich or poor); source of knowledge (e.g., technical vs. personal); or any general characteristic that would be common to both issues.

These advantages should lead to a better solution than if the brainstorming method were applied directly. Thus, the output is of higher quality, which by definition indicates a higher efficiency. Brainstorming on the synectics topic lessens the stress in generating ideas, which reduces the input part of the efficiency equation. Thus, the numerator increases with the denominator decreasing, both of which cause the efficiency to increase.

4.6   INDIVIDUAL SYNECTICS

It may seem that synectics can only be applied in a group format, as the real problem is assumed to be hidden to the group and only known to the facilitator. However, an individual could capture the advantages of the synectics method by developing one or more synectics topics and brainwriting on them rather than the real problem. This practice would encourage broader, more imaginative thinking than just the practice of brainwriting on the real problem. Even though the person knows the real problem, he or she would still be more likely to generate novel ideas if the synectics topic is used to develop the list. If a person finds the synectics topic selected does not provide good ideas, then a second synectics topic should be selected to continue the idea generation. Individual synectics should be more efficient and effective than brainstorming because it approaches problem-solving with a broader perspective.

4.7   OVERT SYNECTICS

The distinguishing characteristic of synectics is that the brainstorming is performed on a seemingly unrelated problem, i.e., the synectics topic. The participants of the group do not know the real problem. This characteristic has numerous benefits and is the primary reason for using synectics.

In cases where the problem-solving group knows the essence of the real problem, the synectics topic may actually be distracting, but in a beneficial way. It may be better to openly hide the real problem such that the group knows the nature of the real problem, but still adopts a more emotional state and generates ideas on a synectics topic. Thus, the overt synectics problem-solving approach attempts to force the group's attention away from the real problem, the essence of which they recognize. Overt synectics, which is intended to encourage the generation of ideas to understand the problem, involves three phases rather than the two typically used with an application of the synectics method.

The following summarizes the three phases of the overt synectics method:

- *Phase I*
  - a. Identify the true problem.
  - b. Using only wild-and-crazy ideas, brainstorm a list of solutions to the problem by saying:
    The [*problem*] is like _____.
    where the statement of the real problem is included in the statement brackets, and the overt synectics topic is entered on the line.
- *Phase II*
  - a. Select the most bizarre response in Step Ib.
  - b. Identify the characteristics of the components/elements of the response of Step IIa.
- *Phase III*
  - a. Relate each characteristic of Step IIb to the problem statement of Step Ia.
  - b. Identify the solutions to the original problem by coordinating the characteristics of Step IIIa.

A possible follow-up activity that can be used to enhance the solution would be to select the second most bizarre idea of Step 1b and perform Phases II and III. Then merge the new ideas with the ideas from the earlier application of phase III.

The overt synectics procedure can be placed in a three-column table, with a column for each phase. The headings for the columns would be: (1) "The problem is like"; (2) "The characteristics of the [insert the phase I item] are"; and (3) "the characteristics relate to the [real problem] by." Table 4.2 provides an example of an overt synectics exercise where the real problem is to identify components of board games to use in developing a new board game. Obviously, an alternative to the overt synectics exercise would be to just list common board games and try to identify the fundamental components of each. Note the variation of the craziness of the six items of Column 1. The first entry ranks low on the craziness scale, while the second and third entries have less obvious connections to the problem of board game design, which can be a beneficial trait.

Overt synectics is nothing more than a synectics application where the group knows the essence of the real problem and the facilitator fears that this may prevent imaginative thinking if the group brainstormed on the real problem. To circumvent this concern, a very bizarre synectics topic is selected in an attempt to distract the participants from thinking about the real problem. By overtly indicating the real problem and developing a bizarre synectics topic, the group has to focus on an unusual problem. Basically, Column 1 is the development of the synectics topic, while Columns 2 and 3 correspond to Phases I and II of the regular synectics method.

TABLE 4.2
Example of an Overt Synectics Problem

Designing a New Board Game Is Like ...	The Characteristics/ Components of [a Snowman Are ...]	The Components of Building a Snowman Relate to Game Board Design By
• Designing a building	1. Have a solid foundation	1. Educational venue; risk of failure; fun activities; good competition
• Making a snowman	2. Plan on its size	2. Number of players; length of play time
• Exploring Mars on a robot	3. Pack it tightly so that it does not fall apart	3. Maintain player interest; challenging game board; ensure rules are not conflicting
• Selecting a new coach for your college football team	4. Add accessories (e.g., carrot for a nose)	4. Use hotels and houses as game pieces; specify types of prizes; include weapons as clues
• Designing an itinerary for a round-the-world trip.	5. Use a good snow/water ratio	5. Well-designed board; fun materials and game pieces; appropriate complexity; will it melt (can it be played multiple times)?
• Developing a recipe for the use of road-killed animals	6. Make it aesthetically pleasing	6. Colorful game board; good box for marketing; easy to read directions

The use of Phase I of the overt synectics method forces the group to engage in divergent thinking.

4.8 CONCLUDING COMMENTS

Some problem-solving methods, such as synectics and brainstorming, are more often associated with creative thinking than with the facts-based thinking methods such as logical thinking. Synectics almost forces the thinking away from the current knowledge and encourages the use of imaginative thinking to solve a problem. For some types of problem-solving, this divergence from logical thinking may lead to a better result. Allowing imaginative thinking to dominate the divergent phase of the analysis is best when the participants have attitudes that are open to imaginative problem-solving. Individuals who are self-inhibitors should not be invited to participate. Therefore, when selecting the participants who will make up a brainstorming or synectics group, the thinking orientation of the candidates who will be solving a specific problem should be considered. Creative problem-solving can be successful only when the participants have attitude characteristics that are inclined toward the imagination. Lacking this characteristic can act as a creativity inhibitor.

Problems experienced by researchers are viewed by some as matters that can be solved using just logical thinking. Emotional involvement in developing a solution is often viewed as not helpful. This anti-affective belief can actually suppress good ideas. While any final solution may have a strong cognitive basis, the identification of alternative solutions may be more productive when affective thinking processes are included with the logical thinking process in the Phase I idea-generating activity. To stress the importance of imaginative thinking, the facilitator may start the session by proposing a few wild-and-crazy ideas.

Synectics is an idea generation method in which emotional and experiential diversity are central to the problem-solving. In synectics, the participants are provided with an analogical activity that encourages psychological processes based on emotion to operate while allowing traditional logical thinking to continue. An emotions-based atmosphere can be very conducive to imaginative thinking and idea innovation. Such an atmosphere enables a variety of thinking methods to be employed in the generation of new ideas. Thus, the facilitator should attempt to obtain facilities that are conducive to imaginative idea generation. Poor facilities can actually act as an inhibitor to the session.

4.9 EXERCISES

4.1. Look up the prefix *syn* in a dictionary. How is synectics like the use of synonyms as alternative words?

4.2. Provide definitions for the words *metaphor* and *analogy*. Discuss their connections to the synectics method.

4.3. Create a detailed list of instructions for holding a synectics session including the selection of both the facilitator and the synectics topic. This would likely involve multiple phases.

4.4. The objective of this problem is to identify synectics topics for the problem of storing flood waters within a watershed. List five wild-and-crazy synectics topics that could be used to generate ideas for the stated problem.

4.5. Identify five imaginative synectics topics to use for the problem of preventing people from slipping on icy walkaways.

4.6. Identify five imaginative synectics topics to use for eliminating a cockroach problem in a college dorm.

4.7. Develop five wild-and-crazy synectics topics for the real problem of preventing trash from blowing around the neighborhood.

4.8. Develop three synectics topics for the real problem of getting people to get vaccinated for COVID.

4.9. Develop an assessment score sheet for use in rating the performance of a facilitator of a synectics session.

4.10. How would assessing the facilitator's performance of a synectics activity differ from the assessment of the performance of a facilitator of a brainstorming session?

4.11. Compare/contrast synectics and divergent-convergent thinking (see Section 13.6).

4.12. Identify actions that would inhibit the effectiveness of a synectics session.

4.13. Discuss the meaning of the numerator and denominator of Eq. 2.4 for a synectics activity.

4.14. What are the advantages and disadvantages of synectics?

4.15. Why is it easier to brainstorm on a topic based on an everyday activity than to brainstorm on a complex technical topic that has some connection to the everyday activity?

4.16. Use the sixth entry "reduce air pollution" of Column 1 of Table 4.2 to complete the analysis for designing a new board game. Compare the resulting characteristics to those identified in Column 3 of Table 4.2.

4.17. Brainstorm a list of ten ideas on the baptism of a cat (see Section 4.2). Then select the wildest idea from the list and develop a solution to the organizational planning problem.

4.18. Develop a list of ten synectics topics if the problem is the need to hire creative people into a company.

4.10 ACTIVITIES

4.10.1 ACTIVITY 4A: LEMONADE SALES

Part I: DO NOT READ PART 2 UNTIL PART 1 HAS BEEN COMPLETED
This activity provides the group with the opportunity to experience the two phases of synectics. Generate a list of about 15 ways that an eight-year-old could increase sales at his or her lemonade stand.

Part II: After generating the list of Part 1 for the eight-year-old's case apply each item on the list to the problem of increasing activity on an electronic sports betting site. After completing Phase II, list ways that the two activities are similar and ways that they differ.

4.10.2   ACTIVITY 4B: IS TWO BETTER THAN ONE?

For one important project, an organization assembles a very large group to hold
a synectics session. The problem to be solved is complex with a wide variety of
disciplines needed. The facilitator decides that the group is too large to operate an
effective synectics session, so the decision is made to separate the group and hold two
sessions. Develop a plan for conducting this double synectics activity. Also identify
the advantages and disadvantages of the multi-group plan when compared to the
original plan of having one large group. This activity provides experience in creating.

4.10.3   ACTIVITY 4C: OVER-POPULATION

Some people believe that increases in the human population during the next 25 years
will cause significant environmental damage to the Earth. You are selected to be the
facilitator to a synectics group that will try to identify a solution to the problem. You
need to develop a synectics topic on which the group can brainstorm. Brainwrite a
list of potential synectics topics that could be used for the actual problem of over-
population. If this is a group activity to identify the Phase I synectics topic, then
brainstorm rather than brainwrite. After generating the list, develop a performance
criterion to evaluate each one; use a quantitative criterion.

4.10.4   ACTIVITY 4D: FLOATING DOWN A RIVER WITHOUT A BOAT

An environmental safety group in your region has found excessive concentrations
of pharmaceutical drugs in local streams and water bodies. The group assumes that
this is the result of people discarding unused drugs. Since studies show that excessive
amounts will cause damage to aquatic life, they are going to search for a solution.
You are going to be a facilitator for a synectics topic on which the group can brainstorm. Brainwrite a list of potential synectics topics that could be used for the actual
problem of excessive concentrations of pharmaceutical drugs in streams. If this is
a group activity to identify the Phase I synectics topic, then brainstorm rather than
brainwriting. After generating the list, develop a performance criterion to evaluate
each one; use a quantitative criterion.

4.10.5   ACTIVITY 4E: SMOKEY THE BEAR

Generate a list of ten synectics topics that could be used for the actual problem of
preventing forest fires. Develop a performance criterion that can be used to assess
the quality of the list. Select five from the list and evaluate the potential value as a
solution to the original problem.

4.10.6   ACTIVITY 4F: THE STAR SPANGLED BANNER NO MORE

With respect to the national anthem, the problem is that the words are difficult for
young people to memorize or even understand. Assume that you are part of a con-
gressional committee with the task of developing a new national anthem. Identify ten

ideas that would be appropriate synectics topics to use in place of the real problem. Imaginative ideas are encouraged. The intent of this activity is to provide experience with imaginative thinking and creating.

4.10.7   Activity 4G: Oil May Be Messy

Use the overt synectics method to identify characteristics of methods for cleaning up oil-spill material deposited on ocean beaches. This activity illustrates imaginative thinking.

4.10.8   Activity 4H: Another Type of Test

A teacher of gifted students wants to develop a new type of test specifically for gifted students. Use the overt synectics method to identify the characteristics of the new test. This activity illustrates imaginative thinking.

4.10.9   Activity 4I: BINGO! We Have a Winner!

Part I: DO NOT READ PART II UNTIL PART I HAS BEEN COMPLETED.
   Assume that a company is having the problem of retaining corporate leadership and the board of directors believes that the loss of personnel is causing a decline in profits. Create a problem statement. Develop a list of synectics topics for the real problem.

Part II: The group decided that the wildest-and-craziest idea (Column 1 entry) was to overcome the problem of a decline in the number of people returning to the weekly BINGO night. Use the list to generate a list of possible solutions to the corporate leadership problem (which would be entries into Column 2).

Part III: Use the Column 2 entries of Part II and connect them with the original problem of Part I.

5 The Delphi Method

CHAPTER GOAL

To detail a procedure that extends the use of brainstorming to conditions where face-to-face sessions of group members are not necessary.

CHAPTER OBJECTIVES

1. To introduce the Delphi method for problem-solving.
2. To list the advantages and disadvantages of the Delphi method.
3. To show alternatives than can increase the efficiency of the Delphi method.
4. To outline the situations where the Delphi method is most useful.
5. To discuss the use of the Delphi method in research.

5.1 INTRODUCTION

Some modern businesses are reducing costs by having employees work remotely, which eliminates the need to occupy and pay for a business office. Simultaneously, the employees can reduce costs for travel and business clothing. Instead, the company can use a short-term rental office when a meeting is needed. However, business decisions are still necessary. If the leadership style is more democratic, rather than autocratic, the CEO may need or want a consensus of relevant employees on any decision. The Delphi method is an approach to decision-making where the employees are located in remote locations such as where relevant employees are in different cities or where the employees work from home. The Delphi method enables problems to be solved using all employees whose input is needed even though they are not on-site with each other. The Delphi method can also be used in cases where the employer has subgroups of employees who do not interact well because of conflicts; in such a case, face-to-face meetings may not be in the best interests of the organization even if the ZOOM option is available.

The Delphi method is a decision tool that uses concepts of both brainwriting and brainstorming. Additionally, it is an iterative method that integrates the idea generation and idea evaluation phases of brainstorming. The Delphi method is used to develop a consensus on an issue without the necessity of face-to-face meetings. The Delphi method is, in part, a brainwriting activity, but it differs from the traditional brainwriting approach in that it is a group activity.

DOI: 10.1201/9781003380443-5

5.2   DEFINITIONS

The Delphi method involves three groups: the stakeholders, the facilitator, and the participants. The stakeholders are the group of people or organization who recognize that a problem exists and that they will be responsible for acting on the final decision. The participants are the people with the socio-technical expertise that can generate and evaluate the alternative options. The facilitator manages the three phases and steps of the Delphi method and is responsible for coordinating the ideas generated by the participants. The facilitator is also responsible for interacting with stakeholders about concerns with the problem statement. Once the group reaches a consensus, the facilitator submits a report with the recommended decision to the stakeholders.

5.3   THE DELPHI METHOD PROCEDURE

Since the Delphi method is an iterative process, the steps are separated into the following four phases:

Phase I: The stakeholders compose a problem statement, select a facilitator, and provide the facilitator with the opportunity to evaluate the problem statement before it is distributed to the participants.

Phase II: The facilitator selects the participants and interacts with the participants to identify any concerns that they have with the problem statement.

Phase III: The facilitator and the participants iteratively work to develop a list of alternative solutions to the problem statement in order to develop a consensus on the best solution.

Phase IV: The facilitator submits the recommended solution to the stakeholders and files all necessary reports on the activity.

5.3.1   PHASE I

a. The first step is for the stakeholders to draft a precise statement of the problem. The issue on which a decision is needed must be stated in a very precise way in order to ensure that each participant interprets the problem statement in the same way and has a clear understanding of the needs of the stakeholders. Any restraints, limitations, or requirements should be included as part of the problem statement. Since the Delphi process does not involve participant face-to-face interactions, failure to have a very clear, precise statement of the problem can introduce inefficiency and lead to a less than optimum decision as the participants will essentially be responding to different problems due to their different interpretations of the problem statement.

b. The stakeholder(s) should identify the facilitator.

c. The facilitator has the option of returning the problem statement to the stakeholders and asking for clarifications. The facilitator must clearly indicate the concerns that the stakeholder(s) needs to address before the participants are asked to solve the problem.

d. The facilitator will establish the plan including setting both the performance criteria and the decision criteria.

5.3.2 PHASE II

a. The facilitator identifies the participants. As a group, the participants should collectively have knowledge of all dimensions of the problem, including knowledge related to the technical, economical, and societal dimensions.

b. The facilitator sends each participant a copy of the problem statement along with all supplementary documentation. The facilitator wants assurance that everyone has the same understanding of the problem. The facilitator requests from each participant a brainwritten list of issues with the problem statement that needs to be clarified prior to initiating the brainstorming in Phase III; this step is essentially an individual brainwriting activity for each participant for the purpose of ensuring that everyone truly understands the problem statement. A deadline for responding must be specified by the facilitator. At this stage, the participants are not responding with potential solutions to the problem, only a request for any needed clarification of the problem statement.

c. The participants have the opportunity to ask for clarification about the problem statement or comment on concerns with any stated constraints or requirements. The participants should brainwrite on the issue and return their comprehensive lists of points that need clarification to the facilitator by the deadline.

d. For the facilitator, the objective of this initial activity would be to generate a comprehensive list based on responses from all of the participants of all aspects of the problem statement that will need to be clarified. The facilitator submits the list and asks the stakeholders to respond to the concerns of the participants. All interactions in this phase are only intended to ensure that the problem statement is clearly understood by both the facilitator and all of the participants. If more than one round of this phase is needed, the facilitator will address the concerns or get additional feedback from the stakeholders.

e. Once all of the participants fully understand and accept the problem statement such that the facilitator does not need to get additional clarifications from the stakeholders, then Phase II is complete.

5.3.3 PHASE III

a. The facilitator distributes the most up-to-date statement of the problem to all participants and requests a set of responses from each participant. The participants are asked to brainwrite a comprehensive list of solutions to the problem and submit their lists by the due date. The facilitator must clearly specify the due date. This is the first round of Phase III.

b. When the participants have returned their individual lists of possible solutions, the facilitator should compile a complete list of all generated ideas received from all participants. Before sending the complete list of possible solutions to the participants, the facilitator may need to organize the collective responses, such as placing similar ideas near to each other in the list of all responses. The facilitator should send the compiled list to each participant and request an evaluation of each entry. If the group is small, the evaluations provided by the participants could be in prose form as a statement of reasons. The facilitator might want an objective assessment, possibly in the form of a Likert scale response for each of the solutions on the list. The facilitator would have established the performance criterion in Phase Id.

c. The participants evaluate all of the suggested ideas that are included on the list, prepare their evaluations, and return their responses to the facilitator in a timely manner. If wide disagreement on the evaluations exists, this phase may need to be repeated. The facilitator should have decided on the format in Phase Id.

d. After receiving the responses from the participants, the facilitator summarizes the responses and based on some pre-determined performance criterion reduces the list to those ideas that would potentially have strong support of a good number of the participants. The reduction in the length of the list is not an inconsequential step and can influence the final decision. The facilitator will need to justify the criterion that is used for eliminating options.

e. The facilitator then distributes a revised evaluation form that includes a reduced list of the current recommended solutions.

f. The participants are requested to rate the entries preferably using an objective rating system and return their ratings and comments to the facilitator.

g. The process of Steps IIIc–IIIf can be repeated, as necessary, until a consensus on the best solution is reached.

5.3.4 PHASE IV

a. When the facilitator decides that a consensus has been reached, which would be based on a pre-specified decision criterion, the facilitator writes a report and transmits the recommended decision to the stakeholders.

b. Shortly after submitting the consensus decision to this stakeholder, the facilitator must submit all required reports on the session. This might include assessment reports on the group as a whole and a separate report on each participant. A report that summarizes the details of the process should also be submitted.

5.4 FACILITATING A DELPHI SESSION

The facilitator is usually selected by the stakeholders. The facilitator should be someone who:

1. Is knowledgeable about all relevant dimensions, including the technical, economic, value, and resource issues.

2. Is knowledgeable about a wide array of methods of thinking, including creative thinking methods.
3. Is generally confident in his or her ability, and avoids counter-skills such as procrastination and pessimism.
4. Would act in an unbiased manner.
5. Is responsible for completing the process in a timely manner and will submit all necessary reports.
6. Will be able to serve without appearing to have a conflict of interest.

The facilitator is generally responsible for selecting the participants and should be someone who is respected by both the stakeholders and the participants. The facilitator sets the decision criteria and the due-dates for the submissions of the participants.

5.5 DECISION-MAKING WITH THE DELPHI METHOD

The speed with which a consensus will be reached will depend on the diversity of stakeholders' interests, the complexity of the problem, the number of participants, the ability and the professionalism of the facilitator, the pre-specified length of time allotted by the stakeholders to make a decision, and the performance criteria used to remove items from the list. For example, the facilitator may decide to remove all entries that have not received a score that exceeds the mean, i.e., 2.5 for a 1–4 Likert scale or 3.0 for a 1–5 Likert scale. The facilitator will also need to pre-specify the decision criterion, such as 90% of the participants agreeing to one of the final ideas. This criterion should be specified by the stakeholders in discussion with the facilitator in advance of the application of the Delphi method; it would be part of Phase I.

5.6 ADVANTAGES AND DISADVANTAGES OF THE DELPHI METHOD

As a group decision-making process, the Delphi method has the following advantages:

1. Face-to-face meetings are not required.
2. This is easily done by email, with the facilitator being the only person who knows the authors of the submittals.
3. If conducted by written ballot, the responses are anonymous.
4. Subjective opinions can be transformed to an objective scale.
5. Each individual's opinion has equal weight.
6. Participants have the time between submittals to think about the issues and possibly gain knowledge that would help in their decision-making.
7. Peer pressure that could occur in a brainstorming session is avoided.
8. Many inhibitors to creative idea generation are not relevant.
9. Introverts are able to respond.
10. Participants who have in the past been antagonistic to each other do not need to interact.
11. A large group can be accommodated.

12. It could be used to develop priorities before convening a smaller-scaled face-to-face meeting.

The Delphi method has the following disadvantages:

1. Piggybacking between individuals is difficult.
2. Voting is done without in-depth discussion.
3. The criterion on when to declare consensus may be subjective.
4. Multiple rounds, even by email, can take considerable time.
5. A facilitator may need to contend with participant procrastination when individual participants do not respond in a timely manner.
6. It is necessary for the facilitator to set rules for participants to contact each other.
7. Anonymity may not encourage broad thinking.
8. Greater effort required of the facilitator.

Like all methods of problem-solving, the Delphi method is most effective when it is applied in situations where its advantages are matched with its characteristics. For example, the Delphi method would likely be more effective in cases where unproductive conflicts would develop among the participants of a face-to-face session; this issue could occur when participants are from competing divisions of an organization. It is also a good method when face-to-face meetings are inconvenient, such as when participants are located in different parts of the country. If one person in the group tends to dominate face-to-face brainstorming sessions, use of the Delphi method allows naturally introverted types to actively participate without in-person interaction with the other participants. The facilitator has a responsibility to adhere to the rules. If the final decision influences the level of resources allocated to the groups affiliated with some of the participants, then face-to-face meetings could be contentious and hurt organizational morale; the avoidance of conflict makes the Delphi method a good option.

5.7   USE OF THE DELPHI METHOD FOR PROBLEM-SOLVING

Obviously, the Delphi method would be useful to research teams, rather than to individual researchers. However, it would be very applicable to teams even as small as two people, e.g., a student researcher and the advisor. In this case, the funding agency would essentially act in the role of the stakeholder, as it defined the problem. Also, the advisor would act as the facilitator, and the student would serve in the role of a participant. The student would be responsible for most of the day-to-day work and would submit a list of problems that arise to the advisor, as the facilitator. In an academic setting, use of the Delphi method forces the student to be more creative and independent, as the student should be proposing options for the solutions to the concerns, with the merits of each option identified. Then the advisor would provide the guidance needed to develop a solution.

The important point here is that in addition to any face-to-face meetings, much of the research progress can be documented by the written communications, possibly via email, between the student and the advisor, i.e., the facilitator and the participant, respectively. The student would need to have a very good understanding of the stakeholder's need, and all communication between the facilitator and the participant must be precisely stated and timely, and not subject to unnecessary delay.

For multi-site projects with numerous participants, such as a professional society committee that undertakes a research project, the Delphi method could be very useful. The method stresses good organization. A responsible facilitator should be assigned such as the chair of the committee. The nature of the research can be an inhibitor, as research often needs flexible deadlines. When it is not convenient to use options such as ZOOM, the Delphi method may be a productive alternative.

5.8   CONCLUDING COMMENTS

The Delphi method makes brainstorming more flexible by removing the constraint of face-to-face meeting of the participants. While face-to-face meetings have numerous advantages, they may require more resources than would be needed to use the Delphi method. Even where face-to-face meetings are practical, the Delphi method may still be preferable where a face-to-face meeting would likely result in conflict among the participants. Stakeholders and facilitators need to recognize the advantages and disadvantages of all creative thinking methods so that the most appropriate method can be adopted.

One advantage of face-to-face activities is the potential for significant interactions among the participants. For example, piggybacking can increase productivity. This advantage can be most important at the time of the final decision. Complex problems often involve diverse technical issues, so without interactions at the time when the decision is being made, the process may not lead to the best decision. Also, when the group is not physically present together, the participants may feel less involved and unwilling to indicate even minor dissatisfaction with the announced decision.

5.9   EXERCISES

5.1.   What is the origin of the word Delphi in the Delphi method?

5.2.   How would the assessment of the facilitator of the Delphi activity differ from the assessment of a facilitator of a brainstorming group?

5.3.   How can a facilitator prevent procrastination of group members when each member of the team is remote to everyone else?

5.4.   In Phase III of the Delphi method, identify alternative ways of ranking the responses for the purpose of making a decision. Justify each and discuss their advantages and disadvantages.

5.5.   In Step III(b), what are the advantages and disadvantages of submitting qualitative rankings versus a quantitative rating for each reason?

5.6.   What factors would influence the number of potential solutions to include in Phase IIId of the Delphi procedure? Identify the criterion if this step must be executed several times.

5.7.   For each factor identified in Section 5.5 that influences the speed to consensus, identify why the speed should be relevant.

5.8. How does the Delphi method differ from brainstorming? From synectics?

5.9. Identify the five major inhibitors to a successful application of the Delphi method.

5.10. The number of ideas that the facilitator includes on a distributed list can influence the result. What factors should the facilitator consider when selecting the number of items in any list distributed?

5.11. Why is diligence an important value relative to the Delphi method?

5.12. A faculty member has a research contract and hires three graduate students to work on related parts of the project. Describe a situation where the Delphi method would be an appropriate problem-solving method to use.

5.13. Critique the following practice: the facilitator initially asks for two lists from each participant. One list includes items based solely on logical thinking. The second list will be reserved for imaginative ideas. Compared to the practice of having all ideas included in one list, would you expect this practice to provide better solutions? Explain.

5.14. What factors contribute to the efficiency of the Delphi method? Use Eq. 2.2 in the discussion.

5.15. How could synectics be implemented into the Delphi method?

5.16. Why are Apollo, Athena, Pythia, and Zeus relevant to Delphi?

5.10 ACTIVITIES

5.10.1 ACTIVITY 5A: SMOKE SIGNALS

Several missionaries are working in different parts of a jungle. They need to communicate about problems. Transform concepts basic to the Delphi method into a means of communication using smoke signals. Also, discuss efficiency issues and inhibitors to smoke signal communication. The intent of this activity is to demonstrate how the principles of the Delphi method could apply to a different form of decision-making where face-to-face meetings are inconvenient.

5.10.2 ACTIVITY 5B: SALES PITCH

The objective of this activity is for a small group to demonstrate the steps of the Delphi method beyond Step 1. The activity is to identify ideas that could be used in developing a company brochure for marketing the talents of the company for both clients and potential candidates for employment. One group member needs to act as the facilitator. Assume that the Phase I activity identified the following ten ideas:

A. Company mission statement.
B. Services and products supplied.
C. Pictures of completed projects.
D. Pay scale and benefits.
E. Sustainability goals of the company.
F. Company activities in the community.
G. Cutting-edge nature of the company's work.

H. Stock price history of the company.
I. Awards that the company and employees have received.
J. Pictures of the office work spaces.

Each person should rate the top six ideas; the players can indicate the idea number in one of the six spaces, with the best option rated #1 and the least of the six best rated #6.

1. ___ 2. ___ 3. ___ 4. ___ 5. ___ 6. ___

Ultimately, the brochure will center on the best three ideas. The task is to reduce the list of ten ideas to the best three ideas. For subsequent iterations, the number of items could be reduced from six to three. Use as many rounds as necessary to find the best three ideas. Document and discuss the process as a tool for decision-making.

5.10.3 ACTIVITY 5C: COVID LOCKDOWN

How would a good knowledge of the Delphi method have enabled a researcher to have more efficient and more effective communications with graduate students if the university policy restricted all face-to-face meetings? Establish specific guidelines for conducting such a practice. Specifically, identify ways to maximize efficiency as it relates to the guideline. Identify specific practices for the facilitator, i.e., the research advisor.

5.10.4 ACTIVITY 5D: COMPANY BROCHURE

Part I: The company for which you work wants to develop a brochure that can be used for recruitment of new employees and to attract new customers/clients (users of company services). What items/entries would be most effective to include (use short phrases in the replies); place your responses in Box 1?

Part II: The facilitator can create a list of all entries and re-submit a sheet with the options and a place for the participants to list the top three. The facilitator can identify the best options.

BOX 1

1. _____
2. _____
3. _____
4. _____
5. _____
6. _____

5.10.5 ACTIVITY 5E: UNRULY AND INSANE

Identify ways that the steps of the Delphi procedure can be modified to increase the likelihood that participant recommendations will include more wild-and-crazy ideas.

5.10.6 ACTIVITY 5F: HITCHHIKING TO DELPHI

Since the group participants for a Delphi activity are not in the same room, pig-gybacking or hitchhiking of ideas would be difficult. However, piggybacking does have some benefits and can contribute novel ideas. Revise the steps of the Delphi method in some way so that piggybacking can be included in the development of solutions using the Delphi method. The intent of this activity is to provide users with the opportunity to essentially perform an innovation of an existing method.

5.10.7 ACTIVITY 5G: EDUCATIONAL IMPROVEMENT

Box 1 should be distributed to each participant, and they should be asked to identify up to six ways that education could be improved. The best option is rated six and the least of the options is rated as one. The responses are submitted to the facilitator. The facilitator compiles the numerical scores and selects the top five based on the sums of the scores. The facilitator uses Box 2 to list the top six options and distributes the revised list to the participants, with the six options listed in random order. Box 2 is distributed to each participant and used by the participants to indicate their top three choices. An allotted time for completing Box 2 should be specified. Then the facilitator identifies the top three options and announces these selections. Another copy of Box 2 is distributed to each participant, and they were asked to identify the order of the options. Then the participants return their Box 2 and the facilitator identifies the top reason. It may be of interest for the group to discuss the value of the activity.

BOX 1

My education would have been better if the following changes had been made:

1. Eliminated the general education requirement.
2. The number of non-technical courses had been more.
3. Computers would have been used more.
4. The number of required credit hours had been less.
5. The freshman math and science requirements were fewer.
6. An internship with academic credit given was required.
7. The amount of technical writing was increased.
8. A research course had been required.
9. Business courses were required.
10. A study abroad semester was required.
11. The design content had been increased.
12. The number of technical electives was increased.
13. A course on climate change was required.

BOX 2

The facilitator lists the top six or the top three options in Box 2. A blank line is provided top left of each option for the participants to rate their selections.

A. _____ B. _____ C. _____ D. _____ E. _____ F. _____

6 Checklist Development and Applications

CHAPTER GOAL

To present checklisting as an applied tool for introducing both breadth and depth into one's creative thinking.

CHAPTER OBJECTIVES

1. To define checklisting as a method of creative thinking.
2. To provide examples of checklists.
3. To discuss the development and application of checklists.
4. To show the benefits of using checklists.

6.1 INTRODUCTION

Do you use a TO-DO list to ensure that you remember to complete your daily responsibilities? Many people use a daily TO-DO list because they believe that it improves their efficiency, i.e., they complete more work in the allotted time. Some people use a TO-DO list because they get an elated feeling when they cross items off the list; it is sort of a reward for being responsible and completing a task on time. TO-DO list users recognize the many benefits of using a list.

Drafting the conclusion section of a technical report is often difficult because the writer is not learning anything new when composing the section, and therefore, little thought or effort goes into this section. Thus, the conclusions sections of reports are often weak, as the authors only present a brief summary of the body of work. A conclusions section of a report is often the last part read by a reader, so the reader is more likely to retain knowledge of the work that is reported in the conclusions section but only if it is well written and complete. If a reader of a report learns little from the conclusions sections, then he or she may feel that the entire report had little substance and was of minimal value. Given that the conclusions section of a technical report is important, but its writing is often not given the serious consideration that it deserves, a critical thinking method could be used to encourage authors to take the drafting of the conclusions more seriously. Whatever approach is used, we should want the conclusions sections of a report to be both effective and efficiently written.

A common criticism of research articles submitted for review to top professional journals is that the authors do not adequately discuss the extended value of the work, i.e., the implications of the conclusions. Conscientious reviewers expect the

DOI: 10.1201/9781003380443-6

authors to show that the research has broad implications, those beyond the immediate focus of the work. Detailing the broad implications of the work would show that the research results have value beyond the specific scope of the paper. It is important to show the implications of research to such factors as public safety, the potential effects on the environment, and the risks to the community. Additionally, a critical review of the implications of someone else's research can be a very valuable source of new research ideas (see Chapter 15). Ways of encouraging the reporting of the broad implications of research are needed.

A checklist is like a TO-DO list. Given the importance of emphasizing the broad implications of novel work, the checklist method is presented herein. Checklists can be classified as imaginative methods because they are intended to extend thinking beyond the norm and allow research results to appear more imaginative. They assist in providing breadth to completed works. An imaginative person can apply a checklist to situations that are well beyond the condition for which the checklist was developed; however, a checklist must be well designed in order to be an effective aide.

6.2 DEFINITION OF A CHECKLIST

What is a checklist? A checklist is *a series of very general statements or questions that can be applied to any one of a number of topics.* A checklist has two parts. It starts with a set of general factors, which can be referred to as attributes, and then a series of questions or brief statements follow and give detail to each of the attributes; these can be referred to as extensions, as they attempt to provide depth to the attribute to which they are aligned. A good checklist usually includes six to twelve attributes and maybe two to five extensions per attribute. A checklist provides a way of ensuring that a breadth of ideas is considered. The attributes provide breadth while the extensions provide depth. In many activities, the inclusion of a wide range of issues is important. For example, the implications of research should be broadly discussed in a research report. Checklists promote breadth at a time when researchers are emphasizing depth; both breadth and depth are important.

A checklist is also a way of improving on an idea. It serves as a prompt to help the checklist user keep an open, broad mind. A checklist can be viewed as a TO-DO list for the topic of interest. For example, in research, the checklist may focus on research implications. Just as we check off items on our daily TO-DO list, an author can use a checklist to identify issues that need to be considered when authoring a technical report. A checklist is nothing more than a "cattle prod" for the person's mind, as the items in the list should encourage the user to view the topic of interest more broadly by thinking about the attributes on the checklist.

6.3 EXAMPLES OF CHECKLISTS

The construction of a checklist will influence its usefulness. A poorly structured checklist would be much like a disorganized TO-DO list. Examples of checklists can be used for demonstrating the structure and content of checklists. Three applied checklists are presented in Figures 6.1–6.3. These examples illustrate a broad range of issues from improving an oral presentation to modifying a commercially marketable product. Note that the primary attributes of each checklist generally suggest breadth. For example, the checklist for the innovation of a commercial product (see Figure 6.1) includes attributes such as size (e.g., minify) and blend (e.g., combine ideas). Also, checklists can provide ideas that can be used beyond the original use. For example, many of the entries for the checklist for oral communication (see Figure 6.3) could be applied directly to improving a written document. Similarly, the checklist intended to be used for discussing the implications of storm water detention basins (see Figure 6.2) could be used to identify the implications of research on the design of any engineering infrastructure, including facilities for green roofs. Even the checklist on the innovation of commercial products (see Figure 6.1) could be used in developing a checklist for an oral presentation.

6.4 APPLICATION OF A CHECKLIST

Before technical papers can be published, they are subjected to a peer review, which is generally completed by two or three anonymous reviewers. The reviewers of journal

MINIFY: Tinier? Delete? Divide up? Decrease dimensions? Constrict?
CHANGE: Purpose? Color? Motion? Add sound? Shape?
EXPAND: Add? Change length or width? Add a component? Exaggerate?
BLEND: Aims? Components? Combine ideas?
REPLACE: Other inputs? What (who) else instead? Other tone of voice?
REVISE: Similar to? Copy? New uses? Slow down? Speed up?
REVERSE: Cancel? Inverse? Make extreme?
REORDER: Rearrange? Interchange? Change sequence? New pattern?
ADDITIONS: Electronic components? Visual? Memory?

FIGURE 6.1   A checklist for product development.

SAFETY: Designed for? Maintenance? Downstreameffects? Failure? Child protection?
POLICY: National? State? Local? Public vs. private? Risk level? Sensitive areas?
COST: Benefit/cost? Individual vs. public? Savings? Return period? Damage assessment?
HYDROLOGY: Small vs. large area? Urban vs. rural? Coastal? Return period?
MAINTENANCE: Access? Cost? Designed for? Equipment required? Clean-out frequency?
ACCURACY: Comparison with existing facilities? Uncertainty of inputs?
ENVIRONMENT: Erosion? Trash accumulation? Effect of climate change?
APPLICATION: Location dependent? Required input? Required technical competency?

FIGURE 6.2   A checklist for storm water management.

TITLE: Shorten? Lengthen? Add humor? More specific? Broaden the scope?
ABSTRACT: Quantitativeresults? Outcomes stressed? Implications noted?
INTRODUCTION: Quotation? Make it personal? Give quiz? Report case studies? Statistics?
OBJECTIVES: General goal? Specific objectives? Novelty stressed? Breadth?
CONCLUSIONS: Controversial recommendations? Relate to intro? Implications? Added work?
AUDIENCE INVOLVEMENT: Eye contact? Ask questions? Needs of audience?
VISUAL AIDS: Sequence lists? Proper orientation? Cartoons? Color? Descriptive titles?
ORGANIZATION: Outlining? Time limit? Suspense built?
CONTROL NERVOUSNESS: Sufficient practice? Good visuals? Planned ahead?
REHEARSAL: Vary location? Concentrate on intro and conclusions? Each visual separate?
VOICE: Control speed? Control loudness? Body language? Inflection? Good pronunciation?

FIGURE 6.3   A checklist for communication activities.

papers are often familiar with the general topic of the paper, but they often lack direct experience with a specific topic. Ultimately, reviewers want to know whether or not a paper may have relevance to their specific area of interest. Therefore, reviewers are especially sensitive to statements of the implications of the work in ways that the methods are relevant to their specific interests. They expect important research to have broad implications. Issues peripherally relevant to the topic can often be included in the conclusions section of a paper. A good conclusion section should be more than a brief summary of the work; this section should demonstrate that the work has broad implications to the state of knowledge in other areas. A conclusion section of a technical report that does nothing more than provide a bullet list of points made in the report is considered to be a very weak effort. Using a checklist can help ensure that the broad implications of the research are identified and discussed. When a report is being drafted, an appropriate checklist can be reviewed by the authors of the paper and the broad implications of the work that are based on the checklist attributes can be added to the closing discussion. For example, the checklist in Figure 6.2 would suggest that the conclusions section of a paper on some aspect of storm water runoff should discuss the ways that their research could improve state policies, influence maintenance costs, or reduce erosion.

6.5   DEVELOPMENT OF A CHECKLIST

The first draft of a checklist should be developed when the problem is initiated, not after completing the research. Additional attributes and the corresponding extensions can be added to a checklist as the research progresses. When developing a checklist, the entries should not be made specific to one problem. The entries should be applicable to a wide array of topics that are similar to the issue of current interest. Then the checklist can be used in the future to provide ideas for different but related problems. Breadth is important. All types of critical thinking can be applied in developing a checklist. A checklist should encourage both breadth of thinking and new relations. For example, the checklist for computer model development (Figure 6.4) would be

COMPONENTS: time dependency; spatial dependency; stochastic
STRUCTURE: linear; nonlinear; exponential; logarithmic; composite.
VARIABLES: empirical; theoretical; interactive.
CALIBRATION: analytical; numerical; subjective; least squares.
VERIFICATION: rationality; split-sample; jackknifing; independent test data.
GOODNESS-OF-FIT: Correlation coefficient; RMSE; RRMSE; bias; relative bias.
SENSITIVITY: absolute; relative; deviation; component.
UNCERTAINTY ANALYSIS: confidence errors; tolerance limits; error analysis.
DATA: record length; missing data; historical events; outliers.

FIGURE 6.4 A checklist for mathematical model development.

applicable to models in engineering, geology, natural resources, medical sciences, etc., so the entries in the checklist should cover issues that go beyond the specific objectives of the research topic currently of interest. The attributes are used to ensure breadth, while the extensions are included to provide depth to the corresponding attribute.

6.6 CONCLUDING COMMENTS

A reader of a technical report judges the quality of the work partly on the basis of the knowledge gained from reading the report. Generally, a reader wants to recognize the applicability of the research to his or her own interests, but they also want to see that the work will benefit society. Yet, authors of technical reports often fail to consider this broad interest. Instead, the authors often provide a very superficial summary, often just a bullet list of points made in the paper without any thought of breadth of relevance. Such a summary reduces the educational value of the total work. A brief summary cannot convey much knowledge and is, thus, of potential less value to the reader. Checklisting is one tool that can be used to assist the author of the report, not just the reader. A checklist includes a broad array of topics that the authors need to consider in establishing the experimental designs for the data analyses as well as drafting the conclusions section of their report.

Checklists are valuable tools for procrastinators. A checklist should include all items that relate to the parts of the work about which the procrastinator has the most difficulty getting started. When the procrastinator is directly faced with the troublesome task of starting the job, he or she only needs to use the checklist as motivation for attacking the most problematic part of the task. A person could include a separate column in the checklist that is used to identify rewards when the item is complete. Another column could be used to identify penalties for not completing a task on time. In any case, the checklist serves as a crutch and can help the procrastinator avoid negative assessments from a superior for not completing a work assignment on time.

Checklists are certainly related to the value dimension of critical problem-solving. Values such as punctuality, industriousness, accountability, dependability, and reliability have obvious connections to completing responsibilities in a timely fashion.

Efficiency is especially relevant to checklists and procrastination. Completing responsibilities on time leads to higher quality work. Since the quality of the work is important, using a checklist can certainly increase the numerator of the efficiency equation.

6.7   EXERCISES

6.1.   Find the definition for the word *checklist*. Compare it to other words based on the word *check* (e.g., check in, check off, check point, checkup, checkbook).

6.2.   Add three new attributes to the Figure 6.1 checklist and three or four extensions for each. Discuss how this improves the checklist.

6.3.   Develop a checklist for a maintenance person who is in charge of snow removal from a college campus. Keep in mind that a good checklist can be applied to other needs.

6.4.   Discuss the ways that a checklist on the innovation of commercial products (Figure 6.1) could be applied in developing an oral presentation.

6.5.   Develop a checklist that could be used for preparing for a job interview. Discuss whether or not the list would help a person who is running to win a political office?

6.6.   Develop a checklist for preparing for a blind date.

6.7.   What changes would you make to the checklist of Figure 6.3 so that the checklist would be applicable for use with oral presentation assignments?

6.8.   How can creative thinking be helpful in the development of a checklist?

6.9.   Develop a checklist that you could use to prepare a two-week-long vacation. Include all items that you might need. For example, everything from remembering to take the airline tickets to a book to read on the plane. Would the checklist be applicable for preparing for a business trip? Would the checklist be applicable for a six-month around-the-world cruise?

6.10. Revise Figure 6.3 by adding one extension to each attribute.

6.11. Compare and contrast daily TO-DO lists and the checklist approach.

6.12. What factors could cause a checklist to be minimally effective?

6.13. How does the concept of efficiency apply to a checklist?

6.14. Efficiency is quantitatively assessed using the ratio of output to input. With regard to a checklist, what factors influence the input and output and how could these factors be quantified to be able to assess the efficiency of a checklist?

6.15. How would a checklist used for short-term activities differ from a checklist developed for long-term activities?

6.16. Develop a set of criteria for assessing the quality of a checklist. Use the criteria to assess the checklist of Figure 6.3.

6.8   ACTIVITIES

6.8.1   Activity 6A: Checklist for Technical Writing

Without reviewing the checklists in Figures 6.1–6.3, develop a checklist to use when writing research papers about any technical topic. Then try to apply it to evaluate a research paper written by a religious studies major.

6.8.2   Activity 6B: Checklist Assessment

The assessment of creative activities is important. Develop a checklist for the general assessment of all checklists.

6.8.3   Activity 6C: Leadership Checklist

Compose a checklist for a president of a student chapter who is responsible for holding weekly chapter meetings. Then discuss the way that the checklist could be used by an office manager in a large company.

6.8.4   Activity 6D: Flood Runoff to Drought

Using the checklist of Figure 6.2, discuss ways that the entries could be applied to a technical paper that describes a model for predicting drought.

6.8.5   Activity 6E: Kinesics

Modify the checklist of Figure 6.3 to cover aspects of kinesics, i.e., body language, for a person who will be making an oral presentation.

6.8.6   Activity 6F: Job Interview

Develop a checklist that a person could use to prepare for an in-person job interview. Discuss whether or not it would be adequate for a ZOOM interview.

6.8.7   Activity 6G: Food Bowl for Fido

Develop a checklist that could be used for the design of small household items, such as small tools, kitchenware, and food bowls for pets.

6.8.8   Activity 6H: Isn't She Pretty!

Develop a checklist that could be used to help a 7-year-old child learn all of the tasks needed to groom his or her pet dog. For what other imaginative activities could the same checklist be used?

6.8.9   Activity 6I: I Can't Hear You!

Develop a checklist that a leader could use during one-on-one interviews to ensure that he or she is a good listener to the concerns of the subordinates.

6.8.10   Activity 6J: Don't Be Late!

Develop a checklist that a procrastinator could use for meeting his or her responsibilities on time.

6.8.11 ACTIVITY 6K: BREADTH VERSUS DEPTH

Assume that a checklist for a topic has ten attributes and five extensions per attribute. Discuss how breadth and depth are relevant to a checklist. What factors enter into establishing limits on the breadth and depth of a checklist?

7 Inhibitors of Creativity and Innovation

CHAPTER GOAL

To make users of imaginative thinking aware of the inhibitors that can reduce the efficiency of their applications of creative methods.

CHAPTER OBJECTIVES

1. To define the concept of creativity inhibition.
2. To show that creativity inhibitors decrease creative efficiency.
3. To present common inhibitors: organizational, peers, and self.
4. To identify pessimism as an inhibitor.

7.1 INTRODUCTION

Jails inhibit freedom! In some ways, gravity is an inhibitor. Aging certainly limits, i.e., inhibits, some physical activity. COVID, the pandemic, limited education, happiness, economic growth worldwide, and personal movement. Inhibitors, such as this variety of constraints, can severely limit many actions and attitudes that we value, and we suffer because of these losses and limitations. In most activities, overcoming inhibitors is necessary for success and for optimizing efficiency.

Over the years, myths about creative problem-solving have surfaced. Many of the myths downgrade the value of creative thinking to any problem-solving activity. The most common myth is that creative problem-solving is no more effective than the standard practices, usually based on logical thinking; yet, creative problem-solving requires greater resources, especially time. Some individuals believe that this myth leads to the conclusion that the use of creative idea generation methods reduces efficiency when compared with the use of logical thinking to make decisions and solve problems. When myths are told and retold, some people begin to believe them even though they lack proof that the myth is true. Since creative problem-solving methods may involve fantasy and imaginative thinking, another inhibiting myth has been promulgated; specifically, this second myth portrays these methods as being immature, maybe even unprofessional. Since many instances of creative thinking take place in the private sector, which is known for its confidentiality and secretiveness, examples of successes of creative problem-solving are minimal and not in a sufficient number to fully discredit the myths. Thus, the myths function as severe inhibitors, as they

DOI: 10.1201/9781003380443-7

discourage potential users of creative thinking from using the methods to solve complex problems.

Numerous factors inhibit the effective application of creative thinking methods. Myths, attitudes, and oneself are a few of the culprits. The inhibition of creative thinking has serious consequences. Solutions to problems can be delayed. Less than optimum solutions may get adopted. Inhibitors can act as constraints on careers, thus, inhibiting prestige as well as paychecks. We cannot overcome these inhibitors unless we first acknowledge them and subsequently take the actions that are necessary to eliminate or at least minimize their effects. Knowing common inhibitors to the solution of complex problems is the first step toward eliminating them. Only when inhibitors are recognized can they be overcome.

7.2   DEFINITIONS

Creative inhibitors can be defined as *attitudes, feelings, or organizational policies that act to limit idea generation and creative problem-solving.* A creative inhibitor is *any factor that acts to suppress creative thinking.* Inhibitors can be organizational policies or practices, colleagues who voice negative beliefs about the intellectual worth of creativity, or the person himself or herself in times when he or she purposely avoids imaginative thinking. One fundamental element of a creative attitude is having the confidence to suppress negative thinking by oneself or allowing the negativity of others to suppress creative thinking. Having to work with colleagues who discourage creative thinking, especially overt discouragement, can decrease a person's willingness to think in a creative way.

7.3   EFFECTS OF INHIBITORS ON EFFICIENCY

Inhibitors can reduce creative efficiency. With respect to Eq. 2.2, inhibitors increase the resources needed to solve a problem, i.e., the input, but more significantly reduce the value of the numerator, i.e., the creative output. Inhibitors are like friction, as greater effort (i.e., input) is needed in terms of time and resources, resulting in an increase in the denominator. The extent of the inhibitions can determine the actual effect on the magnitude of creativity efficiency.

It is important to distinguish between *disposition* and *action*, i.e., attitude versus thinking. Having a creative attitude can enhance a person's likelihood of being successful at idea innovation; however, it is not just the person's attitude that determines success at idea innovation. The person must function in an environment that is essentially free of creativity inhibitors. Unfortunately, creativity is sometimes constrained by organizational policies and the attitudes of personnel, i.e., peers. Leaders of groups must ensure that both organizational policies and other individuals do not restrict subordinates from exercising their creative abilities. The attitudes, personalities, and fears of colleagues may constrain an individual's use of his or her creative abilities. The attitudes of colleagues toward creative thinking can produce an environment that is similar to the effect of corporate restraints on creativity. A person's own negative attitude toward creativity is often the most significant restraint on imaginative idea generation and a reducer of creative efficiency.

7.4   CREATIVITY INHIBITORS: ORGANIZATIONAL

Some organizations have a philosophy that the tried-and-proven problem-solving methods of the past will provide the best solution to every problem. Therefore, they discourage imaginative problem-solving and try to force each new problem into the same problem-solving mold that has been used to solve problems in the past, even if the current problem has characteristics that are quite different from the characteristics of the problems of the past. Leaders who do not reward the creative efforts of their subordinates may discover that the lack of a creativity reward system inhibits future acts of creative problem-solving. The failure to reward creativity and critical thinking reduces problem-solving efficiency and leads to the adoption of less than optimal decisions. Every display of creative thinking should be acknowledged in a positive way by an organization, regardless of the success of the outcome.

Creative problem-solving often requires more time than would the use of logical thinking alone, as more alternatives are proposed and each one must be fully evaluated. This time constraint can be viewed as an inhibitor of creative problem-solving; however, the effect of this negative aspect can be mitigated by more efficient evaluation. If creative solutions are to be viewed as positive outcomes, then the organization must allow more time for problem-solving. The benefit will be better solutions to problems. Not allowing more time can be a significant organizational constraint on productivity, on employee job satisfaction, and on profits. Creative thinking will be suppressed in the future if proper credit is not acknowledged by the organization for current successful uses of creative thinking. Table 7.1 identifies other restraining factors due to organizational policies and practices.

7.5   CREATIVE INHIBITORS: ASSOCIATES

Some employees believe strongly that the use of creative problem-solving methods does not improve problem-solving. They often snidely berate peers who promote creative problem-solving. Criticism made by associates can be a very constraining

---

TABLE 7.1

Organizational Factors That Inhibit Creative Problem-Solving

- Narrow mindedness of management.
- Distrust of benefits of creative thinking.
- Excessive routine work.
- Unwillingness to devote and fund the time.
- No public acknowledgement of creators.
- Belief in myths that are pessimistic about imaginative thinking.
- Arguments that creativity lacks practicality.
- No offer of incentives for creative thinking.
- Creative ability not considered in hiring.
- Lack of breadth of objectives.
- Excessive routine problem-solving.
- Lack of experience with creative thinking.

TABLE 7.2

Restraints Imposed by Associates

- Lack of peer support on new approaches to problem-solving.
- Inflexibility of associates.
- Adverse criticism.
- No encouragement to apply critical thinking.
- Resistance to innovation.
- Pressure against using imaginative thinking.
- Arguments against its practicality.
- Ridicule by peers.
- A stress on logical thinking.

factor with respect to idea generation and problem-solving. If a person expects criticism from an associate, then he or she may be less inclined to even propose innovative ideas, especially when the associate is at a higher rank. Some individuals often suppress wild-and-crazy ideas out of fear of being labeled immature. Table 7.2 includes examples of ways that associates can constrain creative thinking. While criticism of generated ideas may occasionally be warranted, imaginative ideas should not be criticized as they can lead to novel solutions. While creative thinking activities may not always provide the best solutions, this does not imply that imaginative ideas should not be generated and evaluated. It does mean that any criticism should be unbiased and based on a valid assessment.

The responses of peers to efforts to act creatively are important. The motivation to creatively solve a complex problem in a unique way should always be viewed as a positive factor, especially when organizational leadership recommends the application of creativity stimulators, such as brainstorming or synectics. Getting credit for past creative efforts, even recognition of such efforts by peers, will encourage future creative thinking, and thus suppress an inhibiting environment. A lack of positive recognition acts as an inhibitor. When associates outwardly display disdain for the use of creative problem-solving methods, creative individuals may suppress future creative efforts at problem-solving. An inhibiting atmosphere in a workplace is especially detrimental to junior-level employees, as they may not be willing to oppose the attitudes of the senior associates.

7.6 CREATIVITY INHIBITORS: ONESELF

A lack of knowledge of skills that are used in creative activities can be a primary inhibitor for anyone. Any lack of courage to approach problem-solving using imaginative thinking will prevent the person from using skills such as critiquing and questioning. A person who lacks experience at critiquing and is not curious would likely avoid consideration of using methods of creative idea generation. Any failure to recognize the uncertainty of data represents a lack of a skeptical nature, which can be a serious inhibitor. Inadequate development of critical thinking skills is possibly the most significant source of inhibition, as the solutions to complex, important problems

---

TABLE 7.3
Self-Inhibiting Factors

- Failure to learn prior to needing the ability.
- Conformity in thinking.
- Lack of confidence.
- Fear of failure.
- Mental laziness.
- Timidity.
- Tiredness.
- A pessimistic attitude.
- Fear of ridicule.
- Undeveloped imagination.
- Self-criticism.

---

depend on having a functioning critical thinking mindset. A flexible mindset can be a valuable asset to a problem solver. Table 7.3 summarizes just a few self-inhibitors to creative idea development. Overcoming these inhibitors will lead to improved problem-solving efficiency and greater recognition for producing novel outcomes.

An individual's attitude about creativity can be the best stimulator or a significant inhibitor during creative problem-solving efforts. The primary factors of a person's mindset that can restrain creative ability are a lack of self-confidence, a pessimistic attitude, or a fear of ridicule. To be creative, a person must generally be confident in his or her own abilities to solve unique problems. Also, the person must be confident in the benefits of creative stimulators, such as synectics and the Delphi method, and recognize that they are tools for increasing the fluency of generated ideas and improving problem-solving efficiency. Overcoming the effects of these self-inhibiting factors will enhance one's self-confidence.

Fear generally connotes the thought of pain or danger. With respect to the application of critical thinking to complex problem-solving, fear implies that the application of imaginative thinking may cause the person to lose face with peers or within the organization because of the failure of creativity to produce a novel solution to every problem. Any hesitancy in one's faith in critical problem-solving will significantly limit effective actions. Pessimism and the lack of confidence lead to such fear; note the interactions between values and attitudes.

7.7 PESSIMISM: A MINDSET INHIBITOR

Any tendency to have negative thoughts about past or future events prevents future success; therefore, such an attitude can be viewed as a personality or mindset problem. Consistently believing that the outcomes of future events will not be to your liking is indicative of a negative attitude. Such an attitude would cast you as being a pessimist. Negativism or pessimism can be defined as *a disposition that dwells on a gloomy future even without evidence—a disposition that generally assumes that action will contribute to the least favorable outcome*. A pessimistic attitude is often accompanied by the belief that the outcomes to other people are generally much better that the outcomes that you would experience under the same circumstances. The causes of consistently having a negative attitude are probably unknown, although some innate roots are likely, which have probably been reinforced by some unfavorable early experiences. Negative consequences in the past can be interpreted by a pessimist as omens for future negative consequences. Thus, pessimism negatively influences both current outcomes and future actions.

Pessimism is usually interpreted as being counterproductive to organizational goals. Those individuals with negative attitudes are often avoided, even rejected, by colleagues, while most people appreciate those who have positive attitudes. Pessimists may be less likely to be promoted and will likely have slower career growth. Pessimists are often viewed as lacking in self-confidence. Pessimism is rarely beneficial and almost always has a negative influence on success.

Statements like *"It won't work!"* or *"I can't do it!"* reflect both a pessimistic attitude and a lack of confidence in both oneself and the effectiveness of imaginative problem-solving methods. These attitudes will certainly restrain an individual's productivity and create a fear of being a critical thinker. With respect to critical problem-solving, pessimism is the attitude that knowledge of and use of critical thinking will not improve the outcome. A pessimist always believes that creative solutions to a problem will require more time. A pessimist believes that the evils of creative thinking outweigh its benefits. A generally pessimistic person will lack confidence in critical thinking. It is not clear which one, pessimism or lack confidence, is the initiator of the problem. The two attitudes form a cycle, as each one of the two attitudes sustains the other attitude. Before a person will have the confidence to embrace creative thinking, the individual must overcome any pessimistic attitude. For an individual to overcome pessimism requires confidence in either the facilitator of the group or organizational policies that strongly support the value of critical problem-solving.

Overcoming a pessimistic attitude is not easy, but it is definitely possible. First, the person must acknowledge the problem and set about creating a plan for developing a more positive outlook. Second, the person should review past experiences where he or she believes that the outcomes to decisions were not favorable. For each of these, the cause of the negative outcome should be identified and rated on its level of negativity. Then the person needs to identify the action that he or she now believes would have produced a more positive outcome. This practice would represent counterfactual thinking. Third, similar situations that could occur in the future should be imagined, and for each situation the person should develop a plan that is more likely to lead to positive outcomes. This exercise should be repeated as frequently as is practical in order to instill an outlook that is directed toward more positive thinking. Fourth, the person should identify the specific effects of negativity and contrast these outcomes with outcomes expected for an optimistic thinker. Fifth, the person should evaluate friends who they believe seem to always experience positive outcomes and compare these individuals to themselves; identifying the differences should lead to a set of changes to the person's belief system. Actions such as these can initiate the conversion of a pessimistic attitude to one of optimism. Hopefully, the change in attitude will begin to yield more positive outcomes to efforts of complex problem-solving.

7.8   EFFECTS OF INHIBITORS ON VALUES

In the previous section, the point that a person's attitudes can act as a creativity inhibitor was made. Similarly, inhibitors to critical thinking can have consequences to a person's value system. Given the importance of critical thinking to the solution of complex problems, it is understandable that value issues associated with these complex problems could potentially be compromised by the occurrence of inhibitors. In terms of efficiency, it is reasonable to argue that creativity inhibitors reduce problem-solving efficiency and, therefore, any values positively associated with efficiency. Organizational inhibitors could lessen the loyalty of an employee who favors the use of critical thinking. Also, both the person's personal happiness and professional satisfaction might be negatively influenced, and the employee's reputation could be diminished because the organizational policies do not provide the employee with experiences where he or she can experience growth in his or her creative ability. If inhibitors compromise knowledge, then important values, such as public welfare, could be at risk.

In addition to the negative economic consequences of creativity inhibition, the growth of knowledge will suffer when creativity is inhibited. Knowledge increases as problems are solved in the best possible way. Gaining new knowledge of problem-solving can bring happiness and enjoyment, so inhibiting creativity can limit these feelings. Excellence is a prized value, and when factors limit the likelihood of finding the best solution to a problem, excellence cannot be achieved. Duty and competency can also be constrained by creativity inhibitors. The bottom line is: the lack of values can act as an inhibitor to problem-solving, but conversely, organizational inhibitors can adversely influence an employee's value system.

7.9   CONCLUDING COMMENTS

Progress is enhanced through critical and creative thinking, but progress can be stymied by inhibitors to thinking. Inhibitors act much like roadblocks, as they delay or prevent progress. The entries in Tables 7.1–7.3 indicate that three general categories of inhibitors can be identified for creative thinking: organization, associates, and self. Within each of these categories, we could identify subcategories, such as constraints on, restrictions to, lack of, and fear of. Subdivisions can be useful when actually identifying a specific reason for a lack of progress in problem-solving. A list of creativity stimulators, which is the opposite of inhibitors, could be developed. Such a list may be just as beneficial as would be a list of inhibitors. A list of stimulators would suggest positive actions rather than the negative factors found in a list of inhibitors. Optimistic feelings act as stimulators and are usually more productive than those feelings based on a pessimistic attitude. Actions that are based on pessimistic thinking are usually more evident to others than are actions that are based on optimistic thinking, but the pessimistic thinking is accompanied by a negative opinion of the thinker.

Some believe that laws and regulations passed by lawmakers act to suppress creative solutions to business problems. This may be true, but constraints are often due primarily by actions of the organization, including the individuals themselves and the peers with whom they work. Organizational policies and practices also constrain creativity and therefore progress. Identifying the inhibitors is the first step in overcoming the constraints that are placed on progress. Therefore, overcoming inhibitors should be viewed as a starting point for the improvement of both progress and efficiency.

7.10   EXERCISES

7.1.   Why do some believe that they cannot learn to be creative? Does such a belief limit a person's ability to be more creative? Explain.

7.2.   Why do children lose much of their expression of fantasy as they mature?

7.3.   Does the ability to think creatively have innate roots? Explain your position.

7.4.   How could you convince people who do not believe in the benefits of creative thinking that they would benefit from learning to be creative?

7.5.   Provide a general definition of the word inhibit. Modify the definition so that it specifically applies to creative problem-solving.

7.6.   Why do some companies discourage creative thinking?

7.7.   Identify ways that a company leader can encourage creative idea development.

7.8.   Why would adverse criticism by associates at work about creativity inhibit creativity in the workplace?

7.9.   In what way would a person's pessimistic nature act as a creativity inhibitor?

7.10. Oneself is often the most significant creativity inhibitor. Identify ways that a person can inhibit his or her own creative problem-solving and identify a way that the person can overcome each inhibitor.

7.11. Using Eq. 2.2, discuss reasons that creativity inhibitors reduce creative efficiency.

7.12. How do creativity inhibitors act to constrain the growth of knowledge?

7.13. Develop a list of creativity stimulators by transforming each negative inhibitor in Tables 7.1–7.3 into a positive factor.

7.14. How can organizational policies on creative thinking negatively influence an employee's personal value system?

7.15. Can governmental policies act as inhibitors to creative idea development? Discuss.

7.16. Why is it important to have practical knowledge of creative thinking methods prior to the need to use them?

7.17. Why does an emphasis on logical thinking by peers adversely influence a person's creativity?

7.18. Provide specific ways that a pessimistic mindset influences creative efficiency of Eq. 2.2.

7.11   ACTIVITIES

7.11.1   ACTIVITY 7A: A TABLE FOR MARLON AND MARILYN

Tables 7.1–7.3 could be combined into a single table with three columns. The table would represent a table of inhibiting factors that limit a person's ability to be an effective creator of ideas. Create a new table of inhibitors for developing skills and attitudes as an actor like Marlon B. or an actress like Marilyn M. The purpose of this activity is twofold: create something new through innovation and show that inhibitors can limit advancement in any career path.

7.11.2 ACTIVITY 7B: BUSINESS INHIBITING ACTIONS

Business managers often use steps such as the following as a general model of problem-solving: (1) identify the problem; (2) generate alternative solutions; (3) collect and analyze relevant knowledge, data, and information; and (4) evaluate and select the best alternative. Using these steps as an act of innovation, develop a general model for overcoming creativity inhibitors specifically for solving business problems. The model should be sufficiently general that it could be applied to both the problem solver and the environment, which includes both organizational constraints and comments by peers. Brainwrite or group brainstorm on this issue, as the intent is to show that the different types of inhibitors are more prevalent in the different steps of the problem-solving process.

7.11.3 ACTIVITY 7C: CREATING A POLLUTED ENVIRONMENT FOR CREATIVITY

For each of the following three inhibitors, provide reasons why they could inhibit the development of an environment for professional problem-solving and other creative activities in an organizational setting: (1) pessimism of the employees; (2) restraint of unenlightened committees; and (3) failure to hire new employees who have a different background than those hired in the past.

7.11.4 ACTIVITY 7D: + VERSUS –

Part I: Create a table, with optimistic as the heading for Column 1 and pessimistic as the heading for Column 2. For each of the following situations, give a response that a positive thinking person would give and a response that a negativist would give:

- Going on a blind date tomorrow evening.
- Your school expected to win the league championship at the upcoming tournament.
- Your expectation for your performance at tomorrow's job interview.
- Your presentation at a business meeting next week.
- When you see that a police officer is about to pull you over for speeding.
- When you know that you will not make the curfew deadline that your parents set.
- When you are expecting new neighbors to move into the vacant house next door to you.

Tomorrow's weather report calls for rain when the picnic you planned is to occur. BEFORE READING PART II, COMPLETE PART I.

Part II: Now evaluate yourself, *honestly*. Would you be likely to give the positive response or the negative response? Discuss your results and discuss whether it suggests that you are an optimist or a pessimist. Discuss the potential value of this list for assessing the accuracy of a person's rating.

7.11.5 ACTIVITY 7E: THE OPTIMIST VERSUS THE PESSIMIST

Create a table with three columns and seven rows. In Row 1, insert column headings of case, optimist, and pessimist. In the other six rows, insert one of the following statements in Column 1: (1) the person is turned down for a job; (2) the person hits the game winning home run; (3) the person finds out that the car repair will be costly; (4) the person wins a $1,000 lottery; (5) the person gets a $750 speeding ticket; and (6) the person gets a promotion over a peer who is disliked. In each case, identify the reaction of the optimist and the reaction of the pessimist. From the responses, develop general conclusions about the ways that optimists and pessimists react to events in their lives.

7.11.6 ACTIVITY 7F: BAD VERSUS BADDER

Table 7.1 includes 12 items. Rank them in order of importance, with a rank of one for the entry that is the most detrimental inhibitor and a rank of 12 for the least detrimental inhibitor. Repeat the procedure for the nine entries in Table 7.2 and the 11 entries in Table 7.3. Using the top three inhibitors from each group, place the nine items in order of importance; again, rank 1 is rated as the most inhibiting. Does the result suggest that one of the three inhibitors (organizational, peers, self) is generally the most constraining? Discuss the results.

7.11.7 ACTIVITY 7G: DAM UP THE CRITICISM

Part I: Brainstorm a list of objects or ideas that can act as barriers, e.g., firewalls, fences, even exhaust fans that prevent aromas and odors from getting into areas. COMPLETE PART I BEFORE READING THE PART II TASK.

Part II: Barriers inhibit! Use each of the ideas on the brainstormed list from Part I to develop a way to prevent inhibiting comments or thoughts by peers or oneself.

7.11.8 ACTIVITY 7H: THIS IS NOT SANTA'S LIST

Part I: Brainstorm a list of ways that prevent individuals from developing self-confidence. This is a list of inhibitors.

Part II: For each inhibiting attitude or belief of Part I, identify a way of overcoming the inhibitor.

8 Assessment of Critical Thinking Activities

CHAPTER GOAL

To provide guidelines on the assessment of critical and creative thinking activities.

CHAPTER OBJECTIVES

1. To define assessment and discuss its objectives.
2. To provide criteria for assessing facilitators, groups, and individuals who are involved in critical and creative thinking activities.
3. To summarize the content of assessment reports.
4. To summarize the importance of values in assessment activities.

8.1 INTRODUCTION

Before going to see a just-released movie, we might decide to listen to a movie critic discuss the strong and weak points of the movie. Before attending an art show to view the works of up-and-coming artists, we may read the local art critic's assessment of the talent on display. We may use critical assessments such as these that are authored by experienced experts in the appropriate subject matter to decide whether or not to eat at a popular restaurant or attend a performance at the Kennedy Center. These critiques could enable us to make better decisions on the best ways to spend our time. They might also provide expert knowledge that would help us enjoy other such experiences in the future. We may be more enlightened to understand the content of the movie or artworks if we think about the ideas presented in the assessments by the established critics. Considering the general structures of the movie and art critics' analyses may actually improve our own abilities to critique a broad array of issues. Assessments made by well-known critics are not always indicative of our own personal assessments; however, the facts and knowledge that the experts provide may be useful in appreciating the art forms and identifying general factors that we should consider in our own assessments. General knowledge that is gained from the knowledge of experienced critics may also be of value in assessing future movies or artwork. Many factors enter into the decision process about seeing a specific movie or attending an art show, but critiques that have been provided by others can be educational and helpful in decision-making for both the short term and the long term. Such critiques can even be viewed as an indirect form of mentoring.

DOI: 10.1201/9781003380443-8

We make assessments every day, and the assessments lead to decisions about the ways that we spend money, use our time, interact with friends, and take actions. The outcomes of our current decisions and actions will generally be followed by assessments, which are our decisions and actions in the future. Sometimes, our assessments are made with considerable thought, while others are based on thoughts that are very superficial. The decisions that were based on knowledge-based critical analyses are probably the ones that will have a more positive influence on future actions and decisions. It is the quality of self-assessments that lead to positive changes in our value systems. Consider the consequences of not making any assessments of past actions. Then each new action would be nothing more than a random decision, which would likely not lead to quality experiences. Our current assessments of our recent actions can be very important to our futures, including our happiness. Unfortunately, sometimes the results of decisions are regretted because the knowledge that led to the decision was assessed properly.

An assessment of any activity can provide useful knowledge, especially when the assessment is conducted by someone with proven expertise in the subject matter. However, assessments can be, in part, opinions that do not logically follow from quality experiences and rational thought. The quality of any assessment depends on many factors, including both the knowledge and the mood of the assessor. Assessments can be biased, and if the reviewer of an assessment report is not aware of the assessor's bias, then the assessment report can mislead the reader by instilling poor knowledge.

Assessments can be of value in many ways. The ability to assess can influence a person's success, and it is especially important to the advancement of a person's critical thinking ability. Assessments allow the person to gain the maximum benefit from a problem-solving experience. In this chapter, emphasis is placed on understanding the criteria that can be used in the assessment of activities associated with creative thinking and critical problem-solving.

8.2   DEFINITIONS

The word *assessment* is commonly associated with a financial evaluation, such as for a tax that can be levied. This interpretation does not apply in reference to critical analysis. Herein, assessment is used more in the sense of an evaluation that relates to the knowledge of some activity. Where possible, a quantitative assessment is generally of greater value than a qualitative assessment, which may allow too much opportunity for subjectivity or uneducated opinion. Quantitative assessments are more useful in making comparisons as long as the quantities are not very uncertain.

A few definitions can be useful to the understanding of assessment:

- *Critic*: One who forms and expresses judgment of the merits and faults of anything.
- *Critical*: Characterized by careful evaluation and judgment.
- *Criticism*: The art, skill, or processing of making discriminating judgments and evaluations.
- *Assess*: To evaluate; appraise.
- *Evaluate*: To ascertain the value or worth of; to examine and judge.

Many factors influence the compilation and quality of an assessment. Knowledge and experience are the two most general and important factors. Knowledge of both the problem and the implications of the outcome is important. To provide a high-quality assessment, the assessor will need to have had extensive experience with the type of problem. The assessor must also have experiences that provide knowledge of problem-solving of complex problems. Experiences in thinking that go beyond logical thinking and basic idea generation are helpful. Past experiences that are related to the actual problem type will be helpful in developing an assessment based on both proper depth and an appropriate breadth.

The range of experiences in creative problem-solving activities will influence the quality and the value of critiques and assessments. Both breadth and depth of experiences should be sought and properly balanced. Depth of experience should not be sacrificed to breadth, and breadth should not be sacrificed to depth. If possible, cases should be reviewed prior to agreeing to the task to determine whether or not it will contribute to knowledge representing breadth or depth.

In some cases, critical or creative activities may not produce the desired results; these activities may labeled as failures. Of course, the success-failure continuum suggests that a dichotomous failure may not actually be a total failure. Assessment reports of such activities are especially important. Failures can contribute to growth in knowledge of assessment only when the failure is followed by a self-introspective analysis. Reasons for the failure should be acknowledged and used to revise the decision-making strategy. The new strategy will then undergo assessment in the future.

Assessment involves the important skill of critiquing. In Chapters 9 and 13, critiquing is identified as one of the skills used to evaluate the works of others. It can be used to identify new research topics or the quality of completed activities. Critiquing is an assessment of the deficiencies of past research, with new research used to overcome the deficiencies. The end product of research-oriented critical assessment is a list of ideas, not a numerical value that indicates quality. Critiquing can also be used to rate the quality of the work, even assigning a grade to the work. This use of the critiquing skill enables the assessor to understand deficiencies that prevented success. While these two uses of critiquing have different objectives, the assessment uses many of the same thinking skills.

8.3 ASSESSMENT CRITERIA AND RELATIONS WITH CRITICAL THINKING

Just as decision criteria are used to judge results of experimental analyses, assessment criteria are metrics used to judge the results of some activity. Table 8.1 identifies assessment criteria that can be used to judge or rate the overall effectiveness of an idea generation activity, such as a synectics or a brainstorming session. The first six criteria can be used to assess the group as a whole. Items 7–11 in Table 8.1 are primarily oriented toward the assessment of a facilitator. The remaining criteria, Items 12–19, can be used to assess the performance of individual participants. Criteria are often correlated with each other. Some criteria are more appropriate for assessing one type of activity, such as the Delphi method, while others would be better for an alternative method, such as the synectics approach. It is important to select criteria

TABLE 8.1

Criteria for Assessment of Critical and Creative Thinking Activities

	Criterion	Definition	Relation to Critical Thinking
1	Diversity	Extent to which ideas differ; variety.	The set of ideas include a wide array of seemingly unrelated entries.
2	Freedom	Independent; free of restraint.	The fraction of participants who seemed uninhibited to provide ideas.
3	Novelty	Both new and unusual ideas.	The freshness of the ideas is unusual.
4	Comprehensive	The group's ideas show breadth in scope.	The collection of the group's ideas shows breadth.
5	Complexity	The condition of an idea being intricate.	Ideas are detailed and allow for direct connections to decisions.
6	Uniformity	Unvarying; consistency; steady pace at generating ideas.	The similarity across all participants in the fluency and quality of the ideas generated.
7	Disorderliness	Amount of confusion; extent of disarrangement of activities.	The problem was not stated clearly to the group, which hindered idea generation.
8	Elaborate	Amount of detail provided; develop thoroughly.	The facilitator rephrased ideas to be specific, which enable decisions to be made.
9	Flexibility	Capable of being modified; responsive to change.	Extent to which lateral thinking was encourage by the facilitator.
10	Fluency	Flowing smoothly; effortlessness; the number of ideas generated.	Little downtime between the generation of ideas due to the facilitator's control of the process.
11	Originality	Being first; authentic; highly distinctive; independent in thought; unconventional thinking.	The extent to which the generated ideas seem to be independent of the problem/issue.
12	Usefulness	Capable of being used in a beneficial way.	The person's ideas proved to be beneficial.
13	Breadth	Freedom from narrowness; an individual's ideas have a wide scope.	The person's ideas result in a variety of decision alternatives.
14	Quality	Excellence of a characteristic; superiority.	The proven usefulness of an idea or set of ideas.
15	Metamorphological	Transformation by magic; alter in structure; a change in function.	Gave imaginative ideas that could be modified to help with decisions.
16	Imaginative	Indulge in the fanciful creation in the mind; to frame a mental picture.	Generation of ideas that do not seem to be real-world creations; other worldly.
17	Multi-interpretive	Can be interpreted in many ways; not a specific idea.	Individual gives ideas that are difficult to transform to decisions.
18	Uniqueness	Unparalleled; only one of its kind.	Individual generated ideas unlike other ideas in the list.
19	Courage	Willingness to explore new ideas.	The confidence to present ideas; a willingness to overcome inhibitors.

that are most appropriate for the idea generation method being applied, as well as the nature of the problem. Other criteria may be developed and included in assessments of creative activities.

8.4   ASSESSMENT OF IDEA GENERATION SESSIONS

Brainstorming was presented in Chapter 3 as a two-phase activity: idea generation and idea evaluation. It really should be a three-phase activity, with assessment being the third phase. The extent of learning will be minimal unless an idea or activity undergoes assessment. The goal of assessment for a brainstorming session is to identify issues or problems about which lessons can be learned from the session. Such knowledge can make future brainstorming sessions more effective and efficient.

After completion of the two phases of the brainstorming session, i.e., idea generation and idea evaluation, the creative activity should be assessed. The objective of any assessment should be to identify ways that improvements could be made when the techniques are applied to future problem-solving efforts. Assessments of the facilitator, the individual participants, and the group as a whole can provide valuable input to the planning of future idea-generating activities. Other assessments, such as the value of piggybacking, can contribute to learning. If organizational efforts at creative thinking and idea innovation are to continue to be productive and of value in meeting personal and professional responsibilities, then assessment efforts at such activities need to be conducted properly. Assessments should be comprehensive and not limited to just a statement that the outcome of the effort was successful or unsuccessful.

8.4.1   OBJECTIVES OF ASSESSMENT

The following aspects of idea-generating sessions need to be assessed:

1. The performance of the facilitator during the session.
2. The assessment reports completed by the facilitator.
3. The facilitator's control of dead time.
4. The group of participants collectively.
5. Each individual participant.
6. The extent of ideas generated: imaginative versus piggybacked.
7. The effects of conflicts between individuals.
8. The appropriateness of the allotted time for each phase.
9. The quality of the ideas generated.

A thorough assessment of each of these components will enable improvements to be made so that future idea generation sessions are both more effective and more efficient.

At a minimum, assessment should be made of the facilitator, each individual involved in a session, the group as a whole, and the final outcome. Additionally, if piggybacking was involved as part of an idea generation activity, this activity should also be independently evaluated. Piggybacking should be assessed, as it could have implications to the overall effectiveness of the group's productivity. Excessive piggybacking may suggest too little divergent thinking. The extent of divergent thinking relative to lateral thinking should also be a central focus of assessment. Like piggybacking, excessive logical thinking may limit the number of imaginative ideas generated. Too little piggybacking might suggest a narrowness of thinking. The extent of divergent thinking relative to lateral and creative thinking should also be a central focus of the assessment. Like piggybacking, excessive lateral thinking may limit the inclusion of imaginative ideas. If the number of lateral ideas is considered excessive, then the facilitator needs to identify activities that divert the ideation away from lateral thinking and more toward imaginative thinking. The facilitator needs to plan in advance for such a time when piggybacking exceeds the level that it should.

8.4.2 Assessment of the Facilitator

The activity is led by a facilitator, who acts as the leader of the activity. A primary objective of a facilitator is to maximize the efficiency of the idea generation session and to direct the development of a unique solution to the problem. Therefore, the performance of the facilitator should be critically assessed. Criteria for the assessment of a facilitator should answer the following questions:

- Was the session outcome of value above an outcome that would have been achieved using just logical thinking?
- Did the facilitator provide the participants with continuous encouragement?
- Did the facilitator ensure that ideas were not criticized during the idea generation phase?
- Was the amount of piggybacking optimum—not excessive, yet adequate to minimize downtime?
- During periods of downtime, did the facilitator change the format to encourage revitalization of the effort?
- Did the facilitator make timely reports to summarize the effort?

The stakeholders should specify the format for all assessment activities.

In addition to the facilitator's leadership qualities, the effectiveness of a session will influence the appearance of the facilitator's effectiveness. The assessment of the preparation and conduct of a brainstorming activity is also important. Factors that influence session effectiveness and can be used in the assessment of the facilitator could include the following: (1) the clarity of the problem statement provided by the stakeholders, (2) the quality of the physical environment in which the exercise was held, (3) the correctness of the technical makeup of the participants, and (4) the attitude of specific individuals who acted as inhibitors to the group's productivity. Before future group efforts are held, any problems that occurred should be addressed so that they do not re-occur at future sessions.

8.4.3 Assessment of Piggybacking

Piggybacking is a means of increasing the number of generated ideas and simultaneously maintaining creative enthusiasm among the participants. Some critics may argue that piggybacked ideas may not be as valuable as new imaginative ideas. Piggybacked ideas can be very valuable to a brainstorming or synectics session, as they can be used to minimize downtime and maintain the flow of ideas. However, if too many of the ideas reflect piggybacking, then the overall success of the session can be jeopardized, as imaginative ideation may be suppressed by the group. Statistics on the optimum amount of piggybacking are not known, and it would certainly vary with the complexity of the problem. However, piggybacking needs to be controlled by the facilitator, as emphasis should be placed on the generation of divergent ideas rather than ideas based on logical and lateral thinking.

In assessment reports that are drafted following a creative session, the extent of piggybacking can be an important part of the report. The report should discuss the reasons that it was effective or the reasons that the piggybacked ideas did not contribute to the quality of the session. The facilitator should make recommendations for ways to control or encourage piggybacking.

8.4.4   ASSESSMENT OF THE GROUP

A facilitator should also make an assessment of the group as a whole, more importantly, their productivity and their collective attitude. The ultimate criterion for evaluation of a group is the extent to which the problem was solved; however, the efficiency of the group's effort should also be part of the assessment. The generation of ideas should not be dominated by one or two individuals. The uniformity of the number of responses across the individuals of the group would be an important criterion for assessing group performance and should be documented in the report.

If the group's overall productivity was below expectations, then the facilitator should identify reasons for the lack of productivity. Identifying conflicts between participants should be noted, so the selection of participants for future sessions would be sensitive to the importance of including conflicting individuals. The facilitator could comment on the diversity of the technical expertise of the group, as well as its effect on the quality of the results. If the interests and experiences of the participants produced a group that lacked adequate expertise, this deficiency should be noted with specifics provided. Reasons for any assessment deficiency are very important sources of knowledge, which is valuable knowledge for improvement of future activities. Also, the adequacy of the facilities can influence the productivity of the group. Specific problems and positives of the facilities should be noted. The environmental conditions should be assessed and documented; for example, a room with poor temperature control can lead to poor group performance. Other factors include the size of the room, the facilities for recording the ideas, and even the appropriateness of the time of the day.

8.4.5   ASSESSMENT OF INDIVIDUALS

The success of an idea generation session is very much dependent on the facilitator, but also on the attitudes and abilities of the individuals. All participants should be individually assessed, and the extent to which each person met the responsibilities documented. A participant has the following responsibilities during the first phase of a critical or creative thinking activity:

1. To provide his or her fair share of creative ideas.
2. To avoid criticizing the ideas of the other participants.
3. To avoid conflicts with the other individuals.
4. To assist the facilitator by preventing downtimes.
5. To contribute piggybacked ideas.
6. To display body language that indicates that the activity is taken seriously.

The facilitator can use these responsibilities as the basis for assessing each participant. Ultimately, did the participant make positive contributions to the session? One person should not be allowed to dominate the generation of ideas, nor should any individual not contribute ideas. Individuals who do not actively participate in a critical or creative thinking activity are essentially acting as inhibitors, as the other participants may think that the non-contributor believes the exercise is of little value. Therefore, a facilitator should attempt to motivate those who are not actively participating during the session. If a person is not responsive during one session, then it may be best to avoid including the person as a participant in future sessions unless the reason for the non-contributing attitude is justified. Success is more likely to be achieved when all participants are uniformly and actively involved in generating ideas; thus, each participant should be assessed on his or her level of participation, with the assessment discussed in the facilitator's final report.

To the extent possible, the assessments of individuals should be based on objective criteria. Participants should be judged on their fluency, the quality of their ideas, and the extent to which their contributions are based on novel ideas rather than logical thinking. Too few novel ideas would suggest that the person was not an important contributor. The objectivity of each of these criteria varies. The final outcome of the session is likely to be due to significant interactions between group members, as members can serve as either motivators or inhibitors of others. While it may be difficult to identify a single individual as being responsible for proposing the best idea, the relative contributions of different individuals to the optimum solution need to be assessed and recognized; this is true even though assessments of individual contributions can be quite subjective. The best idea may actually be the result of the enthusiasm of the group during the idea generation phase rather than the knowledge of the person who recommended the idea that led to the final decision.

8.5 DEVELOPMENT OF ASSESSMENT REPORTS

Assessment reports are valuable summaries of idea generation activities. They can provide stakeholders with information that can be used to evaluate the quality of the group's decision, the conduct of the current session, and its usefulness in planning future problem-solving activities. Incomplete or poor reports deny the stakeholders with knowledge of quality of the effort. Timeliness is also important. Assessment of the work for which the stakeholder contracted can be used to justify the use of the resources. The facilitator is often the party responsible for compiling most of the reports.

While job-to-job assessment reports might be based on similar assessment criteria, it is preferable to submit unique reports for each activity. In developing an assessment report, the facilitator should list the evaluation criteria that were believed to be important. The actual criteria used may be specified by the stakeholders who initiated the work. The facilitator may choose to list and discuss them in order of importance. Most reports will be qualitative in nature, but quantitative data should be used when possible. The criteria listed in Table 8.1 identify possible criteria to include in the assessment reports. While the criteria may appear to be qualitative in nature, it may be possible to establish quantitative indices for some of them. The stakeholders may ask for additional criteria to be included in a report.

Ordinal scale ratings could be the primary format for the report. Likert scale ratings are common. For some criteria, other formats may be used, such as a pass-fail index or essay-type, open-ended reports. Four- and five-rank Likert scales may be better, as they show some measure of dispersion. These summaries can be important, but the prose part of the detailed report is often the most informative section of the overall assessment. Reports should be submitted as early as possible to avoid the loss of memory of specific events.

8.6 THE ROLE OF VALUES IN ASSESSMENTS

Values are important in decision-making and can be determinants of the success of a creative activity. Unbiasedness was mentioned as an important decision criterion. A facilitator of a brainstorming session must not show bias in his or her assessment of the participants. Individual assessments of the participants are essentially reports on the accountability of each participant, where accountability can be defined as *being answerable for duties*. The facilitator needs to provide accurate assessments because future decisions may depend on the assessments. Promptness is certainly a factor in the success of a session. For example, in the Delphi method, a lack of promptness by one or two participants can delay progress. Self-discipline, reliability, variety, and industriousness are other values that can be used in the assessment of critical thinking exercises. Variety, or the diversity of the responses, is important as it is likely correlated with the number of responses that can be considered novel.

Biasedness is a major issue in any critical analysis or assessment. Every person has biases, some minor but others more significant to the point where critiques and assessments can be adversely influenced by the biases. When a reviewer does not appreciate the extent of his or her own bias, biases can distort critiques and assessments. Thus, the reviewer of the assessment report may develop beliefs or attitudes that are counter to those that would have resulted from unbiased reports. An author of a critique or an assessment who knowing allows his or her bias to distort the work of another person is actually acting unethically unless he or she acknowledges the bias as part of the assessment report. Readers of the reports must be aware of the potential effects of biases on any knowledge gleaned from someone else's work.

Assessments can influence the thoughts and actions of reviewers of the assessment. Therefore, a critical assessment should be unbiased, which requires the assessor to have a good value system, and knowledge of both the subject matter and critiquing. The assessor or critic, therefore, must know the best way to critique the work and to provide a fair assessment to the reviewer.

8.7 CONCLUDING COMMENTS

Brainstorming and synectics were presented as two-phase activities, idea generation followed by idea evaluation. The need for the two phases is obvious. The assessment of sessions is essential, and thus, it represents a third phase. Assessments are important because they provide the stakeholders with some measure of the quality of the work and the degree of certainty that can be placed on the final decision. An assessment that indicates that the group provided a poor effort might concern the stakeholders as to the strength of support for the group's decision.

How can critiques that are provided by the art and movie critics help a person be a better assessor of creative activities in his or her own field of interest? A movie critic may criticize movies on the basic components, such as plot, character development, special effects, and acting. The art critic would use quite different criteria, such as the use of color, the location of objects, and the use of shadows. Therefore, some assessment criteria are very topic dependent. A person who is not in the arts and entertainment fields should still recognize that the art forms have general characteristics on which the specific art form can be evaluated. Therefore, when a person has a responsibility to submit an assessment, it is important to ensure that the most relevant decision criteria are selected. A person who provides a critical assessment of a professional publication to identify potential ideas for research would need to have a different set of criteria than a person who is interested in assessing the same article but to evaluate it for novelty. The point is: the effectiveness of any assessment depends on using the set of decision criteria that will provide the most useful knowledge.

Promptness is a value and should be applied in any assessment. The facilitator should provide the stakeholders with an assessment summary shortly after the decision is submitted. A delayed submittal may result in the adoption of the stakeholder's decision to implement the finding and then later realize that the group's effort was sub-par, which places uncertainty in the value of the action.

Assessments can provide useful knowledge, but assessments can be somewhat subjective, as a portion of any assessment can be nothing more than opinions. An assessment that is based on a person's biased opinion may be of little value. In many cases, a bias is legitimate because the person may have responsibilities beyond the immediate concern. A bias is wrong when it exceeds its legitimacy. A bias that places an unusual constraint on decision-making can result from selfishness. Therefore, the quality of an assessment depends on the assessor's ability to properly critique an activity.

Critics are often accused of being pessimistic, as many of their comments tend toward the negativity, i.e., they stress the poor side of the issue. Yet, if a person digests the work of a pessimistic analysis, knowledge can still be gained. This thought suggests that one purpose of assessment is the accumulation of knowledge.

Understanding can arise from the results of a critical assessment, whether it is positive or negative. Thus, assessments should be viewed as a valuable source of knowledge about an issue, so they should be sought, not avoided. Negative critics can provide important knowledge.

8.8 EXERCISES

8.1. Define assessment in general terms. Develop a definition that could specifically apply to the assessment of research.

8.2. Define the word *bias*. Discuss how it can be applied to the assessment of a brainstorming session. Identify ways that bias can be detected.

8.3. Explain why assessments of creative activities are of value. What can be learned from someone else's assessment?

8.4. Why might quantitative assessments be more useful than qualitative assessment? Why might qualitative assessments be more useful that quantitative assessments? How can quantitative measures be developed for the qualitative indicators?

8.5. How does a person accumulate sufficient experience to become a critic of creative activities?

8.6. What criteria could be used to indicate that a creative activity was a failure?

8.7. Develop a quantitative metric to assess the performance of a facilitator in directing a brainstorming session.

8.8. Develop one or more criterion that would measure the extent to which piggybacking helped or hindered the quality of a brainstorming session. Develop quantitative metrics for each criterion.

8.9. How can a facilitator control a group that is spending too much time generating piggybacked ideas?

8.10. Develop a new method that a facilitator could use to combat downtime during a brainstorming session.

8.11. Outline a format for a report that a facilitator can use to summarize the assessment of individuals who participated in a brainstorming session. Include all of the important factors.

8.12. Define the value of accountability and discuss the ways that it applies to the assessment of creative idea generation activities.

8.13. What assessment criteria should be used to rate the quality of a checklist.

8.14. Would it be appropriate to make assessments for intermediate rounds of the Delphi method? Provide the criteria that would be used. Explain.

8.15. Outline a format for a report that a stakeholder could use to assess the performance of a facilitator.

8.16. Outline a format for a report that a facilitator could use to summarize the assessment of a group.

8.17. Outline a format for a report that a facilitator could use to summarize the assessment of the use of piggybacking during the session.

8.18. How can a person accumulate sufficient experience to become a critic of creative activities?

8.19. How might a reviewer of a critique identify a writer's bias?

8.9 ACTIVITIES

8.9.1 ACTIVITY 8A: ASSESSMENT OF INHIBITORS

Chapter 7 is devoted to creativity inhibitors and their effects. Develop metrics for assessing the effects of inhibitors on the creative productivity of an idea generation activity.

8.9.2 ACTIVITY 8B: DEVELOPMENT OF ASSESSMENT REPORTS

Provide a format for a facilitator's assessment of a brainstorming session for (a) the group as a whole, (b) an individual in the group, and (c) the degree of piggybacking versus novel ideas.

8.9.3 ACTIVITY 8C: ASSESSMENT OF UNDERGRADUATE RESEARCH

Identify and define assessment criteria that could be used in the assessment of undergraduate research. Also, discuss ways that the criteria could be ranked to provide an overall measure of research quality.

8.9.4 ACTIVITY 8D: NOVELTY

Novelty is often used as a measure of the quality of research output. How could the ideas generated during a brainstorming session be assessed on the basis of novelty? What quantitative measure could be used to rate the novelty?

8.9.5 ACTIVITY 8E: ATTITUDES AND ABILITIES

Six responsibilities for participants were identified in Section 8.3. Identify a quantitative criterion that could be used to assess each one of the six, including ways of ranking each of the six.

8.9.6 ACTIVITY 8F: RATINGS OF 1, 2, 3, AND 4

For each of the following, develop assessment metrics: (a) college lecturers; (b) mystery novels; and (c) rock concerts. Discuss the development of a single-valued assessment metric.

8.9.7 ACTIVITY 8G: PASS, RUN, OR PUNT

Develop a list of five or six criteria that you use to rate the experience of watching your favorite football team. Transform each of the criteria to a criterion that could be used to rate the quality of a brainstorming session.

8.9.8 ACTIVITY 8H: IF THE GLOVE FITS...

You are owner of an advertising agency. You were hired by a company that produces gloves to wear while washing dishes. Your top person has finalized the selection to five possible advertisements to market the gloves. What decision criteria would you use to pick one of the five alternatives?

8.9.9   ACTIVITY 8I: PROTECT SMOKEY

You are hired by the state to develop a method of assessing the quality of small parcels of land of about 40,000 m^2 for wildlife. The wildlife might include deer, raccoons, and fox, among others. What criteria would you include in your assessment?

9 A New Perspective on Critical Thinking

CHAPTER GOAL

To place critical thinking into a broader framework, one based on the diverse dimensions of values, mindset, skills, and types of thinking.

CHAPTER OBJECTIVES

1. To define critical thinking and present it as a process.
2. To summarize important attitudes and characteristics of critical thinkers.
3. To present ways of improving and providing assessments of critical thinking abilities.
4. To contrast critical thinking and creative thinking.
5. To identify the factors that influence problem complexity and show the ways that problem complexity influences problem-solving.
6. To introduce group (organizational) critical thinking.

9.1 INTRODUCTION

Was Nicolaus Copernicus (1473–1543) a critical thinker or was his heliocentric theory just an innovation to explain the ideas of heliocentric thinkers of the past dating back as far as Aristarchtus in the third century BC? Did Francis Bacon (1561–1626) really develop the Scientific Method or did he just codify da Vinci's method of solving problems? Was Charles Darwin (1809–1882) a critical thinker or just an innovator of Wallace's work? Did critical thinking exist during the scientific revolution of the fifteenth and sixteenth centuries? How about creative thinking? When did people start to use brainstorming? Were Leonardo da Vinci's (1452–1519) sketches evidence that brainsketching existed as a common creative thinking practice in his time or was it just da Vinci's informal way of recording random thoughts, i.e., doodling? Before questions like these can be answered, the concept of critical thinking must itself be explained.

The word critical can be defined as either *inclined to judge earnestly* or *characterized by exact judgment*. So, it appears that the word *critical* refers to something that is being evaluated or judged thoroughly and unbiasedly. The word *critical* is often associated with a person's health, i.e., a hospitalized person is in critical condition, which again emphasizes a condition that is of special note. With respect to critical thinking the word *critical* needs to be extended to use with the word complex, as

DOI: 10.1201/9781003380443-9

problem complexities are the severe conditions to which thinking is applied. The term critical thinking can be interpreted as formulating thoughts about important activities. The term critical thinking could refer to the processes of problem-solving, research, and decision-making; therefore, the term critical thinking may be a pseudonym for critical problem-solving, critical research, or critical decision-making. It is only when the situation specifically points to a complex activity such as one of these that the distinction needs to be made.

The word think means both *to have a thought* and *to formulate in the mind*. The word thinking can be defined as *the mental processes that are used during thought about some event or issue that could be a problem*. A dictionary may provide the following insight to the word *thinking*: (1) to reflect on; (2) to decide using mental processes; (3) to exercise the power of reason; (4) to conceive of ideas, draw inferences, and use judgment; and (5) to weigh an idea. These definitions suggest that thinking is a tool for decision-making. With respect to critical thinking the word critical needs to be extended to use with the word complex as problem complexities are the severe conditions to which thinking is applied. These brief insights suggest that thinking itself is a process, and not just a momentary state of mind where an idea is conceived. Definition 2 suggests that thinking involves decision-making, while Definition 4 suggests that thinking starts well before making a decision. Integrating these five definitions clearly suggests that thinking is involved in all phases of the decision process from the identification of a problem to the endpoint of making a decision. The view of thinking as a fundamental element of all phases of the process is a necessary precursor to education about critical thinking. In moving on, thinking can briefly be defined as *mental processes used to reason or reflect on an issue*.

The solutions to complex problems often lead to the development of new policies, significant advances in the state of knowledge, major reorientations of organizations, or beneficial changes to our society. Knowledge is very important to society and knowledge-based problems can be very complex and difficult to solve. For our purposes herein, the intent is to identify ways that thinking and acting improve decision-making that is part of solving complex problems. Critical thinking is one such way.

A person's success in solving complex problems partly depends on his or her critical thinking ability. When a person needs to solve a complex problem, knowledge of critical thinking can be a valuable asset. Answers to questions such as the following can help advance a person's knowledge of problem-solving:

1. What is critical thinking?
2. What special characteristics do critical thinkers possess?
3. How does someone develop or enhance his or her critical thinking ability?
4. How does knowledge of critical thinking influence a person's success?
5. Will strong critical thinking skills improve one's chances of success in solving complex problems?
6. What constitutes a complex problem?
7. What education and experiences are necessary for a person to be a critical thinker?

It is questions like these that this chapter intends to answer.

9.2 INTRODUCTION TO CRITICAL THINKING

What is critical thinking? We should all agree that it is more than just creative thinking, although many critical problem-solving exercises may involve creative thinking activities. Beyond this fact, the scope of critical thinking is largely debated and unfortunately agreement has not been reached on a universal definition.

9.2.1 THE DIMENSIONS OF CRITICAL THINKING

A proposal is presented herein to develop a model of critical thinking that is a multi-dimensional system for solving complex problems. Putting aside some problem-specific dimensions, such as the technical and economic dimensions, critical problem-solving will center about four other dimensions: values, mindset, skills, and thinking types. These four dimensions act as constraints on problem-solving; however, focusing on these four dimensions does not imply that the other dimensions are unimportant, just that dimensions like the technical dimension are too specifically oriented to the problem that needs to be solved. It will be assumed that the critical thinker has the necessary expertise with problem-specific dimensions like the technical and economic dimensions. The four dimensions discussed herein are universally applicable and should be applied with the technical, political, and economic aspects of a problem, which can be quite variable.

A primary focus in this book is in finding solutions to complex problems. Complex problems have quite varied characteristics, and these characteristics dictate the need for a multi-dimensional set of decision criteria. Thinking type is considered to be an important dimension because the type of thinking needs to be matched with the characteristics of the problem; this will optimize the effectiveness of the analysis. The likelihood of successfully solving a complex problem is greater when the thinking type can direct the imaginative thinking to the most difficult aspect of the problem. A person's mood, disposition, beliefs, and principles reflect the mental processes that he or she will use in developing solutions to problems; therefore, the mindset is the second important dimension. Decisions will be a function of the decision maker's mindset. The way that a person approaches the solution to a complex problem depends on the person's state of mind or the time when he or she is confronting the problem; thus, the mindset dimension is important. A person's mood or disposition could distort the person's thinking based on his or her beliefs. Controlling one's mindset is an important characteristic of a critical thinker. Both the decision maker's values and the values that are relevant to the stakeholders of the problem can greatly influence the outcome; therefore, when solving most complex problems, it is important to include a value dimension. Stakeholders often have quite different objectives, with some objectives being value laden. Value conflicts are often a dominant constraint on problem-solving. Mental skills, such as curiosity and skepticism, influence the way that a decision maker approaches problem-solving; therefore, critical thinking must include a skills dimension. Knowing the roles that these four dimensions play in solving complex problems will increase both the effectiveness and the efficiency of the problem solver.

Critical problem-solving is a method for determining the outcomes of important, complex problems. Critical problem-solving goes beyond the creative problem-solving approach, as other factors must be considered. From my perspective, critical problem-solving is a multi-dimensional process for solving complex problems and at a minimum involves the use of the following four dimensions:

1. *Thinking type*: Selects the thinking types that are most appropriate for solving complex problems.
2. *Skill set*: Systematically applies a sequence of skills to extract all relevant knowledge from theory and data.
3. *Value set*: Properly balances all value issues relevant to the making of decisions.
4. *Mind set*: Controls the moods, dispositions, and attitudes that influence actions and decisions.

These four dimensions act as restrictions on the solution of complex problems. Acting as restrictions is actually not a negative or limiting factor. The added dimensions act as restrictions but are positive factors because they ensure decisions will not be made haphazardly, i.e., that aspects of these four dimensions will improve the quality of decisions. Additionally, these four restrictive dimensions should add confidence to decisions, as uncertainties associated with variations in all of the four dimensions would be under greater control than if the factors were not restricted.

9.2.2 DEFINITION: CRITICAL THINKING

The individual definitions for the two words *critical* and *thinking* are not fully able to convey the real meaning of the term critical thinking. In reference to problem-solving, critical thinking can be defined as *the art of analyzing the existing state of knowledge to identify known problems and then use imaginative mental processes that involve multiple types of thinking, multiple values, and multiple skills to identify potential solutions from which an optimum solution to a complex problem can be synthesized.* Thus, the term critical thinking represents *the thought processes used when making a careful examination of an issue, especially a complex problem.* This definition is useful as a starting point for examining ways of more effectively and efficiently investigating a complex problem. Hopefully, critical thinking will yield better decisions than those that result from the narrowly focused one-dimensional thinking. We need to recognize that this previous interpretation of the individual words *critical* and *thinking* must be viewed more broadly when they are combined, i.e., *critical thinking.*

Critical thinking involves both analysis and synthesis. Through methods of analysis and synthesis, novel solutions to complex problems can be identified by systematically considering the multiple dimensions. For example, innovative research ideas can easily be developed by identifying and analyzing assumptions or constraints that the authors of published papers have placed on the reported research; this detection process involves both a challenging mindset and the directed application of a set of skills. Any assumption or constraint can be relaxed and used as the basis for new research. Identification of the problem is the analysis part of the solution while relaxing the constraint and developing a solution is the synthesis part. Variations of the assumptions that were used to develop the existing state of knowledge can be proposed, which represents a method of analysis. Testing of the assumptions would represent the critical synthesis part of critical thinking. The ability to analyze the assumptions and constraints on a research topic requires a critical attitude, one that is founded on the belief that published research is not the end, but instead published research is the starting point for related work that can be both novel and provide advancements to the state of knowledge. Analysis as the input leads to synthesis as the output. Before this skill of identifying new research topics can be learned, it is important to learn the characteristics of those who are critical thinkers and the knowledge that is needed to become a critical thinker.

9.2.3 CRITICAL THINKING AS A PROCESS

As indicated in Chapter 2, many processes can be used to solve problems, such as the Scientific Method and the decisions process. Critical thinking is a process, but in a different framework. Critical problem-solving borrows one of the processes of Chapter 2, and subjects it to the constraints of the relevant dimensions, such as the skill and value dimensions. While the steps of a process are followed, the process is applied with consideration of each dimension. Each relevant dimension is applied, but the importance of each dimension would vary from step to step of the process being applied. Appreciation of the different objectives of the steps of the critical thinking process should lead to a more effective application of the different dimensions. In order to have a full understanding of the term critical thinking, issues such as mindset, values, skills, and thinking types need to be addressed, as a person's understanding of these factors can influence the extent to which he or she will develop the necessary critical thinking philosophy for solving specific complex problems.

9.2.4 DEFINITION: CRITICAL THINKER

A critical thinker is one who applies the principles of critical thinking while attempting to solve complex problems. When someone refers to a person as being a creative thinker, he or she is usually suggesting that the person is knowledgeable about creative thinking methods, such as brainstorming and the Delphi method, but has thinking abilities that are more appropriate for complex problems. While a critical thinker may use creative thinking in problem-solving, he or she views problem-solving more broadly by using the constraints of the other dimensions. In summary, understanding the attitudes and abilities that enable a person to be considered a critical thinker requires an appreciation of the characteristics of a critical thinker, which are discussed in Section 9.6.

It is not possible to present a single definition of a critical thinker as we can for many other words, as a critical thinker is one who possesses broad knowledge of various aspects of problem-solving and uses a variety of mental processes in the solution of the problem. The knowledge of the different dimensions that is required varies from problem to problem. Additionally, the depth of knowledge for any one aspect of critical thinking will vary with the problem. Thus, both the breadth and depth of various aspects of the problem are required and influence a person's level of critical thinking ability.

9.2.5 EXPERIENCES

When I refer herein to the term *critical thinking*, the thinking aspect of the method is possible only when the person has both the characteristics of a critical thinker and sufficient past experiences where it was necessary to apply the critical attitudes and skills. Experiences with complex problems promote learning, and the learning enables the person to participate in solving even more complex problems. The experiences should be broadly based, as complex problems are characterized by quite varied issues. To get a broad base of experiences, the person should seek problems that vary in the type of complexity. Experiences with a narrow set of complexity issues will limit a person's ability to confidently solve problems that have complexity characteristics about which the person is not familiar. The topic of experience is discussed in Section 9.12.

9.2.6 ADDITIONAL DIMENSIONS

The four previously mentioned dimensions of the critical problem-solving process do not include any hint of the technical details of the problem; however, the technical dimension will likely be a very important part of the problem. Other dimensions, such as economics, environmental, cultural, and political, may also be very constraining and important; however, the proposed four-dimension model should be applicable to any technical discipline, such as engineering, sociology, or music. Other general dimensions may need to be part of the growth in knowledge that is required to solve a specific problem. For any one case, dimensions such as one that emphasizes the effects of risk or one that promotes social issues such as public health may also be needed. Development in knowledge of these additional dimensions would follow the paradigm used to advance the state of knowledge of the four dimensions stressed herein.

9.3 CRITICAL THINKING VERSUS CREATIVE THINKING

Does critical thinking differ from creative thinking? If it does, how do the two methods differ? When the term creative thinking is mentioned, most people would narrowly think of brainstorming, i.e., the imaginative generation of ideas. Furthermore, Phase I, idea generation, of brainstorming would be emphasized. Brainstorming involves two phases, with a decision always made at the end of Phase II. Thus, creative thinking is really a problem-solving method, but it is not critical problem-solving, as the latter has greater breadth in terms of its multiple dimensions and the complexity of the problem that it is used in solving. Creative problem-solving focuses on generating ideas on a narrowly specified concern. Critical thinking focuses on solving a much broader problem, one that is complex in many ways. In summary, the two methods differ in the complexity of the problem for which a solution is needed.

One objective of this chapter is to create a broad framework for a new perspective on the term critical thinking, as it is really more than just thinking. It is thinking in the broader realm of applying constraints on the problem-solving process to ensure that the decision meets important decision criteria for multiple dimensions. When compared to creative thinking, critical thinking is more of a process. The steps of creative problem-solving focus on a statement of the problem (Step 1), the synthesis of ideas (Step 5), and a decision (Step 6). Creative thinking places minimal emphasis on resources collection (Step 2), experimental planning (Step 3), and analysis (Step 4). The use of the complete process is not necessary in creative thinking activities because of the relative simplicity of the task.

While a creative problem solver may consider aspects of dimensions beyond the technical aspect of the problem statement, they are generally not recognized as fundamental metrics of creative problem-solving. Critical problem-solving will generally consider a broader array of dimensions than would be needed to just manage a brainstorming or synectics session. With critical problem-solving, in contrast to creative problem-solving, the use of multiple thinking types is most often a requirement for solving complex problems. Thus, the values, skills, and mindset dimensions of critical problem-solving are additional distinguishing factors in making the comparison. With creative thinking activities, it is best to have an open-ended mindset for the duration of the activities. Critical thinking involves a broader perspective on the mindset, as principles and variations in mood are more relevant. The creative thinker needs to be in the mood to generate ideas, with emotional thinking being the central element of the thinking. Critical thinking uses the sequence of the skill set, while creative thinking focuses primarily on the skills of courage and critiquing, where the courage skill focuses on the mindset to let the emotional thinking dominate the mindset dimension. Critical and creative thinking have the same general objective, which specifically is to develop a novel solution for a problem, but the two methods approach a solution with quite different philosophies, with critical thinking formally tied to multi-dimensionality.

A primary difference between creative and critical thinking is the emphasis that is placed on different characteristics of the thinker. Critical thinkers are constrained by many dimensions, while creative thinking is less constrained. Critical thinking focuses more on skills like curiosity and questioning at each step of the problem-solving process. Being highly proficient at the sequence of problem-solving skills is ideal. Creative thinking places greater emphasis on the imaginativeness associated with the generated ideas and less emphasis on the entire skill sequence. The two methods also differ in their purposes. Critical thinking seeks more to identify and solve a problem using an array of skills and thinking types while being constrained by value and mindset issues. Creative thinking is less restricted and aimed more at unconstrained idea generation, but for the purpose of solving a relatively narrowly focused problem. Creative thinking is not as constrained as critical thinking, but the constraints placed on critical thinking should be viewed as positives.

9.4   A FRAMEWORK FOR CRITICAL THINKING

As previously indicated, critical thinking does not have a unique series of steps. Instead, it adapts the steps of another process to the problem under investigation. In other words, the critical thinking process builds in the characteristics of the relevant dimension into an existing process, such as the research or decision process.

9.4.1 STEPS OF THE CRITICAL THINKING PROCESS

The steps of the critical problem-solving process and their unique factors can be summarized as follows:

1. *Problem identification*: The objectives of this step are: (a) to help identify and firmly establish the underlying problem; (b) to translate the problem into a general goal and a set of specific objectives; and (c) to ensure that the stated goal and objectives will adequately address the problem. Given that critical thinking is used to solve complex problems, often with multiple stakeholders with conflicting objectives, the problem statement and objectives will need to reflect the complexity and these conflicts, as well as the ways that are needed to develop a solution that is appropriately sensitive to all of them. The courage skill is applied to have the confidence to approach solving the problem. The problem solver will need a mindset attitude that is based on principles of knowledge.

2. *Resource collection:* The objective of Step 2 is to obtain a broad array of sources of knowledge, including all relevant theory and existing empirical findings, and to assess the potential adequacy of the required data. It is also necessary to identify the dimensions that will need to be considered. The stakeholders and their demands and biases are part of the existing knowledge base and so they must be identified. The skill set capacity and values need to be known. Since the existing state of knowledge must be established, the critical thinker will need to fully appreciate the knowledge level and value constraints of each stakeholder. As an auxiliary skill, persistence is required to search through the array of sources of knowledge. Just finding all of the sources of knowledge and data requires persistence. The perception skill is also important, as it is necessary to decide on the importance of the various elements of the available knowledge base. Mindset constraints must be overcome, especially procrastination and pessimism, as they usually delay the completion of this step. When solving complex problems, the problem solver needs to be applying the most appropriate mindset; therefore, mindset control is an important ability of a critical thinker.

3. *Experimental plan development:* The objectives of this step are: (a) to establish all aspects of the experimental designs that will be necessary to test the specific objectives of Step 1 and (b) to ensure that the experimental plans meet all requirements for the general principles of experimental analysis. The results of each experimental plan are expected to be sufficient to fully answer the corresponding problem statement of Step 1. For complex problems, the experimental designs will need to be sensitive to the multidimensional nature of the case. Since problem characteristics often establish needs for different thinking types, the thinking dimension must be given special consideration. Imaginative thinking types may be needed to ensure that the experimental plans will enable all of the implications to be identified.

4. *Analysis and generation of results*: The objectives of this step are: (a) to collect and assemble the necessary data and use the data for all analyses

of the experimental designs, (b) to perform all statistical and/or modeling analyses, and (c) to generate results from the analyses and ensure that the results properly follow from the given extent and limitations of the data. For complex problems, results will need to be sensitive to stakeholder concerns. A critical thinker must ensure that the results are not distorted by stakeholder biases; thus, the value dimension is important. An attitude of open mindedness will enable the results to be fully identified from the analyses. Values of unbiasedness and truth should form a value foundation for the interpretation of the results.

5. *Synthesis of outcomes*: The objectives of this step are: (a) to critique the extent of the results of Step 4 and develop a set of conclusions, (b) to ensure that the results do not suggest outcomes that are unrealistic, and (c) to ensure that all potential benefits of the effort are identified. The mindset dimension is important here to ensure the mood and disposition during decision-making. A critical thinker should not distort the outcomes in a way that suggests a value dimension is being dominated by honesty and truth. The skill set dimension is a central concern here because the results will need to be verified.

6. *Make and report and decision*: The objectives of this step are: (a) to use the decision criteria and present the final decision and (b) to ensure that the outcomes selected are extended to show the implications of the problem-solving to issues beyond the basic technical conclusions. This task may be factors such as risks to society, application of the conclusions to governmental policies, and the implications to environmental risk.

This example of the critical solving process included six steps, with each step having a direct correspondence to the steps of the research and problem-solving processes. In each case, all four of the dimensions are relevant to the steps.

9.4.2 GROUP CRITICAL THINKING

While it is common to think of critical problem-solving as an individualistic activity, the term critical thinking can easily be extended to group problem-solving. The term *group think* generally has a bad connotation. Group think is often viewed as problem-solving by a group with nefarious intentions. For example, the decades-long cover-up of the Manville asbestos case has been cited as a case where immoral group-think could not be overcome. Just as group think can occur, group critical problem-solving can also occur, but group critical thinking usually results in positive consequences, especially where the value dimension is important. Very often, the group will consist of personnel of an organization and consultants who have unique knowledge and abilities. If critical problem-solving by an individual is considered a positive act, then group critical problem-solving should be afforded the same respect.

Group critical problem-solving is nothing more than a group of people who collectively have the same characteristics as an individual who has the characteristics of a critical thinker (see Section 9.6). These characteristics include specific criteria that are related to the dimensions of critical problem-solving. It is possible that no one person in a group can be independently considered a critical thinker as each may have at least one deficiency; however, the group would collectively have all of the necessary characteristics as would an individual who is a critical thinker. The group would need to appoint a leader or a facilitator who has most characteristics of a critical thinker and can manage the critical thinking characteristics to ensure that the internal actions meet the demands of critical problem-solving.

Collectively, the group must be sensitive to all dimensions of the critical problem-solving. In order to be considered as a critical thinking group, collectively the group should be relatively free of biases, i.e., they should not allow biases to inhibit their thinking and decisions. If a person lacks sufficient experience and knowledge of an issue, he or she should withdraw from the decision-making on that issue. The group should be open to using all types of thinking, not just be limited to logical thinking. When it is appropriate, the group should be capable of practicing imaginative thinking. When confronting a problem, the group should use all of the sequence of skills, including critiquing the issue and asking questions that will clarify everything starting from the problem statement to the final decision. Group curiosity that possibly involves imaginative thinking should be a dominant skill at meetings. The group should be optimistic and confident in evaluating alternative solutions and their actions and decisions.

Before being fully engrossed in the solution of a problem, the group should select a facilitator, as leadership will be vital to the success of the group critical thinking activity. The person who is selected to facilitate the activity should have many of the characteristics identified for a critical thinker. He or she will also be responsible for performing all elements of leadership. The facilitator should use a democratic style of leadership, as an autocratic style would likely be ineffective in a group where each individual has many critical thinking abilities. The chosen leader should be comfortable in facilitating imaginative problem-solving activities. While we commonly think of the mindset dimension in terms of a single person, the mindset of a group will be an amalgam of the mindsets of the individuals in the group, as the facilitator will need to ensure that the group mindset is always functioning properly. For many complex problems group critical thinking should be more effective than a single person, but differences in experiences and knowledge can introduce a source of conflict that will need to be addressed by the group facilitator.

9.5 CRITICAL ANALYSIS AND SYNTHESIS

Recall that analysis implies *to take apart* while synthesis implies *to put together.* Each step of the critical thinking process has an outcome that serves as an input to later tasks and the final decision. Thus, analysis and synthesis are part of each of the six steps of the problem-solving process. In the analysis task of each step of the process, the relevant information is collected and assembled into a form that can be analyzed. The intent of the analysis task for each of the six steps is to identify the primary unknown element of the step. In the synthesis phase of each step, the results of the analysis provide material that can be assembled, i.e., put together, into an end product of that step. The synthesis task for each step focuses on identifying the output for that step, which may serve as the basis for the input to the next step. Thus, the analysis-synthesis duo is central to each step, with the synthesis structuring an output for each of the six steps. Table 9.1 summarizes the intent of the analysis and synthesis at each stage of the problem-solving process.

9.5.1 CRITICAL ANALYSIS

Analytical skills are needed at each step of the problem-solving process. Table 9.1 provides a summary of the use of critical analysis in each step of the process. In Step 1, a person's ability to fully analyze the problem statement is necessary, as it is used to develop the problem statement, the general goal, and the specific objectives. This statement will influence the knowledge requirements of Step 2 and establish the research hypotheses of Step 3. Step 4 involves the actual analyses of data for each experimental hypothesis of Step 3. In Step 5, the experimental plans of Step 4 are analyzed to develop a set of conclusions that reflect a solution to the problem

---

TABLE 9.1

The Intent of Analysis and Synthesis for Each Step of the Problem-Solving Process

Step	Analysis	Synthesis
Problem identification	To identify the basic elements of the problem, the factors involved, and the potential implications of success.	To develop a final statement of the problem, the research goal, and the specific research objectives.
Resource collection	To analyze the current state of knowledge and to identify the available database.	To create a detailed statement of the state of knowledge; to identify deficiencies in the existing database.
Research plan development	For each objective of Step 1, develop an experimental plan; to analyze the data requirements for each objective.	To finalize the experimental plans, including establishing data and theory needs.
Analysis and generation of results	To analyze all data using such tools as statistical analyses, simulation, and qualitative analyses; to summarize any apparent deficiencies that may need additional analysis.	A willingness to question an inexplicable finding; to provide the experimental results of each experimental analysis.
Synthesis of outcomes	To evaluate each alternative using decision criteria and quantify or rank the criteria that will be used to assess the outcomes.	To identify the most feasible solution and to assess the extent to which the corresponding objectives were met.
Make and report a decision	To analyze the outcomes; to analyze whether or not the goal of Step 1 was achieved.	To write reports that identify the new state of knowledge, and distribute them to stakeholders and other interested parties; identify the implications of the effort.

---

statement of Step 1. In Step 6, the conclusions of Step 5 are placed in the context of a broad framework to identify the implications of the work to society or the stakeholders for which the work was completed.

The critical thinking process is better understood by identifying its objectives, knowledge requirements, and the attitudes that are best promoted through analyses. When conducting a critical analysis, the motivation is to identify both the weaknesses of the current state of knowledge and any restrictions on the application of the knowledge. It is the weaknesses of the existing state of knowledge that provide the motivation for developing potentially new knowledge that can be uncovered using a more in-depth study and used to form a better understanding of the true system. The skill and value dimensions enable the weaknesses to be identified, but they will only be effective when the person has the proper mindset. Problem statements can be formulated from identified weaknesses. Having the confidence to develop a statement of the problem is a necessary ability. Having the courage to act on the identified deficiency in knowledge is the first skill needed. For critical thinking to be most effective, the researcher has to have both narrow and broad knowledge of the issue. The application of the principles of critical thinking will be necessary to execute each of the six steps.

For a novice to develop the ability to make critical analyses, it may be best to start by developing the skills using everyday events rather than starting with a complex technical issue about which he or she knows very little. Research into topics about which the novice is more knowledgeable may serve as a better experience for learning problem-solving than using a complex topic. Issues such as solving a campus parking problem or finding solutions to problems with educational facilities may produce better learning experiences than trying to solve a complex technical problem about which the novice has little knowledge. A person can review day-to-day activities and identify real problems that will provide good problem-solving experiences. A novice should not select a topic to analyze about which he or she has recently read articles that included possible solutions; this type of problem will not provide sufficient challenge from which the experience would be beneficial. While situations that involve conflicting values or needs may not be complex to an experienced problem solver, they may seem complex to a novice critical thinker. Concepts of critical analysis and synthesis can be used to solve such problems in a reasonable period of time and produce maximum learning. After solving a few cases, the process can be assessed from a general standpoint and a general model of the learning process developed. These activities may provide the necessary preparation for using critical thinking skills for actual complex problem-solving situations.

9.5.2 CRITICAL SYNTHESIS

While synthesis is usually associated with Step 5 of the problem-solving process, synthesis is actually part of every step, as each step has an outcome developed via synthesis. The analysis part of Step 1 established the background of the problem. The objective of synthesis is to develop the final statement of the problem, which includes statements of the general goal and the set of specific objectives; the problem statement is important because the best solution cannot be found if the problem statement is not accurately formulated. In Step 2, the analysis involves discarding irrelevant knowledge and data, and the synthesis actively involves assembling the relevant knowledge and data that would be available and needed for Steps 3 and 4; this task may also include synthesizing a list of additional data that may not be currently available but would be needed for the analyses of Step 4. In Step 3, the objective for synthesis would be to ensure that the experimental designs that resulted from the Step 3 analyses will be adequate to test the study objectives stated in Step 1, as the collection of objectives must be able to fulfill the goal stated in Step 1. In Step 4, a set of verifiable experimental results will be synthesized into a form from which general conclusions can be identified using all of the possible alternative parts of Step 5. The synthesis of Step 6 is to identify the broad implications of the conclusions and communicate the results in the most effective way of disseminating the advancements in knowledge. In Step 6, a final assessment is also made as to whether or not the goal and objectives have been met; this step also involves identifying the implication of the effort. Critical synthesis has a role in each step of the problem-solving process, but the focus of critical synthesis centers on assembling a set of meaningful conclusions that show the problem statement of Step 1 has been solved.

9.6 FUNDAMENTAL CHARACTERISTICS OF A CRITICAL THINKER

The characteristics of a problem will influence the characteristics that the critical thinker will need to solve the problem, but the degree to which a person can achieve success depends on factors that are independent of the problem. A critical thinker requires knowledge and abilities that go beyond the norm. Both innate-based abilities and learned knowledge influence the potential capacity of a person to have the necessary capacity to think critically. Experience is a good teacher, so part of the capacity to be a critical thinker will be the result of quality experiences.

9.6.1 ESSENTIAL CHARACTERISTICS

A person's principles and beliefs are an integral reflection of various personal characteristics. The ability to think at a level that enables a person to solve complex problems depends on those principles. While any one person may not have command of all of the important characteristics, the critical thinker generally possesses many of them to a level that enabled him or her to make good decisions. Critical thinkers actively seek to improve their position on each characteristic. The following ten characteristics are central to becoming a critical thinker:

- **Imaginative:** Automatically and willingly applies his or her power of imagination to the solution of problems that seem unsolvable. Generating a large number of ideas seems natural to an imaginative creator and the challenge to identify novel options provides personal satisfaction.
- **Honesty:** When critically evaluating ideas and solving problems, the decisions should be impartial and unbiased.
- **Inquisitive:** Intellectually curious, interrogative, and given to asking insightful questions, an individual is willing to change the pattern of thinking to ensure that all ideas and alternative solutions are recognized and evaluated.

- **Receptive:** Willing to entertain new ideas and seek knowledge and experiences that will lead to self-improvement; however, the individual is willing to suspend judgment until the time is appropriate for making a decision or making a change.
- **Optimistic:** Takes a positive view into all problem-solving situations, with the belief that the best possible solution will be found by habitually expecting a favorable outcome.
- **Affective:** Willing to use emotions, especially to distort ideas in order to obtain a fresh viewpoint on a problem or issue.
- **Confident:** Has a self-assured feeling that he/she can contribute to finding the best solution; does not sway in the face of inhibiting attitudes or criticism believed to be unfounded.
- **Perseverant:** Steadily persistent in finding a solution and is not discouraged by temporary setbacks; indefatigable in the effort to find the best solution.
- **Tolerant:** Thrives on uncertainty; venturous and disposed to new ideas; enthusiastically entertains unconventional ideas; understands the uncertainties of decisions.
- **Initiator:** Takes the leadership to propose new ideas; does not wait for others to propose new solutions; the initiator's ideas often generate novel solutions.

Obviously, many of these characteristics are correlated with each other to some degree; for example, confident individuals are often optimistic. A person does not have to possess all of these characteristics in order to be considered a critical thinker. For example, a timid person who often shies from contributing imaginative ideas may still be a strong critical thinker because his or her experiences have led to the development of other characteristics that are important to a critical thinker. If a person believes that any one of these characteristics is especially important, but it is a personal deficiency, then the person needs to develop a self-improvement plan for that characteristic. Improving any of these ten characteristics should improve one's problem-solving efficiency, as it will ensure that the numerator of Eq. 2.2 increases at a greater rate than any increase in the denominator. Guidelines for improvement are provided in Section 9.10.

9.6.2 CHARACTERISTICS IN CONFLICT

The characteristics that were identified in the previous section should not be viewed as an either-or dichotomy. Instead, each one represents the upper end of a continuum. For example, a person does not have to feel either confident or not confident. The person might feel mildly confident, which is an intermediate position within the continuum. Also, a person may feel confident under one set of conditions, but not confident when other conditions prevail. Thus, we need to acknowledge that attitudes can vary both over a range and over time. We can loosely define ends of the spectrum for each characteristic and accept that at any point in time the influence of the characteristic on problem-solving can vary.

For each of the positive characteristics, a contrasting negative characteristic can be identified, with the positive and negative aspects serving as ends of a continuum for that characteristic. A person will usually not function at one of the two extremes, but will act at some intermediate level. A person's goal should be to move toward the positive end of each spectrum. If circumstances tend to suppress a positive feeling at any point in time, then a usually critical thinking person may not be creatively efficient at that point in time. It is worthwhile assessing the competing traits by showing both the positive and negative ends of each spectrum; adopting the viewpoint of characteristics varying over a range will make it easier to assess the importance of a weakness in any of the characteristics. The following provides a perspective on the spectrum ends of each of the ten characteristics:

Imaginative vs. Non-whimsical: Some people limit their thinking pattern to logical thinking; they generally are unwilling to recognize the merits of unconventional ideas, affective thinking, or taking actions that deviate from existing knowledge.

Honesty vs. Deceitfulness: Individuals may approach problem-solving in a way where their thinking prevents honest appraisals of imaginative ideas and creative solutions to problems. Their actions usually lean toward those that favor themselves at the expense of legitimate claims by others, which indicates a bias.

Inquisitive vs. Unquestioning: The unquestioning person generally lacks important skills such as curiosity trait and, therefore, does not recognize the merits of asking questions when confronting a problem.

Receptive vs. Narrow-minded: A narrow-minded person consistently lacks the vision to recognize the value of practicing alternative methods of thinking; having a breadth of ideas is discouraged and viewed as being noncommittal.

Optimistic vs. Pessimistic: A pessimistic person takes the view that his or her input to a problem will not help in developing a solution to a problem.

Affective vs. Unintuitive: Unintuitive people believe that productive ideas can only result from logical thought processes; they believe that affective or emotional involvement is counterproductive and a sign of immaturity with respect to problem-solving.

Confident vs. Unassured: The unassured person is uncertain about his/her ability to solve problems; he or she believes that other people are better problem solvers.

Perseverant vs. Indolent: A slothful individual is disinclined to work, non-adventurous, and prefers confronting only simple problems—those where solutions are found only by logical thinking and with minimal effort. Because of laziness, complex problems are immediately viewed as being unsolvable.

Tolerant vs. Dogmatic: The dogmatic problem solver is unwilling to believe in and use new ways of problem-solving; the dogmatist attempts to force opinions, ways of problem-solving, and beliefs on others in an authoritative way. The dogmatic individual maintains a loyalty to the traditional way of problem-solving.

Initiator vs. Non-enterprising: Non-enterprising individuals avoid proposing ideas and are generally hesitant to provide leadership in problem-solving.

It is easy to recognize that specific circumstances could influence a person's position on the scale between any of the two endpoints. For example, a person can be very optimistic under one set of circumstances, but much less so under other conditions; an optimistic person may even be pessimistic at times. A person who performs a self-evaluation based on these competing attitudes must recognize and consider the continuum between each set of endpoints, as well as recognize that both experiences and variations in one's mindset can alter the position within the spectrum either momentarily or for an extended period of time.

A critical thinker will usually rank strongly on the positive end of each spectrum identified. Most people have some level of ability with each characteristic. A person's goal for self-improvement should be to enhance the attitudes that are personal weaknesses. At any point in time, circumstances may greatly influence the extent to which a person acts on these important characteristics. Both a person's mood and the type of problem that the person confronts can influence the extent to which the person is able to properly act on any of these characteristics.

9.7 PROBLEM COMPLEXITY AND CRITICAL THINKING

What are the characteristics of complex problems that require the involvement of a critical thinker? Problems are complex for many reasons, and a person is more likely to solve the problem when his or her critical thinking characteristics match the needs of the problem. Just because a problem involves a technically complex issue does not mean that a critical thinker needs to be involved in solving the problem. Truly complex problems involve more than just complex technical issues. Complex problems are most often multifaceted, with demands for expertise in solving value issues, having complex aspects that require the ability to innovate, require breadth in the thinking skills, and have the need to generate novel ideas. Novel solutions generally require the involvement of imaginative thinking and a requirement for experiences with multiple thinking types, as well as the systematic application and control of important mental processes. Mindset control is important. The critical thinker will provide the best solution by integrating his or her knowledge of and experiences with these multi-dimensional characteristics of problem-solving.

When a problem is first identified, it is necessary to identify the factors that contribute to the complexity of the problem, as the importance of these factors will influence the ability of a person to solve the problem. In terms of the characteristics, the closer the problem solver's strengths are to the nature of the problem, the more likely the best solution will be developed. Additionally, the greater the extent of agreement between the problem solver's abilities and the characteristics of the problem, the greater the problem-solving efficiency. Complexity is not only connected to the technical aspects of the problem. Complexity also depends on the ability of the problem solver to address issues related to the problem. If the problem solver lacks the qualities associated with any of the dimensions, then the problem will seem complex. In many cases, complexity is dominated by the lack of the necessary qualifications of the problem solver.

9.7.1   DEFINITION OF COMPLEXITY

Simple problems can be solved with a minimum of problem-solving experience. As problems become more complex, solving the problems requires greater knowledge and experience. What constitutes a simple problem? A complex problem? Simple implies *something that is not elaborate or confusing*. Complex implies *an object or idea that consists of many interconnected parts; the parts are so interconnected as to make the whole difficult to understand*. Complexity can also result from a complicated relationship of the components even though the number of components is small. These two definitions can be applied to ideas and problems, not just systems or objects. Complex problems are those that involve multiple dimensions, especially when the dimensions are complicated by very conflicting factors. For example, problems that involve value conflicts may be subject to only a few dimensions, but the problem can be complicated more by a group of stakeholders who have very different value goals. Conflicts that are centered around potential environmental issues can be quite contentious as the problems involve many stakeholders who have very varied value-laden goals. Even though these two example problems may only involve a few dimensions, the stakeholders have very different goals along with an unwillingness to recognize conflicting viewpoints. The technical issues may be important, but the decision will depend on the ability of the conflicts to be fairly reconciled. These types of problems are best reconciled by a critical thinker.

9.7.2   SELECTION OF A CRITICAL THINKER

The first two steps of the problem-solving process are relevant to deciding the complexity of a problem. The problem is identified using a descriptive statement, a general goal, and set of specific objectives. The second step of the process is to identify the resources that are needed and those that are available. For deciding on the complexity of the problem and the expertise required of the critical thinker, it is necessary to identify the dimensions involved, the stakeholders and their concerns, the existing knowledge base, the data available and the data needed, and the expected life of a project. External dimensions, such as political or economic constraints, might also influence the expertise needed by the critical thinker who will act as the project leader or facilitator.

9.7.3   DIMENSIONALITY AND PROBLEM COMPLEXITY

Complex problems are considered complex because they are multifaceted with quite different, even competing, demands, with increasing complexity making it increasingly difficult to reach a decision. When we think of problem complexity, most project personnel would only think that technical complexity was the primary cause. Critical problem-solving generally involves more dimensions than just the technical issues. While technical complexity is often a primary issue for a project, other issues can be principal causes of complexity. Economic complexity, political constraints, environmental restrictions, and risk issues can contribute to the overall complexity of a project. Additionally, complexities associated with the four dimensions of values, skills, thinking types, and mindset factors must also be considered. While these latter four dimensions may seem less important than technical complexity, they can introduce perplexing issues that are more difficult to resolve than the technical issues. Value conflicts are often a source of complexity. While the technical issues of a problem may involve conflicts, the demands from other dimensions may be more responsible for the overall conflict. The selection of the facilitator will depend on the extent of all sources of conflict. The facilitator may not need to have extensive knowledge of the technical issue if other sources of conflict are more confounding to the goals.

Generally, the necessary breadth of the decision maker's experiences will increase as the complexity of a problem increases. That is, the type and degree of complexity can determine the most appropriate approach to solving the problem. Many factors contribute to the complexity of a problem, including stakeholder characteristics, the number and diversity of problem objectives, the importance of factors that rely on qualitative knowledge, economic considerations, the types and importance of value issues, and even the way that the decision will be made. Value conflicts between stakeholders' demands can contribute to complexity and complicate decisions. Biases of individual stakeholders are a common issue. The time frame over which the stakeholders will be influenced by the decision can also be an important issue. A problem that has long-term consequences is associated with greater uncertainty, which increases the complexity of decision-making. When a final decision is to be made, the number of alternative decisions may make the decision somewhat more difficult because the alternatives may likely reflect the interests and biases of stakeholders who have conflicting goals. The breadth of constraints will likely reflect the diversity of stakeholder interests and, therefore, the decision complexity. As the number and breadth of conflicting issues increases, the need for the decision maker to have a breadth of experiences in making critical decisions will increase.

Any dimensions that will influence a decision can contribute to the overall problem complexity. The four general dimensions discussed herein are potential sources of complexity. The following sections outline ways that these dimensions can contribute to problem complexity.

9.7.4 PROBLEM COMPLEXITY AND MINDSET ISSUES

Problem-solving must contend with mindset issues related to both the problem solver and the stakeholders who are involved in the problem. The mindset dimension can be a prime source of issues that contribute to problem complexity. The problem solver can be either an individual or the collective personnel of an organization. An organization's policies and practices are a reflection of the organization's mindset. Individuals within an organization can interfere with or constrain the completion of tasks; constraints are one type of inhibitor, and these inhibitors would contribute to the apparent complexity of the problem. Organizational policies can be primary inhibitors. Stakeholder mindsets influence the ability to make decisions and, therefore, are a complexity issue. Stakeholder biases are partially mindset constraints, especially when they suggest excessive selfishness; in such cases, the biases can constrain progress as project personnel must devote time to the resolution of the effects of biases. Pessimism by either personnel or stakeholders can act as inhibitors. Procrastination by the facilitator, personnel of an organization, or stakeholders can slow progress, which is a contribution to problem-solving complexity. Mindset issues can be quite diverse and can be significant inhibitors to progress and therefore developing a solution. A critical problem solver must be capable of dealing with mindset issues of himself or herself as well as others who are involved in the decision process.

9.7.5 PROBLEM COMPLEXITY AND VALUE ISSUES

Stakeholder demands are often value sensitive and quite varied, which can contribute to conflict and thus add to the overall complexity of a problem. Thus, the critical thinking model includes the value dimension because value conflicts are common to complex projects. Environmental health and risk issues are just two examples of value issues that contribute to the complexity of decision-making. Critical thinkers must recognize the importance of values, as well as the emotional baggage that is most often associated with obvious differences in values. A critical thinker must be capable of accounting for these differences and uncertainties when making a decision. A critical thinker will need to appreciate both the difficulties in addressing these value conflicts and the uncertainties in trying to weight the competing values. A problem solver will need to incorporate the often imprecise values with the other project objectives into a decision-making framework. A person who lacks either value sensitivity or knowledge of the way to balance conflicting values will have difficulty in solving any complex problems that involves value issues.

Value issues can be a source of conflict, which can add to the complexity of a case. Conflicts of interest, which can occur between project personnel and stakeholders, can add to the complexity of a project. A conflict of interest occurs when a person knowingly takes on new value obligations that compete with responsibilities to other stakeholders. Conflicts of interest may involve values such as loyalty to the organization, honesty, confidentiality, and public safety. For example, personal responsibilities can conflict with the responsibilities that a person has to an organization. Project personnel may have to sacrifice personal responsibilities in order to fulfill professional responsibilities. Those involved in projects often have multiple value responsibilities such as those to society, the profession, the client of the employer, the employer, and those responsibilities to himself or herself. Balancing all of these value responsibilities can cause considerable stress, which could be considered as an inhibiting mindset factor, as well as a value-related problem, both of which are sources of project complexity.

Stakeholder responsibilities and self-interest factors that are in conflict with responsibilities of other stakeholders can contribute, even dominate, problem complexity. Some stakeholder bias is to be expected, as all stakeholders have legitimate responsibilities to their specific causes. A decision maker who allows stakeholder biases to cause unfair weighting of the value issues confounds the assessment of alternatives, which introduces considerable complexity in the decision procedure. Stakeholders can be quite varied in their opinions and needs, value demands, and even in their rationality of thinking. Cultural differences can contribute to problem complexity. Environmental issues such as climate change can introduce stakeholders who have unique, sometimes highly biased demands. It is not that a bias is wrong; it is only that biases can increase the difficulty in evaluating alternative solutions to a problem. As the complexity of a problem increases, so does the demand for greater knowledge of critical problem-solving skills, including the incorporation and balancing of many human values. Complexity even makes the selection of a leader and decision maker more difficult.

9.7.6 Problem Complexity and the Skill Dimension

Problems have different characteristics, and the solutions to these problems may be sensitive to the skills employed. Curiosity, questioning, and critiquing are three important problem-solving skills. A person who is weak in the necessary skills may not be able to identify the real problem and provide the best solution. Thus, the ability to solve these complex problems would depend on having a well-developed skill set. An individual who is deficient in problem-solving skills will struggle to assist in developing solutions to skill-sensitive complex problems. For example, a problem that involves questionable use of advanced statistical methods will require a decision maker who is adept at the critiquing skill, as it would be necessary to identify weaknesses in the experimental analysis. Problem complexity may require the problem solver to be capable of using specific skills (see Chapter 10). A critical thinker who is confronted by complex problems will have to appreciate the importance for systematic application of the full sequence of problem-solving skills. Being qualified with the skill model of Chapter 10 should lead to greater problem-solving efficiency. Knowing the individual skills but lacking knowledge of their systematic organization may reduce the efficiency of any solution.

9.7.7 Problem Complexity and Thinking Type Dimension

The characteristics of complex problems need to be matched with an appropriate type of thinking. Complex problems that require imaginative solutions should be approached using a brainstorming or synectics type of thinking. Time-dependent complex problems may be solved more efficiently using upward-downward thinking. Problems that are dominated by highly uncertain factors may find stochastic thinking to be the most effective. Knowing just one thinking type is not sufficient to solve all complex problems, especially where accuracy is important.

A requirement for multiple thinking-type experience is another dimension of the critical problem-solving process and is relevant to addressing the issue of problem complexity. Many types of thinking are available. Solutions to certain types of problems will be best solved by specific types of thinking. For example, for problems where sales of a product are in decline even though an increasing trend is desired, the upward-downward thinking type would be the most appropriate method; many economic problems can be characterized by a trend that is progressing in the wrong direction. Stochastic thinking may be best for solving problems subject to considerable uncertainty. Counterfactual thinking may be the best choice to use when progress to date indicates partial failures. Thinking type, problem complexity, and problem characteristics are discussed in Chapter 13.

It is important for the critical thinker to be aware of the multiple types of thinking and be able to select the most appropriate types for the demands of the problem at hand. A person with a minimal understanding of thinking types will likely try to force the problem to fit the thinking method that he or she knows best. Obviously, experiences with different problem types will improve decisions for complex problems and increase problem-solving efficiency. More importantly, varied experiences with thinking types will increase a person's confidence in dealing with complex problems.

9.8   KNOWLEDGE REQUIREMENTS OF CRITICAL THINKERS

Critical thinkers have a unique inclination that allows them to identify the basis of a problem. This ability may be the result of mastering the multiple dimensions of critical thinking, including the technical dimension in the field of interest. Critical thinkers have the important talents and skills that are needed to advance knowledge. These skills include primary skills like critiquing and curiosity and auxiliary skills such as self-confidence and perseverance. They should also have experiences with an array of creative thinking methods, such as synectics, brainwriting, and introspective idea innovation. Utilizing alternative approaches to thinking forms an attitude that critical thinkers regularly apply. Knowledge of and past experiences with these tools should enhance the outcome of any critical assessment. When confronted by a complex problem, past experiences with each of the essential dimensions enable the critical thinker to achieve success.

While a few dimensions of critical thinking have been emphasized, being a critical thinker requires an array of attitudes and abilities. While depth of knowledge is certainly important, breadth is also necessary. A critical thinker must have the ability that is needed to manage stakeholders who have very conflicting interests. A critical thinker should have broad leadership skills, such as planning, organization, communication ability, and experience with personnel management. Experience with decision-making is important, and a record of moving the state of knowledge forward is essential. Based on such consideration, a critical thinker will need to have demonstrated the following specific abilities:

1. A critical thinker should have some depth in the state of knowledge of the technical issue relevant to the problem statement. Critical thinkers also need to seek greater breadth of knowledge about related issues. Critical thinkers should be capable of assessing the accuracy and uncertainty of all elements of the knowledge base relevant to all proposed decisions.
2. Critical thinkers must have broad knowledge of the principles of decision-making, including the process, the types of decisions, a leader's responsibilities, and the way to address all associated risks. The subjectivity and uncertainty in the assessment of decisions can complicate a critical thinker's ability to make correct decisions and take relevant actions; the critical thinker must have the knowledge needed to overcome the corresponding problems caused by uncertainty.
3. Conscientious assessment of all actions and decisions can lead to personal and organizational growth. The intelligent selection of appropriate

assessment criteria combined with a sincere belief in the value of introspective analyses of past decisions are essential abilities of critical thinkers.

4. Leaders have broad powers, and they must avoid misusing them. A critical thinker who is in a leadership role and has the responsibility to solve complex problems must be aware of the consequences of all actions and ensure that the powers that are inherent to his or her role of leader are properly used. This responsibility has a basis in the value dimension.

5. Critical thinkers are responsible for evaluating the positive and negative consequences of the positions of all stakeholders. Some stakeholders' biases are legitimate, but other biases stem from selfishness. Critical thinkers should be able to identify all biases of the stakeholders and consider the legitimacy of the biases in making final recommendations and decisions. A critical thinker needs to have the confidence that is necessary to ensure that stakeholder biases do not adversely influence decisions relative to the problem.

6. External parties, such as the media, may misrepresent the facts about a case. A critical thinker should recognize biased decisions that result from external influences and be willing and able to suppress the effects of such distortions of facts.

7. Critical thinkers are often confronted by problems that involve fuzzy factual bases. The fuzziness could be in the knowledge base of the problem or in data used to calibrate or verify the decision. A critical thinker must feel comfortable in dealing with the uncertainties of both data and knowledge and be able to judge the potential effects of uncertainty on any recommendation or decision made.

8. While research is a form of problem-solving, working on the forefront of knowledge imposes special responsibilities. As one responsibility, a critical thinker who is involved in research should ensure that the end product of the research does no harm to society and that all implications of the conclusions are identified.

9. Critical thinking inhibitors can have detrimental outcomes for both people and society, as well as distort any decisions made. A critical thinker accepts the responsibility to be knowledgeable of all inhibitors and be willing to redesign the activities when inhibitors are expected to have a detrimental effect on a decision.

10. Critical thinkers know the fundamental value positions that they personally hold and should not allow these value positions to improperly bias their decisions. Critical thinkers should have a strong background in fundamental human values, which include both personal values, such as honesty and diligence, and societal values, such as the recognition of the importance of public safety and welfare. They also need to have the ability to identify the value positions of all stakeholders who are involved in a case. A critical thinker must know the way to weight and balance values and assign honest, unbiased weights to each value-based issue in order to provide some indication of the importance of the value for decisions about the issues under consideration.

TABLE 9.2

Self-evaluation Criteria for Four Stages and Four Dimensions

Stage	Values Dimension	Skill Set Dimension	Thinking Type Dimension	Mindset Dimension
Stage I: Novice	Only recognizes personal values; does not recognize the application of values.	Minimal experience with most skills; unable to apply skills in solving problems.	Unaware of imaginative thinking approach; basically confined to logical thinking.	Difficult in making own decisions; often procrastinates; is usually pessimistic; little control of personal mindset.
Stage II: Specialist	Recognizes values relevant to the organization; no experience in balancing values.	Understands the skills but only in an individualistic sense; cannot apply them as an organized sequence.	Knows the fundamentals of creative thinking but lacks experience in their application.	Leans toward pessimism; some procrastination; allows moods to control decisions and actions.
Stage III: Master	Some knowledge of balancing values; little experience with stakeholder value biases; minimal ability to apply values to the other dimensions.	Appreciates skill-set model, but lacks actual experience applying the skills in solving complex problems.	Knows many of the types of thinking but not always able to select the most appropriate type to match the demands of the problem.	Leans toward optimistic attitude; able to control moods; minimal practical experience with mindset control.
Stage IV: Expert	Good at balancing values; easily connects values to other dimensions.	Is able to apply the critical thinking skill set based on experiences and education.	Fully able to apply all thinking types in a diverse set of problem types.	Good control of all mindset factors, both personally and in decision-making; extensive experience.

11. Critical thinkers should know and be able to use multiple types of thinking, as such knowledge provides greater flexibility in problem-solving. This ability will enable the critical thinker to select the thinking type that is most appropriate for the characteristics of the problem that they are responsible for solving.

12. Critical thinkers must know both the principles of many skills, such as critiquing, and the best approach to applying them when making decisions; they should recognize the importance of using these skills in making decisions, especially when the problems are open-ended and subject to considerable uncertainty.

13. Knowing the proper questions to ask can be essential to identifying the best decision for a problem. Questioning stakeholders to know their needs will be necessary for the critical thinker to fully understand the problem that

needs a solution. Critical thinkers must feel comfortable asking penetrating questions of stakeholders, as well as questions about all of the data used in making decisions.

14. A critical thinker needs to have self-confidence and have his or her actions display this confidence to stakeholders and subordinates. Self-confidence usually results from successfully completing challenging projects.

While innate factors can influence one's capacity for each of these qualities, experiences are good teachers. Experiences can enhance one's communication skills, provide depth of knowledge of technical issues, enhance one's organizational skills, and provide exposure to complex problem-solving. It is both the depth and breadth of experiences that are important to the development of one's critical thinking ability.

For a person who has not had a strong record of achievement as a critical thinker, self-evaluation can be a useful way of identifying causes for the lack of growth in critical thinking ability. Self-evaluation will also identify the actions that are necessary to overcome weaknesses. Table 9.2 provides some characteristics that could be used to identify both strengths and weaknesses for each of the four dimensions of a critical thinker. In the chapters to follow, more specific assessment criteria will be discussed. Reviewing the items made for stages that are higher than the stage at which a person currently rates can provide guidance for actions that should be taken to move to a higher stage. An individual could develop similar criteria for other dimensions, such as an economic or environmental dimension.

9.9   PROBLEM CHARACTERISTICS AND CRITICAL THINKER NEEDS

The ability to critically think partly depends on a person's characteristics; the ability to apply critical thinking partially depends on several characteristics of the problem. For simple problems, adequate decisions may be made by individuals who have minimal critical thinking experience. As problems become more complex, higher stages of critical thinking ability become more necessary. Problem characteristics that can contribute to complexity are many and include technical, economics, environmental, risk, cultural, societal, aesthetical, image, and temporal change, i.e., time frame. These characteristics focus on the issue. The problem solver should have experience in those characteristics that are applicable to the problem.

Problems where critical thinking is most useful have unique characteristics. Any one problem may not involve all of these characteristics, and a critical thinker may not need to use all of his or her critical thinking ability with every problem. When a problem involves a high degree of conflict, critical thinking ability is most necessary. Conflicts can result from stakeholders who have different values, concerns about the potential implications of the final decision, disagreements on technical issues, or the need to produce outcomes even when the problem involves apparently irreconcilable differences. Problem-solving can be especially troublesome when the outcomes would be considered morally questionable by some stakeholders. Such problems often involve multiple stakeholders whose value needs conflict. The decision maker must eventually adopt weights to indicate the importance of each value; these weights are rarely expressed quantitatively and qualitative weights are difficult to justify to those who have unreasonable biases. Decisions about the weights should be sensitive to the objectives of all stakeholders, including the community perspective rather than just the demands made by individual stakeholders. Assigning numerical values to human values is often very problematic and debatable, which makes it difficult to arrive at a universally accepted set of weights for the competing value issues. In some cases, codes of ethics and legal requirements dictate that emphasis be placed on some of the values issues. Public safety and human welfare are two such values that are often assigned higher weights than those assigned to personal issues, such as economic profit. Again, all of the weights may not be numerical quantities, just a ranking or even a less well-defined system of weighting. Of course, the more difficult task is to establish the priorities of the values. In order to resolve the value dimension of critical problem-solving, the decision maker will need to have broad experience.

In addition to value conflicts, technical and economic issues can be primary causes of discord during the decision process. These two one-dimensional issues are generally easier to resolve than value conflicts because more quantitative analyses are usually possible, even going as far as being capable of computing a value for a quantitative criterion such as the benefit-cost ratio. Stakeholders often place greater weight on qualitative issues such as aesthetics. Many stakeholders reject the idea of placing a monetary value on aspects of the problem that relate specifically to them. Different stakeholders would assign quite different monetary values for these qualitative issues. This disparity reflects the uncertainty that is inherent to qualitatively weighting values. This source of conflict is of special concern when some of the monies are publicly funded because the public would be an unidentified stakeholder.

Open-ended cases refer to problems with broad stakeholder demands on less well-defined issues, especially values. Developing solutions for these cases generally requires the abilities of a critical thinker. Such cases usually involve multiple stakeholders who often are quite biased toward their own issues. Some of these biases are legitimate, while others reflect selfishness. In open-ended cases, it is more difficult to balance the demands of the stakeholders because multiple feasible solutions are possible for these problems, with the degree of conflict depending on the factors that contribute to extreme biasedness. In such cases, stakeholder biases become more distorted and may even be irrational. A critical thinker must resolve such biases before recommending a solution. Open-ended problems generally require greater knowledge, more data, and a wider variety of resources. Balancing the demands and placing them into a less debatable framework confounds the decision-making and the reporting of results.

9.10 IMPROVING CHARACTERISTICS NEEDED FOR CRITICAL THINKING

It is easy to say that an ability to think critically is important; however, it is not so easy to develop critical thinking attitudes and abilities. Self-inhibition is often the constraint that prevents a person from developing a critical attitude. Thus, identifying practices that can help a person overcome the self-inhibiting attitude may be the most fruitful way of guiding a novice in developing a more critical attitude. Such a person must strongly believe that being a critical thinker will be a valuable asset and that developing the attitude will influence his or her future in a positive way. Mentors should encourage their mentees to gain appropriate experiences and act in ways that will help them improve themselves in each of the dimensions of critical problem-solving.

Having a critical attitude is especially important, maybe even necessary, to those who are involved in solving complex problems. The best decisions usually result from situations in which numerous alternatives are proposed and evaluated. Success in creative idea development and subsequent decision-making requires experiences that stem from practice with judgmental assessment. Individuals who have personal weaknesses that limit critical thinking ability will improve their critical thinking ability only after success in efforts toward improvement and completion of critical thinking experiences. Experience is a good teacher. A few general actions that individuals can take to develop and enhance their critical attitude are as follows:

- Review the ten characteristics previously identified (see Section 9.6) and grade yourself on each, such as on a 1–4 Likert rating; then work to improve those that you believe are personal weaknesses, i.e., have the lowest scores.
- Read, digest, and practice material that is contained in books that are relevant to your weaknesses.
- Become more familiar with the Scientific Method and identify general ways that it can be used to assess problem-solving ability; brainstorming is one way to practice this activity.
- Actively critique all books that you read, movies that you attend, experiences at restaurants, and sports games that you watch; before each experience, identify the criteria that you will use to assess the activity. These experiences at critiquing different events can be very beneficial. Learning to be a critic is important. Criticism does not have to be negative. It is important to provide positive criticisms when they are warranted.
- Think back to experiences in your life where the failure to be creative had negative consequences. Perform a self-introspective analysis. Identify a learning outcome from each of the experiences of failure and imagine the way that your actions in similar situations in the future will be different.
- With the objective of overcoming a weakness and developing critical thinking skills, assemble a small discussion group of those who share the same weakness. Discuss ways of overcoming the weakness and perform activities that will contribute to growth in the quality.
- Select an article from a trade magazine and another article from a trade magazine from a totally different discipline. Identify ways that the ideas in the two articles could be combined to provide a new, publishable article. This provides experience at the integration of ideas.

These seven activities provide experience at different steps of the problem-solving process. In addition to these general methods, the following are some specific ways of improving on the ten traits discussed in Section 9.6.

9.10.1 IMAGINATIVENESS

- Identify a currently debated topic and develop an array of possible future directions for the issue, especially directions based on fantasy thoughts.
- Identify an issue that was of interest in the past and identify the alternative futures that existed at the time of its interest. From a *post factum* perspective, discuss the merits and faults of the direction that was taken at that time, especially imagining the facts that led to the decision.
- Read an article in a magazine, such as *Smithsonian* or the *National Geographic*, and develop a list of questions about information that you believe should have been included in the article; try to identify reasons that the issues were not considered.
- Find a copy of a famous painting; then identify its weaknesses from your perspective and identify ways that you would improve upon the artist's work.
- When you are on the non-driver's side of a car, look at the license plates of other cars and use the letters on the plate to make up humorous phrases.
- When you are watching a TV show, identify weaknesses in the plot and identify ways that the scriptwriter could have made improvements.

9.10.2 HONESTY

- Develop a list of human values. Provide a definition for each. Identify ways that each of them may be related to the value of honesty.
- Recognize your own positions on issues, such as gun control or mandatory vaccination; for a moment, view each of these as a personal bias rather than the correct perspective and evaluate the reasons that you hold those beliefs. Be willing to modify all of your positions that are based on biasedness. Self-criticism can lead to self-improvement.
- Obtain a copy of a code of ethics for your area of professional interest. For each section of the code, identify a relevant example of dishonesty.
- Review some of the many dishonest actions of politicians and note the negative outcomes to society (unfortunately, these dishonest actions may have had positive outcomes to the politicians).

9.10.3 INQUISITIVENESS

- Find articles on topics that you know little about and analyze and critique them; be critical of each point made by the author, so that you develop a position that is counter to the views of the author. Then identify potential changes that you believe would be advancements to the article.
- Use a current issue, such as the use of public funds to support charter schools, and list a series of questions that would help uncover the best pathway for resolving the issue.
- Ask questions about any problem that you recognize; try to develop an extensive list of questions about the problem.

- Write an essay on any subject of your interest. Then put it aside for a week or two and then re-read it. Critique the ideas by identifying changes that are needed to improve it; these should not be changes to grammar but to changes in the underlying ideas about the subject matter.
- Critiquing is a method of analyzing and questioning, both of which require a sense of inquisitiveness. Review published critiques of books, movies, plays, and works of art by expert critics to learn the fundamentals of critiquing the works of others; fully analyze each critique to understand the factors that are relevant to critiquing the works of others.

9.10.4 RECEPTIVENESS

- First, identify factors that contribute to effective teaching. Then think of a teacher who you have had and considered to be a poor teacher; specify your reasons. Compare the reasons that you rated the teacher to be ineffective and the qualities that make for an effective teacher. For each factor, provide a critique of your negative view for the purpose of possibly believing that the poorly rated teacher was not as bad as you originally believed.
- Critical analyses can be either positive or negative. Review your position of any issue and critique it from an opposing viewpoint. Then for each of the viewpoints, identify experiences that would cause you to change your attitude.
- Identify a public policy or a school system practice and find faults with it. Identify arguments that you would use to encourage changes in the policy or practice.

9.10.5 OPTIMISM

- Identify an issue about which you have negative feelings. Identify three primary reasons for your negative feelings. For each of these, imagine a person who has strong positive feelings about the issue and the arguments that they would make to counter your negativity. In light of your negative feelings, explain reasons that the other person may have positive feelings.
- Identify an issue that is currently in the news about which you have a negative feeling about its future outcome; identify the reasons that you hold the negative expectation. For each negative feeling, identify a corresponding positive perspective. Discuss the differences in perspectives.
- Identify a person who you believe is a negative thinker i.e., a pessimist, and list the consequences to him or her because of his or her negativity.

9.10.6 AFFECTIVENESS

- Take a scientific concept such as evaporation increasing with temperature and ideate about the relation but using an emotional involvement rather than a scientific or physical reasoning; this distortion of the underlying concept may enable you to view the relation from an emotional perspective.

- Select a painting, maybe something scenic, and distort the image that you see in some emotional way; the distortion will force you to improve your emotive thinking ability.
- Contrast evidence of the emotions of winners and losers in a sports game.
- Review common advertisements in magazines or on TV and evaluate the ways that your emotions influence your assessment of the ads.

9.10.7 SELF-CONFIDENCE

- Identify one of your weaknesses, such as lacking talent in art or in a tendency to procrastinate. Identify activities that you could do to overcome your weakness. Now imagine one of your strengths. What life experiences and thoughts enabled the activity to be a strength? Now compare your responses and identify differences that relate to self-confidence. By generalizing this, you should have a better recognition of ways of gaining confidence in your other weaknesses.
- Identify and study methods of idea innovation and creative thinking; then practice them to develop unique solutions to problems with which you are confronted. Practice provides experience, and the experience helps improve one's self-confidence.
- Explain ways that self-confidence increases one's chance of success in some endeavor, such as in sports, playing in a band, or taking tests in school. Evaluate yourself on the basis of these criteria.
- Think of a person that you know who seems to be very confident. Then think of another person who obviously lacks self-confidence. Contrast the two specifically to identify the factors that contribute to self-confidence and those that limit self-confidence. What are the effects of not having self-confidence?
- Identify ways that you could help a younger person overcome the aspects of self-confidence that you lacked at that age and that you wish you had known to take actions needed to overcome these deficiencies.

9.10.8 PERSEVERANCE

- Prodigies such as those in the arts or sports have spent many hours of their youth practicing their craft. Read a book about one of them and identify the sacrifices that they made. Then relate the sacrifices to the general concept of perseverance and evaluate the sacrifices made by the person.
- Efficiency in any activity is important. Identify connections, both positive and negative, between perseverance and efficiency as they relate to problem-solving.
- Identify two activities: one that you like doing and the other that you dislike. What factors contribute to your willingness to persevere at the activity that you like? Then discuss changes to the later task that would make you more likely to persevere at it.

9.10.9 TOLERANCE

- Make a list of diverse activities, e.g., sports, playing a musical instrument, traveling to distant lands, participating in community service activities. For each one, identify ways that you could get a friend who was reluctant to participate in such activities to be an enthusiastic participant.
- Tolerance can refer to the practice of respecting the beliefs of others. Identify a cultural issue about which you have strong feelings that seem opposite of the feelings of your peers. Identify the changes to your belief system that would be necessary for you to be more tolerant of the cultural norm. Then identify corresponding changes that your peers would need to make in order to be more tolerant of your perspective.
- Recognize that newscasters spin news to fit a viewpoint. Then practice taking the opposing viewpoint of the commentator on a nightly news program. Identify the beliefs that you would need to have to be more tolerant of the newscaster's spin.

9.10.10 INITIATOR

- Identify a problem about which you know little specifics, e.g., medical ethics or an aspect of child development. Imagine that you were asked to be part of a brainstorming group that needed to resolve the problem. Make a list of the ideas that you could contribute. Try to transfer knowledge from your interests to the other area.
- Identify reasons that people do not contribute to brainstorming sessions. Then identify ways that you could encourage and motivate them to contribute at the next brainstorming session.
- A contributor is one who is willing to speak out during a group discussion. To improve this ability, respond to questions posed when in a classroom setting or other type of group activity.
- Review professional journal papers; identify assumptions and constraints that could lead to improvements (i.e., contribute to) in the published results.

9.11 DEVELOPMENT OF CRITICAL THINKING ABILITIES

One of the questions posed in the introduction to this chapter was: what education and experiences are necessary for a person to be identified as a critical thinker? Colleges generally do not have specific programs that prepare students to be critical thinkers. While some courses can provide a background for improving some of the characteristics, it will usually be necessary for individuals to gain the important abilities through experiences rather than through their formal education.

Critical thinking can be taught, and it can be learned. While self-teaching is possible, experience at problem-solving is the best teacher. The effectiveness will partly depend on the person's innate characteristics, as well as factors such as the person's mindset and the levels of curiosity and imaginativeness. While imaginativeness is more closely related to creative thinking, it is also important in critical thinking. A person's interest in the subject matter will also influence the level of motivation toward action and achievement. Learning to be a critical thinker depends on knowledge, values, abilities, and attitudes, but experience may be the best teacher.

9.11.1   EDUCATIONAL CONSIDERATIONS

How does a person learn to hit a 95-mph fast ball? How does a person learn to be a proficient skateboarder? How does a chef become adept at consistently producing the most delicious veal scalopini? Practice! Practice! Practice! How does a researcher become proficient at critical problem-solving? Yes, experience is the answer. Experiences that provide continual practice of solving critical problems are a necessary element of education. For someone who is new to the conduct of research, identifying weaknesses of existing professional literature is not a natural ability. A curious mind is necessary. If a person's skill at curiosity is improved, his or her ability to critical think will improve. Confidence including an I-can-do attitude must be developed, as confidence influences many of the necessary constructs of critical thinking. Becoming a critical thinker comes from practicing the art of critical analysis and synthesis over an extended period of time. The improvement of one's skill at critiquing was discussed in Section 9.10. Experience in applying skills that are important in the problem-solving process is necessary for success in solving complex problems.

Education and experience are essential to the development of a critical attitude. The teaching of critical thinking must begin with the development of the students' attitudes with respect to problem-solving. The ability to think critically is not an innate characteristic but it does have some innate roots. To reach one's innate capacity requires both education and experience. A curriculum for critical attitude development would focus on helping the individual become more inquisitive, more prone to asking questions, and more perseverant in the pursuit of knowledge. Those involved in research should be curious about knowledge and ways to advance it. Note the skills: inquisitiveness, questioning, curiosity, and perseverance. Having a critical attitude will enable a person to use an array of important attitudes, such as optimistic thinking and self-confidence, when solving complex problems or conducting research.

Most of the characteristics of someone who can be considered as a critical thinker are quite general, and unfortunately, they are not amenable to quantitative assessment. For example, how can a numerical rating be placed on someone who is slightly biased or someone who is better able to accurately interpret a problem statement? Is there a metric for rating people on curiosity or on skepticism? Given the subjectivity of these characteristics, it is difficult to set a minimum quantitative value that would enable a person to be labeled as a critical thinker. Even qualitative rankings would be difficult to develop. Given this restriction, it is necessary for a person to accumulate a broad range of experiences that will provide the opportunity to appreciate the knowledge or skill that underlies each of the many characteristics that are important to critical thinking ability. Each individual should develop a personal plan that accounts for his or her strengths and weaknesses.

9.11.2 CRITICAL THINKING CAPACITY

A model of eight primary skills is presented in Chapter 10, with the support of five auxiliary skills; a person can improve his or her abilities of all 13 of these skills. Therefore, students of critical thinking must first acknowledge the importance of these skills and be motivated to improve their appreciation of the roles of the skills in problem-solving. Mentors can introduce novices to the skills and the way that the skills are sequentially applied. The mentor can begin by posing a series of mildly complex problems and have the mentee respond with questions that reflect the various elements of the problem. Initially, the problem can be about day-to-day issues, but with time and experience, more complex problems can be presented to the mentee. The mentee should also be encouraged to make critical analyses on his or her own time. Continually experiencing success will develop the confidence to attempt greater challenges.

The critical thinking process is more than just a coordinated act of reasoning. The process extends from the development and interpretation of a problem statement through the critical assessment of the final decision and the specification of the implications of the conclusions. Attitudes are influential at each step of the problem-solving process. Thus, attitude development is an essential part of the educational process for developing a person's ability to think critically. Characteristics such as curiosity and inquisitiveness are central to a critical thinking attitude, both as a skill and as an attitude. Educational programs to teach creative thinking methods are needed and should be centered on improving students' skills such as imaginativeness. A critical mindset, which is developed through experiences and insight, is defined as *the intellectual capacity to perceive and control the essence of a situation*. The internal thinking is combined with the individual's values and mental capacity to make judgments. The control of moods and dispositions will influence the extent to which experiences lead to learning, i.e., knowledge accumulation. Mind control is very important, as a negative mood at a critical time can lead to failure. This discernment and internal thinking produces the external action of a decision, which is a primary outcome of the critical thinking process. Developing the capacity for critical thinking requires a major intellectual effort over time. The person must evaluate each experience and identify the attitudes that were used at each phase of the problem-solving. In the analysis of failed experiences, the person must determine the reasons that the appropriate attitudes were not applied and develop the attitude to use the correct procedure in future problem-solving responsibilities. It would also be necessary to identify the reason that the inappropriate attitudes were applied at that time.

Thinking is often interpreted as a unitary activity, i.e., mental processes being taken by one individual. The critical thinker is using his or her mental processes to progressively move through all phases of the critical thinking process (see Section 9.4). Very often, it is applied as an iterative process, as the person may start thinking about the problem and intellectually think through to the decision step, but then return to the problem step, i.e., Step 1, to refine the understanding of the problem. Returning to any of the other step is possible. In a sense, this is system feedback. Feedback can actually occur between any two states of the process. A questioning attitude is a positive factor and is more likely to produce a recognition of the need for and benefits of feedback.

9.12 THE INFLUENCE OF EXPERIENCE

With reference to both creative problem-solving and research, experience refers to *the accumulation of knowledge that is generally learned from active participation in relevant activities.* To rate experiences as being worthwhile, the activities need to advance the person's knowledge of any of the dimensions of critical thinking, including the other external dimensions such as technical knowledge or environmental health. Accumulated experiences of high quality yield the development of an attitude that encourages the appreciation of the fundamental tenets of the underlying knowledge in ways that will advance the person's critical thinking ability. Experiences are not skills, but they are necessary to hone the corresponding skills and shape a person's thinking. Experiences can create or change beliefs and attitudes, with the quality of the experiences instrumental in establishing the feeling of the truth of a belief. Experiences that are sufficient to create a change in a belief result only when the person fully appreciates the benefits of the accumulated knowledge and experiences a variety of problem-solving situations. Of course, experiences that involve unusual problems can have a more significant effect on one's beliefs and lead to greater advancement in critical problem-solving ability than experiences of little content.

The benefits of experience are many; therefore, problem solvers should seek a wide variety of experiences, as breadth increases the opportunity to apply and solidify beliefs, attitudes, and skills under a diverse set of conditions. Experiences should be sought in complex projects that emphasize problem-solving, critical thinking, creativity, and research. The projects should involve conflicting values so that experience can be gained in balancing competing values. The projects should be centered on a variety of issues that would enable a person to test or challenge their beliefs under a variety of conditions. Given that problem-solving skills can be an important part of any experience, *a priori* evidence should be sought that suggests an experience would provide for varied, yet important, skill development. The experience should provide opportunities for critiquing the current state of knowledge, acting curious, and questioning ways that knowledge can be advanced. Post-project self-introspection of the experience should always be a priority, as it can be a good learning experience. In all of these facets of critical thinking, it is always best for the experiences to extend the person's knowledge of all dimensions beyond his or her existing level of knowledge.

Research is a special case of problem-solving. The temporal development of experience in research should involve the successful application of all steps of the research process under a variety of circumstances such that the application of the research process becomes the person's standard approach to problem-solving. For research that would require critical thinking, experience is developed through expanding knowledge of methods of imaginative idea generation and evaluation, the application of creative problem-solving methods to challenging problems, the generation of outcomes that have implications to society, and the firm belief that imaginative thinking will likely produce outcomes that would not have been generated through logical thinking analyses. Researchers should seek involvement in creative activities that will advance their personal states of knowledge. To accomplish this, a person will need to have multiple experiences that involve critical problem-solving skills. The researcher may also need experience in auxiliary disciplines such as statistics or modeling concepts for decision-making. Simultaneous use of imaginative thinking as part of research provides a multifaceted experience that will best advance one's critical problem-solving ability.

Each experience should be followed by a self-introspective analysis, which would have the following purposes:

- To assess the performance of relevant skills.
- To recognize the importance of values at each point in the experience.
- To gain experience in balancing qualitative values with quantitative decision criteria.
- To assess the importance of beliefs.
- To understand if change is needed.
- To ensure that negative moods or other pessimistic feelings did not hinder decision-making.
- To assess the effectiveness of any imaginative thinking used.
- To evaluate the quality of the decision-making.

It is always important to discuss any failures or weaknesses with a mentor. Such interactions can place the weaknesses into a better learning context, thus solidifying positive change.

Periodic self-evaluations are generally a good practice regardless of the issue. Religious leaders recommend periodic self-evaluations of one's beliefs. Even companies perform company self-evaluations. They often have annual retreats, which is where planning for the future is a dominant activity; however, this is often preceded by an assessment of progress made on goals identified at previous retreats. Prior to a retreat, the company leaders often meet with the individuals who report directly to them with the thought that the leaders can identify the concerns of their employees and at the same time refresh the employee's commitment to the company. The supervisor can identify actions that may be needed to improve the employee's commitment to the company. Research leaders often hold periodic meetings with research staff to identify the concerns of the individuals whom they supervise. The supervisors and research leaders may have their subordinates do a self-evaluation prior to the meetings.

The experiences of researchers and creators are generally quite different, in spite of the creative roles in both. The values and attitudes can be different, but that is also true from one research project to another research project. While both the creator and the researcher should have experiences in decision-making, the underlying objectives of the decision-making can be quite different. The skills referenced in research, such as self-discipline and curiosity, may be different from those that are emphasized in creative problem-solving within an organization, where unbiasedness and flexibility may be considered to be more important. Organization problem-solving often involves more varied stakeholders that are influential in a research environment. In both cases, experience is necessary, and it is beneficial for both types to have experiences in both realms.

9.13   THE ADVANCEMENT OF CRITICAL THINKING ABILITY

Being a critical thinker is not a case of you-are or you-are-not. Given the multi-dimensional nature of critical problem-solving, the qualifications of each critical thinker are often quite different than the qualifications of other critical thinkers. Recognizing that assessment of a person's qualifications in any one dimension of critical problem-solving is both variable and uncertain makes it difficult to begin assessment of one's overall ranking as a critical thinker; however, a person may want to make a self-assessment so that he or she can act on self-improvement. Therefore, a method of assessing a person's level of being a critical thinker can be of value.

9.13.1   INTELLECTUAL DEVELOPMENT

A taxonomy proposed by Bloom in 1957 was discussed in Chapter 2. Since that time, numerous modifications of the steps have been proposed, including the one method mentioned in Chapter 2. Bloom's six-step taxonomy provides a framework for a systematic advancement of knowledge about a subject of interest. A significant modification of Bloom's model is proposed herein. The intent of this model is to show the way for a person to improve his or her knowledge of critical problem-solving. The model will be referred to as the intellectual development model, or the ID model. A person's intellectual development, as described by the ID model, may be better appreciated if its five stages are viewed in light of the multi-dimensional scales of the critical problem-solving model. Note that I am not referring to Bloom's original model, but to a Bloom-like model that classifies the educational goals of the multi-dimensional critical problem-solving model. Advancing from one stage of the ID model to the next higher stage reflects a collective improvement of knowledge of the critical thinking dimensions of skills, thinking types, values, and the elements of a person's mindset. While it may not strictly require simultaneous growth in all four critical thinking dimensions in order to advance one stage in the ID model, some growth in each of the four dimensions should be expected to advance a stage in the ID model.

The five-stage ID model provides a general pathway of assessing educational advancement in critical thinking. The underlying principles of the ID model could be used to assess any dimension. The intent of the method is to provide specificity to the various aspects of knowledge development. Growth in any dimension of knowledge should provide a broader understanding of other dimensions. This is also true of external dimensions such as the technical and economic dimensions; however, the intent of this integration of the concepts of the ID model with the multi-dimensional critical problem-solving model is to provide guidance for advancing both a person's level of knowledge of critical problem-solving and the ability of a person to advance from one stage to the next higher stage.

9.13.2 Stages of the Intellectual Development (ID) Model

The five-stage ID model is intended to describe both the growth of a person's knowledge related to an issue and the way that the accumulation of knowledge of the issue can be advanced. One outcome of any worthwhile educational experience should be the advancement of a person's level of knowledge from one stage to the next higher stage such that the educational experiences provide a more complete understanding of the relevant knowledge, which in this case is critical problem-solving. Experiences can change attitudes and beliefs, enhance the ability to weigh values, and improve skills.

In terms of knowledge of a topic of interest, the five stages of the ID model for the advancement of knowledge of critical problem-solving are as follows:

I. *Comprehend*: To establish the state of knowledge using the current facts of record about an issue; this might include the identification and definitions of variables and processes that are the foundation of the current knowledge, as well as an understanding of existing relationships under the current state of knowledge. For critical problem-solving, this would be knowledge of the relevant dimensions and their objectives.

II. *Experience*: To experience the way that the existing system functions under the current conditions including an understanding of the influence of each of the system components. The application allows the underlying effects to be assessed for a variety of conditions and provides experience in critical problem-solving; past cases could be reviewed to compile this knowledge. In the case of critical thinking, identifying the most appropriate idea generation method or the best type of thinking for the specific experience might provide the necessary knowledge.

III. *Analyze*: To critically analyze the existing state of knowledge, which yields results from which measures of the weaknesses in the state of knowledge can be identified.

IV. *Synthesize*: To synthesize the results of the analyses into a set of general, broadly applicable conclusions and integrate them into an advanced state of knowledge of the issue.

V. *Create*: To verify that the state of knowledge has been advanced and now the state of knowledge better reflects the functioning of the real system. Greater prediction accuracy could be achieved and an improved understanding of effects should be possible under the new state of knowledge. The conclusions will enable the development of broad generalities, with the generalities showing the implications of the study. Thus, the conclusions can be induced for conditions beyond those in the measured data.

The important point is that systematic learning that is based on critical problem-solving will provide a new end state of knowledge that is a better description of the functioning of the system of interest than would be the knowledge of the current time. The first two steps assemble a statement of understanding of the system. To create a more accurate model of the subject of interest, the weaknesses of the state of knowledge are identified in Stage III and corrected in Stages IV and V. The most significant advancements are made possible by following the intentions of the five stages of this ID model. The process attempts to ensure the progressive advancement in knowledge. For critical problem-solving the advancement should be in all dimensions, but the discussion here is limited to each of the four dimensions. It would be straightforward to extend the practice to other dimensions.

9.13.3   DISCUSSION OF STAGE I

The five-stage model of intellectual development must begin with the recognition of the fundamental elements of knowledge related to an issue. Stage I would begin by identifying all relevant factors and provide definitions for each one. For example, for the value dimension, values such as fairness and loyalty may be relevant. For the skills dimension, definitions of critiquing and curiosity would likely be needed, while for the thinking type dimension, definitions of logical, counterfactual, and lateral thinking might be needed. Recognizing the relevant knowledge basis for these fundamentals would establish a Stage I level of critical thinking.

An understanding of the relationships between the fundamental components of knowledge would also be needed as part of Stage I. This knowledge of both the effects and the interrelationships will need to characterize the state of knowledge both within and between the multiple dimensions. For example, values of fairness and unbiasedness have obvious connections to the advancement of knowledge as do critiquing and unbiasedness. The former pair is within the value dimension while the latter pair reflects important knowledge between the skills and values dimensions. All relevant interrelationships should be identified during Stage I with their connections relevant to the issue of concern clearly identified, which could be a technical dimension subject.

9.13.4   DISCUSSION OF STAGE II

An intent of Stage II is to evaluate the person's ability to apply the knowledge of Stage I. Stage II involves participating in experiences that solidify learning, especially about the current state of knowledge. Analyses of existing case studies provide a good opportunity for novices to gain experience in the potential applications of critical thinking. A person will need to be capable of applying the current state of knowledge of these interrelationships to existing problems. These applications should provide experiences in activities such as problem-solving, leadership, research, and decision-making. The appropriate process will be used, e.g., the research or decision process. They also establish the effects to be understood and maybe even quantified in the case of issues where knowledge is more quantifiable. Experiences with multiple issues or systems enable a person to recognize any limitations on the state of knowledge of that issue.

9.13.5   DISCUSSION OF STAGE III

Stage III is a transition stage, with the existing knowledge being analyzed to identify potential improvements in the state of knowledge of Stage II. Gaining knowledge of weaknesses that exist in the current state of knowledge or constraints that have been placed on the method is an effective way of moving the state of knowledge forward. The intent of Stage III is to identify ways of improving the state of knowledge. Note that the first two stages focus on the existing state of knowledge while the last two stages will focus on advances that are made to the state of knowledge. Stage III is the beginning of a search for new knowledge for the purpose of establishing an advanced state of knowledge about the issue. Thus, Stage III is a transition stage, i.e., transitioning from the current state of knowledge to a more advanced stage. This new state of knowledge should ultimately provide a greater and more accurate under-standing of the system under consideration, but weaknesses in the existing state of knowledge of the issue must first be detected. These weaknesses lead to hypotheses and experimental designs that must be developed as part of Stage III. Weaknesses may result from either inadequate theory or poor data, or both. Restrictions that were placed on the existing relations could also be factors responsible for the current less-than-acceptable accuracy of the Stage II knowledge. Another problem with the current state of knowledge might be that the effect of a causal variable was initially ignored because data on that variable were not collected with the existing model of the system. Another problem might be that the method of quantifying relation-ships in the existing state of knowledge was relatively unsophisticated. Fitting by eye is less sophisticated than some of the analytical and numerical methods of fitting functions. So if the existing state of knowledge depended on approximations to the coefficients of the model rather than the more sophisticated methods of calibration, then the state of knowledge might be improved through the use of better methods of fitting functions. Other deficiencies may have been detected from the results of Stage II. Correcting any simplifications could lead to advancements in the state of knowledge. The output of Stage III is a list of potential changes identified from the analyses. These can identify issues that may suggest inaccuracies in design or predic-tions when the existing state of knowledge was applied. Also, the effects of changes suggested by the existing model may not be reasonable, especially at the extremes. All of the problems that are identified from the efforts of Stage III provide directions for Stage IV efforts.

9.13.6 DISCUSSION OF STAGE IV

The objective of Stage IV is to make syntheses that will advance the state of knowl-edge, which may be adding theory to the existing base of knowledge or improving the database, which could be a larger sample size or a more representative sample of data. Ideally, the improvements in knowledge and/or data will enable the state of knowledge to advance. A hypothesis would have been developed as part of Stage III to reflect each weakness. Stage IV analyses lead to potential advancements. Each Stage III hypothesis should be analyzed in Step IV and should lead to a correspond-ing result. In Stage IV, the role of critical thinking would be to use the four dimen-sions to ensure that the new knowledge from the added resources, i.e., new data and/or theory, is helpful. Of course, good values, such as honesty and unbiasedness, should be used in all of the analyses. The output would be a set of results that would serve as input to Stage V.

9.13.7 DISCUSSION OF STAGE V

To move from Stage IV to Stage V involves the assembling of a new state of knowledge based on the interpretation of the results of Stage IV. The new knowledge may be a new model that can provide quantitative indications of the responses of the system being modeled. The new knowledge could also be a set of rules for decision-making, with the rules providing guidance for making decisions for systems other than those that can model quantitative effects.

As part of Stage V, conclusions are extracted from the results identified in Stage IV. The accuracy of the analyses should be assessed to ensure that the new state of knowledge represents a substantial improvement over the state of knowledge of Stages I and II. The rationality of effects should also be evaluated to ensure that the new knowledge provides reasonable measures of the effects of variables that are part of the model. The output of this step would be a set of general conclusions.

One additional objective of Stage V is to verify the results and summarize the broad implications of the conclusions. Verification of the model can be based on rational analyses. In some cases, statistical analyses are used for verification. To reflect an advancement in knowledge, performance criteria that are selected to measure accuracy should show significant improvements. The improvement is judged based on the method showing significant improvement in terms of the physical system. While it may be worthwhile to show a statistical improvement, the true measure of verification is an improvement in design or prediction accuracy. In addition to prediction accuracy, the new model should yield more rational assessments of effects, again in terms of the physical system.

9.13.8 APPLICATIONS OF THE CRITICAL THINKING MODEL

Table 9.3 summarizes the ID model as it uses the four dimensions of critical problem-solving to advance the state of knowledge on some issue. Each of these will be discussed in more detail in the next four chapters.

9.14 ASSESSMENT OF CRITICAL PROBLEM-SOLVING

The assessment of creative thinking activities is much less complicated than the assessment of critical problem-solving efforts. The former has more specific responsibilities, but each is easier to judge. For example, in a brainstorming session, the number of ideas, the number of wild-and-crazy ideas, and the extent of downtime are quite easy to assess. The end product of an idea generation session is much easier to recognize than the end product of a critical problem-solving session.

Four general dimensions of critical problem-solving are emphasized herein. Thus, the assessment criteria will revolve around these four dimensions. Criteria for assessing other dimensions, such as the technical and economic dimensions, should be appropriately developed and assessed, but they may be less troublesome to quantify than the other four dimensions. Establishing assessment criteria for the values, mindset, skills, and thinking type dimensions will be difficult; however, learning that follows from the assessments of past critical thinking efforts can lead to improvements

TABLE 9.3
The Five-Step Intellectual Development (ID) Model for Assessment of Critical Thinking Ability

Cognitive Process	General Purpose	Skill Set Dimension	Value Dimension	Mindset Dimension	Thinking Type Dimension
I. Comprehend	Identify and assemble basic knowledge. Understand the basic connections between knowledge and decisions/actions.	Identify skills relevant to the type of problem. Understand the general roles that each skill plays in solving problems.	List values related to issue and value responsibilities of the individual. Understand how value responsibilities influence decisions.	Identify mindset vocabulary; understand the effects of pessimism and procrastination.	Identify thinking types relevant to solving different types of problems. Understand the way that thinking type influences decision-making.
II. Experience	Using cases, show relevance of knowledge to decisions/actions subject to uncertainty.	Apply skill set to show the types of outputs that can result from the various skills.	Apply value issues and value responsibilities to cases where the decision is largely dependent on values.	Apply different mindsets to different situations and discuss effects on outcomes.	Apply different types of thinking to situations and evaluate the outcomes.
III. Analyze	Analyze complex cases to show influence of diverse factors on effects and outcomes.	Analyze actual cases to see the types of skills that were applied and the outcomes; note skills associated with failures.	Analyze cases where value conflicts greatly influence the optimum solution.	Analyze situations from the perspective of the state of mind of competing stakeholders.	Analyze complex cases to recognize the influence of thinking type on outcomes.
IV. Synthesize	Synthesize alternative outcomes and effects for variation in inputs to complex problems.	Synthesize a variety of cases to show the importance of various skills.	Synthesize values into situations where values are not the only issues to show the ways that values influence good decision-making.	Synthesize cases to show the array of outcomes due to variations in states of mind and dispositions.	Synthesize cases to show the ways that the thinking type should be matched to the characteristics of the problem.
V. Create	Create new knowledge on the topic; e.g., new plans, new educational programs, ways of dealing with complex problems.	Create plans that will enable novices to learn of skills that they will need and when they need to apply them.	Create standards of conduct and methods of balancing values for complex problems.	Create complex problems and show the effects of lack of control of one's mindset.	Create guidelines on thinking type to ensure that best decisions are made for complex cases.

in critical problem-solving ability. Therefore, after completing a critical problem-solving activity, an introspective analysis of the entire activity should be made to assess the effectiveness and efficiency of actions on each of the four dimensions.

Assessment of the four dimensions is primarily a post-project task; however, efficiency can be improved if the fundamentals of each dimension are reviewed prior to the start of a new project. For complex problems, assessments can be made at selected stages of a project. These intermediate assessments can help to ensure that problems are detected sufficiently early so that corrective actions will prevent the need for post-project collective actions. Intermediate assessments of the dimensions can also improve the overall efficiency of the project, as they will refresh the participants' memories of the important general aspects of each dimension.

9.15   CONCLUDING COMMENTS

Were Copernicus, Bacon, Darwin, and da Vinci excellent innovators or truly critical thinkers? Problem-solving is partially influenced by the knowledge base that exists at the time. Copernicus lived in a knowledge world where astrology was a dominant force and religious beliefs limited thinking. Copernicus needed to be a critical thinker to overcome these inhibiting constraints on thinking. Copernicus imaginary use of spheres suggests that he informally applied synectics when he was developing his theory. He certainly needed to apply many of the skills of the skill dimension, especially the skill of courage when he approached the problem in the face of the religious restrictions that were placed on scientific investigation during that period of time.

Francis Bacon was a strong believer in the benefits of experience and experimentation. While his work to some extent was constrained by religious pressures that were imposed on scientists of his time, Bacon was also constrained by the dominance of an educational philosophy of Aristotelian scholasticism, which promoted deductive investigation. Bacon filled in many of the steps of scientific investigation that da Vinci did not formally address. Bacon combined mathematics and philosophy of science in his development of the fundamentals of the Scientific Method. He indicated that scientific research should involve imaginative thinking, hypothesis formation, and inductive reasoning in an environment of a strong moral code—this in spite of his own moral laxity when he was Chancellor of England (1618–1621).

Years after his five-year voyage to the South Pacific, Charles Darwin wrote a book that fully described a theory of evolution of natural selection. Darwin transferred knowledge from analogous thinking of the ways that farmers selected crops and bred livestock. This thinking went beyond innovation. Darwin was also influenced by the works of Thomas Malthus (1766–1834) who suggested that population, if unchecked, would increase exponentially while food production would increase arithmetically, which would cause starvation. Much of Darwin's data had been collected over a long period of time prior to the publishing of his book, *The Origin of the Species*. Darwin used the data to support and document his theory, which Darwin had proposed decades before his book was published. Darwin's approach to his theory was clearly indicative of critical thinking.

Are the ideas that form the basis for critical thinking accurately represented by the term critical thinking? Is critical thinking critical? Is critical thinking really thinking?

Does the definition of critical really apply to the intent of critical thinking? Does the definition of thinking accurately describe the mental processes that take place when someone is thinking critically? I contend that the term critical thinking does not really describe or explain the mental processes and actions that reflect the multi-dimensionality of problem-solving. Neither the word critical nor the word thinking fully reflects the stresses that take place when making decisions on complex problems.

Unfortunately, the term critical thinking has become a commonly used term instead of the term critical problem-solving. The term critical thinking seems to present a very narrow view of problem-solving and decision-making. I believe that it may be preferable to replace the term *critical thinking* with words that better reflect the most specific objective of problem-solving. As indicated throughout this book, complex problems need solutions that address values, thinking types, and a variety of skills, as well as the control of mental processes. One purpose of this discussion has been to provide a new perspective on critical thinking, one that emphasizes its multi-dimensionality.

Critical thinking is a widely used term; unfortunately, it is often misunderstood because it means different things to different users. The term *critical thinking* is really more than just a one-dimensional activity. Additionally, it is more than creative thinking. Critical thinking is actually a process that involves both analysis and synthesis, and this fact should be fully recognized and appreciated when the term *critical thinking* is used as a synonym for *critical problem-solving*. Even the term critical problem-solving might not fully reflect the uses inherent to the multi-dimensional process presented herein.

A person needs to have critical reasoning ability before he or she can think critically. This think-before-action tenet is central to education related to critical thinking, as well as creativity and idea innovation. Teaching creative analysis depends on the student being willing to depart from the normal mode of thinking, i.e., logical thinking. The person must be willing to entertain critical thinking strategies. The central element of education related to critical analysis is to reorient the students thinking away from the commonly used logical approach to one that encourages systematic use of primary skills, values, thinking types, and a sound mindset. An initial task for education related to critical analysis is to develop the student's attitudes that encourage questioning of the current state of knowledge. Specifically, a person who needs to perform critical analyses must have a sense of curiosity, be inquisitive, and be willing to question the adequacy of existing knowledge. The first requirement is for the person to have the courage to go beyond the current state of knowledge. Developing a critical attitude should be the goal of every person and, more importantly, a component of the curriculum of every educational institution. An ability to critically analyze problems is a necessity when working in a state-of-knowledge environment.

9.16   EXERCISES

9.1.   Find definitions of ability, attitudes, talent, and skill. Discuss how they enter into critical problem-solving.
9.2.   What values conflicts could develop in debates on the issue of global climate change?

9.3.   Define the word critic. Then compare the application of the definition to the responsibilities that you would have as a movie critic and to you as someone responsible for critiquing professional reports.

9.4.   What types of problems can be solved using logical thinking only? What additional restrictions make imaginative problem-solving methods necessary?

9.5.   Summarize the main differences and main commonalities between creative thinking and critical thinking.

9.6.   Why are the values and skill set dimensions not overly relevant to creative thinking? Why are the mindset and thinking dimensions relevant to creative thinking?

9.7.   Apply the six steps of the thinking process of Section 9.4.1 to the situation of a person purchasing a gun.

9.8.   Expand the six-step critical thinking process to a ten-step process.

9.9.   Identify the responsibilities of the leader of a group critical thinking session.

9.10. Based on the concepts of critical thinking, what criteria would you use to judge a science fair contest?

9.11. Assume that you are conducting an experiment to test whether or not the weight of a baseball bat influences the distance a batted ball travels. Briefly outline your experiment and discuss the parts of the experiment that you would consider analysis and synthesis.

9.12. Expand the entries of Table 9.1 by elaborating on the analysis and synthesis responsibilities for the six steps.

9.13. Which three of the ten essential characteristics of a critical thinker (Section 9.6.2) would be the most difficult to master? Briefly defend your selections.

9.14. Create a table that identifies factors that contribute to problem complexity. Associate the factors with the dimensions of critical problem-solving.

9.15. Identify three types of problems where you believe that critical problem-solving would be necessary to provide solutions. Briefly justify your reasons.

9.16. Why are values important in solving complex problems?

9.17. How can someone identify the values that they use in making value decisions?

9.18. Characteristic 11 of Section 9.8 suggests that knowing multiple types of thinking is important. Provide reasons for this requirement.

9.19. Fourteen knowledge requirements of critical thinkers were discussed in Section 9.8. Select three and discuss the types of experiences that will best develop a person's ability at each of the three.

9.20. Discuss the way that a person would conduct an introspective analysis of recent failures in his or her life.

9.21. Identify three inhibitors to critical problem-solving and discuss ways of minimizing each of their effects.

9.22. Separate the comprehend stage of the ID model (see Section 9.13) into two distinct stages. Discuss the reason for the separation. Define each section.

9.23. What would the office climate be like if a tolerant person has a dogmatic person as his or her immediate supervisor?

9.24. Imaginativeness is an important characteristic of creative people. What activities can enhance a person's imaginativeness?

9.25. Develop a metric that could be used to measure a person's confidence in generation ideas. Assume that the confidence versus unassured is a spectrum.

9.17   ACTIVITIES

9.17.1   ACTIVITY 9A: THE ENVIRONMENT HAS A HEADACHE

Assume that you work for a company that is heavily involved in environmental health. Create a dimension that is comparable to the four dimensions that are discussed in Chapters 9–13. Also, briefly discuss the ways that it interacts with the four dimensions.

9.17.2   ACTIVITY 9B: GOOD AND BAD FAILURES

Create an assessment metric for rating of critical thinking activities, such as those in Section 9.17.

9.17.3   ACTIVITY 9C: WILD-AND-CRAZY VALUES?

Incorporate a value dimension into the creative thinking process that emphasizes idea generation. Identify and briefly discuss the values that would be most important.

9.17.4   ACTIVITY 9D: ID-TECH

Select any technical area of interest and develop a seventh column for Table 9.3 to develop the entries of the ID model for the technical dimension.

9.17.5   ACTIVITY 9E: MIGHTY MOUSE VERSUS SQUEAKY THE MOUSE

One of the spectrums in Section 9.6.3 is confident-unassured. Using these as the endpoints of the spectrum, identify three intermediate points with the endpoints and briefly discuss the characteristics that would assign a person to each of the five points on the spectrum. Provide a descriptive title to each of the points.

9.17.6   ACTIVITY 9F: NO GUESSING!

For the following items, identify those that are most descriptive of critical thinking (CT) and most descriptive of creative thinking (C), and those that are equally relevant to both CT and C:

- Generate ideas.
- Curiosity is important.
- Subject to self-inhibition.
- Appropriate to use piggybacking.
- Can be a group activity.
- Suitable for solving complex problems.

- Efficiency is difficult to assess.
- A facilitator is a primary component.
- A multi-dimensional method.
- Follows a process.
- Broad experience is necessary.

After completing your analysis, compare your responses with those of someone else and discuss all differences.

9.17.7   ACTIVITY 9G: A DOLLAR BREAKS INTO CENTS

Analysis means *to break apart*. Efficiency means *producing effectively with a minimum of waste*. Efficient analysis means *to have a minimum waste when breaking something apart*. This latter term could apply to a company that has seen a significant decline in sales over the last six months. How does the term efficient analysis apply to the company's problem?

9.17.8   ACTIVITY 9H: PUPPY TRAINING!

Synthesis means *to put together*. An inhibition is some factor that adversely influences efficiency. A company executive believes that the company receives too many customer complaints, which seems to limit the growth in sales that the company should be experiencing. The executive has traced the problems to the attitudes of the sales force and recognizes the need for a training program for the sales force. Develop definitions of synthesis and inhibition that apply directly to the company's problem and identify factors that would be relevant to solving the problem.

9.17.9   ACTIVITY 9I: THE FIFTH DIMENSION

If we include technical competency as a fifth dimension of the problem-solving process, what characteristics of Section 9.6.2 would we need to include? Discuss your response.

9.17.10   ACTIVITY 9J: GROUP THINK

Identify the responsibilities of a leader or facilitator for each of the four dimensions of the critical problem-solving process. What assessment criteria could be used to judge the leader?

9.17.11   ACTIVITY 9K: METACOGNITIVE THINKING

Develop a plan for improving critical problem-solving, i.e., critical thinking. The ideas should take problem-solving beyond the current critical thinking approach.

10 The Skill Dimension of Critical Problem-Solving

CHAPTER GOAL

To propose a sequence of mental skills that will maximize the efficiency and effectiveness of solutions to complex problems.

CHAPTER OBJECTIVES

1. To propose a model of the skill dimension of critical problem-solving.
2. To identify the purpose of eight primary skills that define the skill sequence.
3. To discuss ways of enhancing important skills.
4. To show the importance of experience in becoming a critical problem solver.
5. To discuss the assessment of one's level of the skill dimension.
6. To outline the process of iterative questioning.

10.1 INTRODUCTION

After more than a millennium, applied science seemed to blossom about the time of the Renaissance. For the past centuries, improvements in technology were minimal, as those responsible for the advances were artisans who needed advances primarily to improve the efficiencies of hunting and farming. During the prior millennium, scientist-philosophers directed their effort toward advances in knowledge about life, i.e., a universe that was difficult to understand. Neither of the two groups were focusing on advances in the toils of daily life. The know-how artisans did not interact with the know-why scientist-philosophers until science became a stronger force in the fifteenth and sixteenth centuries. The skills of the artisans were integrated with the new knowledge generated by the scientist-philosophers. This integration of intention and scholarship established an intellectual environment for the types of thinking needed for the rapid advancements that occurred. A focus on knowledge related to experimental analysis was a primary advancement in thinking. Evidences of innovation emerged from the darkness.

Critical thinking, which has its roots in the Scientific Method, requires more from its practitioners than the thinking abilities of the sixteenth-century scientist-philosopher. Critical thinkers need the science-based knowledge of the philosophers and the skills of the artisans, but they need these attributes in a more advanced

DOI: 10.1201/9781003380443-10

way. The ways of the current critical thinkers are more systematic, yet with greater breadth than the way that propelled applied science six centuries ago. The problems that need solutions today are more complex than the problems of that time, so the problem-solving process needs to incorporate a greater number of dimensions.

When we hire people to do a job, we expect them to have a certain level of expertise obtained from worthwhile experiences. Without a mastery of the appropriate skills, the person's efforts may be ineffective and completely inefficient. The type of skills required for a job varies with the job responsibilities. Expert cesspool cleaners need skills that are quite different from the skill requirements of a radiologist. We could not then expect experts from the two jobs to trade places and immediately provide the same quality of work. Skills vary from job to job, and achieving expertise in one job does not easily translate to expertise in unrelated types of work. This is true for critical thinkers. Critical thinkers who are experts in engineering research would not be expected to be experts in open-heart surgery or cesspool cleaning. Critical thinkers in different disciplines may have some common skills, but the individuals generally lack the complete array of skills to be critical thinkers regardless of the discipline. This observation refers specifically to the technical dimension, but it is likely partially true for other dimensions, including the four dimensions focused on herein.

It is likely safe to assume that some skills would be common to most problem-solving disciplines, and it is these skills that need to be identified and organized into a framework for solving complex problems. Research is one subdivision of problem-solving. Since complex problem-solving in general and research specifically are two specific interests herein, it seems reasonable to identify the specific skills needed for problem-solving, whether it is for business practices or research in any discipline. The goal then is to propose a sequential set of abilities that provide a framework for acting in the search for a solution to a complex problem.

10.2   DEFINITIONS

Skill can be defined as *an ability or proficiency at some activity*. This definition does not address the issue of the level of expertise of the skill, and it is reasonable to assume that variations in ability are expected. This variation can occur from person to person or from time to time for any one person. Some knowledge of skill application can be gained from reading; however, to be considered an expert at a specific skill requires a significant level of expertise with evidence of both experience and some knowledge acquired from a source of educational development.

The term critical was defined as *the inclination to judge severely*. The word critical applies to the nature of the problem complexity, with specific reference to the consequences of the problem. For complex problem-solving, the ability to critique requires knowledge and experience, both of which serve as motivation for improving skills. Thinking can be defined as *the use of mental processes to reason or reflect on an issue usually for the generation of an idea*. These definitions of critical and thinking do not by themselves adequately summarize the idea of critical thinking. The difficulty in making decisions partly depends on the complexity of the problem. The skills required to overcome this difficulty can be identified, but the quality of a solution to the problem partly depends on the proper application of the skills. The mental processes or mindset influence the application of the skills. Thus, the term *critical thinking* or the term *critical problem-solving*, by definition, requires knowledge of various skills and the way that the skills are applied.

The term *critical thinking* is not isolated to its relation to the application of the skills, as the skills both influence and depend on other dimensions. A critical thinker must think to identify alternatives and then think about weighting the consequences of conflicting values and outcomes. Additionally, for solving complex problems, critical thinkers will need to incorporate various types of thinking beyond just logical thinking. A person's mindset at the time of decision-making greatly influences his or her application of the skills. Finally, a critical thinker will need to have developed an array of values that interact with the different skills. Note the four dimensions: mindset, values, skills, and thinking types. These dimensional interactions must be understood in order to be a critical thinker.

Creative and critical thinkers have the ability and inclination to analyze and judge intellectual ideas, as well as the willingness to uncover without bias both the merits and faults of important ideas. With respect to analyzing complex problems, a creator can be defined as *someone who approaches problem-solving with both an orientation toward critical thinking and knowledge of and the motivation to use methods of idea innovation.* Such a creator is one who firmly believes that integrating the creative thought process into the problem-solving process will lead to the most efficient and effective solutions. Successful creators have the ability to use logical thinking, but they approach problems with a more mature attitude and broader thinking skills. They achieved their added skills through both education and experience.

10.3   OUTLINE OF A SKILL MODEL

Regardless of the technical discipline, problem solvers follow many of the same steps. This was evident from the similarity of the processes outlined in Chapter 2. Efficient problem-solving involves many of the same mental processes and actions. For example, different mindset attitudes influence the ways that skills are applied. The mental processes lead to actions that influence the effectiveness of problem-solving. These actions are the basic skills needed to solve a problem. The overall efficiency of problem-solving will depend on the efficiency at which each individual skill is applied and the point in the decision process where each skill is applied. The following is an outline of a possible sequence of skill application during the steps of the problem-solving process.

10.3.1   DEFINITIONS OF SKILLS

The skill dimension includes a series of eight primary skills; five auxiliary skills are also identified. The eight primary skills are best applied as a sequence. The model also includes two counter-skills, which are skills that tend to reduce, rather than improve, the efficiency of problem-solving. Specifically, the counter-skills are procrastination and pessimism. The following provides brief definitions of these skills in a way that the skills apply to critical problem-solving:

10.3.1.1   Primary Skills

Courage: To actively pursue difficult tasks with little fear of failure even when the odds suggest some doubt should exist. Courage as a skill would reflect the willingness to risk criticism for adopting a position that conflicts with common feelings.

Perception: To recognize the true nature of and achieve significant understanding of even incomplete knowledge through reasoning, with reasoning implying the ability to detect important underlying characteristics.

Skepticism: To doubt the truthfulness of an existing state of knowledge, thus demonstrating a willingness to seek the truth, to advance the state of knowledge even if the complete truth is unattainable.

Critiquing: To dissect the existing knowledge for identifying the weaknesses in the current state of knowledge.

Questioning: To develop an interrogative expression that summarizes a weakness in knowledge.

Curiosity: To identify and develop specific ways of improving identified weaknesses.

Emergence: To use the results of the analyses to synthesize preliminary conclusions.

Consequence: To broaden a conclusion to a general understanding of the synthesis which enables the implications of the body of the work to be identified.

10.3.1.2   Auxiliary Skills

Self-confidence: To trust in oneself to successfully solve complex problems.

Self-discipline: To control one's own actions especially where a desired activity must be sacrificed in order to perform a less desirable activity in order to meet a responsibility.

Persistence: To refuse to give up on the search for an answer or a solution.

Flexibility: To have the ability to accommodate change.

Unbiasedness: To be fair and impartial when making decisions.

10.3.1.3   Counter-Skills

Procrastination: To needlessly postpone or delay a responsibility.

Pessimism: An inhibiting attitude characterized by a belief that positive outcomes will not result.

The inverse perspective on the auxiliary skills could be viewed as additional counter-skills; this would include a lack of self-confidence, no self-discipline, a tendency to quit, inflexibility, and biasedness.

10.3.2   The Sequence of Skills

In general, a sequence is defined as *a succession of related items* or *a set of items in consecutive order*. All of the items in the series have some obvious connection and the items are systematically arranged based on the intended application of the items.

When the skills are applied in the proper sequence, the highest efficiency will be achieved. The order of the items provided a rational application of the items.

The eight primary skills form a sequence that covers the six steps of the problem-solving process. Any individual who faces a complex problem must first have the courage to approach a solution procedure. He or she must believe that the problem can be solved; this belief is the basis for courage. The extent of the courage or lack of it will influence the efficiency of the problem-solving. A person who has the courage must have the perception to recognize that a solution is needed. Then the person must adopt a skeptical attitude in order to identify any deficiency in the existing knowledge base of the issue, especially any invalidity associated with the uncertainties, and that he or she can significantly advance knowledge beyond the weaknesses. Again, overcoming weaknesses requires courage. Given a sufficient level of skepticism, the person can critique documents that indicate the current state of knowledge and, through his or her ability to critique, identify the weaknesses in the knowledge of the issue. The person must be adept at critiquing the current state of knowledge and be able to argue that a solution is needed to overcome or at least reduce the uncertainties. Specifically, critiquing is a skill that promotes searching for a solution to the problem. Thus, the issue is investigated by critiquing the existing state of knowledge. Self-discipline, self-confidence, flexibility, and perseverance are auxiliary skills that are needed to support the effort to find a solution. Identification is often achieved through questioning to find the specific causes of weaknesses in knowledge. Then the person's curiosity can identify specific changes that may be made to overcome the weaknesses or restrictions. Having a definitive statement of the weaknesses, the problem solver must use reasoning of the results of the analyses to deduce one or more results. From them, a potential solution will emerge. The solution provides a general understanding, i.e., emergence. The conclusions emerge from the results. To make the conclusions more broadly applicable beyond the framework of the existing database, general understandings and their implications must be synthesized with the consequences of the synthesis leading to a set of broader implications of the study. Other attributes of a problem solver can hinder progress. Specifically, procrastination, rationalization, negativism, and biasedness must be overcome for successful solutions to be found.

10.3.3 A SKILL SET FOR PROBLEM IDENTIFICATION

The first phase of the problem-solving process (see Section 2.11) is the identification of the problem. This act of identifying may seem obvious, but it is not! Yet, it may be the most important aspect of the task, as failure to identify the true nature of the problem will limit the ability to advance the state of knowledge. With complex problems, the underlying enigma may not be universally obvious, as different individuals may envision quite different fundamental problems from a given set of underlying experiences and an existing knowledge base that is of minimal quality. These are characteristics of issues with complex problems.

The input to Step 1 of the problem-solving process consists of facts or observations that suggest an important element of knowledge is not known. In terms of the problem-solving process, this inadequate knowledge base is the basis for the problem. The inadequacies must be stated as a problem and then the problem statement must be transformed into a general statement of an overall goal and a set of specific objectives. Unfortunately, the facts may be inaccurate and the knowledge base weak; therefore, different individuals may interpret the knowledge that lacks accuracy into quite different goals and objectives, which would then lead to quite different solutions to the problem. This result is understandable, and the multiple advancements in knowledge are certainly acceptable. Without sound reasoning skills, an erroneous statement of the goal and objectives may make it difficult to advance the state of knowledge. Enhancing one's problem identification skills will enable the best goal and set of objectives to be identified.

The skill set for Step 1 of the problem-solving process consists of two skills: courage and perception. Without the courage to accept the responsibility to seek a solution to a complex problem, success is unlikely. Perception is the act of registering inputs, knowing their meanings, and knowing the way to respond to them. In the framework of complex problem-solving, perception is more than just superficially understand that the existing knowledge is inadequate. The sensing of the inadequacy is the first part and is followed by an understanding of the knowledge that was sensed and then the ability to take action on the existing state of knowledge. Registering implies the mental occurrence of a phenomenon; recognition implies the nature of the understanding of the registered phenomenon, and responding implies using reasoning to take the necessary action on the phenomenon.

Perception herein is not the sensory understanding of objects, but instead the higher ideational process of imagination and reasoning. The recognition needs to be at a higher level than just a basic sensory event. The underlying characteristic of any weakness in knowledge is really unknown at this phase in the solution process. Past experiences will influence the ultimate replacement of the deficient knowledge base and help prepare for a more complete statement of understanding, but only if those experiments had sufficient depth to advance the person's knowledge base and problem-solving ability.

How does a person improve his or her abilities at these skills? While formal education may help, experience and introspective thought are the two primary mechanisms. At the conclusion of any problem-solving activity, the person should reflect on the extent to which the initial goals and objectives fully answered the lack of knowledge that existed when the facts or observations were perceived. If the introspective analysis concludes that the incomplete knowledge was inadequately addressed, then the reasons for the misperception need to be identified and understood, with appropriate changes made to the person's problem-solving philosophy. The cause could have been an incorrect perception, registration, recognition, or response. Analyzing the failure from a multi-source perspective may help ensure that the person's perception improves and that similar failures are avoided in the future.

10.4 PRIMARY SKILL: COURAGE

Courage is usually associated with military personnel who are involved in a war zone where the soldiers face the dangers of warfare. However, with respect to critical thinking, courage can reflect a steadfast determination in facing any adversity in complex problem-solving, but with a certain confidence that overcomes the fear of failure. The soldier has the courage to face an enemy combatant who may feel that it is a him-or-me situation. The critical thinker has the courage to face a different sort of unknown, a need to overcome a deficiency in knowledge with little concern over the thought of failure that the state of knowledge will not be advanced. Copernicus had the courage to promote the heliocentric viewpoint even though religious leaders placed the Earth at the center of the universe. Galileo had the courage to promote ideas that conflicted with the Pope's philosophy of nature; it cost Galileo his freedom over the last years of his life. They both were determined to advance the state of knowledge in spite of the potential adverse consequences.

In the sense of critical thinking, adversity is the fear of not being able to make a significant contribution to advancing the state of knowledge. Someone who works at the edge of knowledge needs the courage to allocate the mental stress, time, and resources that are necessary to complete the required tasks and successfully produce an advancement of knowledge in a timely manner. A failure to solve the problem is always possible, so the person needs to have the courage to overcome this fear of failure. Courage is connected to self-confidence in that the person should approach a critical problem-solving opportunity recognizing the potential for failure, but letting the fear of failure serve as a challenge to his or her skills.

As a person experiences success in increasingly difficult tasks, the fear of failure will lessen and a sense of courage will strengthen. However, risk takers often search out riskier problems, so that the likelihood of failure remains part of the challenge. As the problems become more challenging, the need for knowledge of creative problem-solving methods increases; however, the more significant the challenge, the greater the improvement in this skill dimension of critical thinking.

10.5 PRIMARY SKILL: PERCEPTION

Perception as a skill in the critical problem-solving process is somewhat different from perception as the term is used in psychology and philosophy. In psychology, perception refers to the act of observing inputs and understanding the meaning of each input. The individual who perceives the activity then has a response that is a function of his or her interpretation of the collective inputs. While some similarity exists with respect to critical problem-solving, the use of perception differs quite a bit from that in psychology.

Before categorizing perception with the other skills for critical problem-solving, it is worthwhile to review the elements of perception as a general topic. Perception does not have a universally accepted definition. It involves recognition through sensory activation followed by a response, which for the discussion herein involves the mental response. This loose definition implies several fundamental components: observation, registration, recognition, and response, which can be defined as follows:

- *Observation*: to make a sensory analysis of a fact or an event.
- *Register*: to become mentally cognizant of a fact or an event.
- *Recognition*: to understand the characteristics of a fact or an event.
- *Respond*: to act on the recognition, which can be a mental or physical reaction.

These definitions suggest that perception is a series of steps that a person takes to gain understanding so that action can be taken as a response to the understanding.

10.6 PRIMARY SKILL: SKEPTICISM

During the COVID pandemic, the two dominant political cultures were both claiming that their positions were based on science and science was the justification for their positions on masking, social distancing, vaccinations, and other policies that one side or the other believed necessary or not necessary to control the spread. Obviously, science meant different things to the two sides. The typical man-on-the-street was somewhat skeptical about the positions of the two sides because the scientific basis for COVID seemed to be very uncertain, i.e., the truth was not known and appeared to depend on politics, not knowledge of the risks. The same might be true of the climate change issue. One group indicates a 2.5 degree increase by the year 2200, while another group indicates only a 1.8 degree change. Another group has suggested that there will not be any change. Again, what is the typical man-on-the-street to believe if the disparate parties are claiming that their science is the correct science, i.e., the truth? On a more short-term case, one computer model might predict the landfall of a hurricane to hit at Mobile Bay, AL, while another model might forecast that the same hurricane will come ashore at Panama City, Florida. It seems that the scientific truth for complex problems is unknowable and that uncertainties can be significant, which leads some people to be skeptical of science in its broad sense.

Skepticism is directly connected to courage. The courage to seek the truth needs to be accompanied by a skeptical nature. Skepticism can be defined as *the doubting of the truthfulness of an existing state of knowledge*. The intensity of a person's skepticism is influenced by the person's mindset. The skeptic naturally views the conclusions of past research as a starting point. The skeptical nature invigorates a critical mind.

Skepticism is difficult to define, but it seems to occur when the truth is unknown, maybe unknowable. A skeptical decision maker recognizes the significant uncertainty involved in the situation and approaches the issue favoring a dispassionate attitude, which suggests unbiasedness. Is skepticism bad or can it be recognized as a positive? Skepticism can be an inefficient attitude when the doubting is accompanied by pessimism. Actually, skepticism can be a very positive attribute when it is balanced by a sense of optimism. In such a case, skepticism can initiate a search for truthful knowledge via systematic investigation, which results from a lack of definitive answers to a question.

How can a person become more skeptical? Increasing one's skill at courage, as it was defined in a previous section, is the first step in becoming more skeptical. As we can easily see from the COVID and climate change issues, a person improves his or her skepticism skill by searching for all viewpoints on the issue. The definition of skepticism itself identifies the best way to improve the skill; specifically, just acknowledge that all aspects of knowledge are uncertain. The person should try to define the certainty of the issue and the difficulty in doing this will lead to the recognition of the uncertainty. It is important to identify the uncertainties in every aspect of the problem, and complex problems likely involve considerable levels of uncertainty in quite different aspects of the problem. Accepting one side of the debate without searching for other viewpoints suppresses the development of a skeptical attitude and, therefore, suppresses critical thinking ability. It is important to assess the quality of the work used in developing the existing state of knowledge that is attributed to each of the competing viewpoints. This assessment should include the identity of the uncertainties of all so-called science. If alternative viewpoints on any issue are not immediately evident, a person should try to create an alternative, with such mental exercises serving to provoke a skeptical nature. A lack of skepticism can reduce the breadth and depth of alternative solutions proposed, which likely reduces the quality of the outcome.

10.7 PRIMARY SKILL: CRITIQUING

Imagine that you are visiting an art gallery with two friends. You and your friends are standing side by side, looking at an original Picasso. One friend says that he could do much better than the artwork that appears before them. Your other friend says that she can understand why the painting is worth $8.7 million. Now your two friends ask for your assessment. What criteria would you use as part of your critical assessment of the painting? Similarly, when you leave the theater after watching a movie, you generally make critical comments about the movie, which can be either positive or negative in nature. You decide whether the movie was worth the time and the money spent for admission. You may criticize the movie on the basis of the character development or its plot. You also make assessments of food when you go to a restaurant. Was the taste of the food the primary decision criterion? The service? The atmosphere? If you have analyzed any of your past experiences, then you are a critic and have used some level of critical thinking. But, could you be a critical thinker for a complex problem?

Just as you evaluate works of art, movies, and food, researchers make critical assessments of problems that they face and creative works that they review, such as articles published in journals. Your experience at the restaurant will influence your future dining-out decisions. Similarly, assessments of journal articles influence decisions made by researchers. All such evaluations depend on a person's knowledge of the subject matter, their mindset and mood at the time, and relevant personal characteristics, especially those that influence their talent for critical analysis.

10.7.1 DEFINITIONS FOR THE CRITIQUING SKILL

A few definitions may be beneficial in understanding the role that critiquing plays in the skill dimension of critical problem-solving:

- *Critic*: one who forms and communicates judgments of both the positives and faults of an issue.
- *Critical*: characterized by careful evaluation and judgment.
- *Criticism*: the art and skill of making discriminating judgments and evaluations.

Learning to be a critic can be a valuable lesson, if only to be able to critique one's own work. But criticism should also be viewed as being helpful to others. A good critic is one who has a good value system and performs all critical assessments in an unbiased fashion. A good critic has learned to control his or her mindset when making an assessment. The output of a critical analysis is a statement of the strengths and weaknesses in the state of knowledge as it existed at that time.

Critiquing may not be a word that is found in a dictionary. The closest word to critiquing would be criticize, but the word criticize generally conveys a negative impression of the topic being reviewed. I wanted a word that was related to criticize, but a word that conveyed a neutral feeling of assessment. So in this book, the word critiquing is used to suggest an unbiased critical analysis that yields both positive and negative results.

10.7.2 CRITICAL ANALYSES

Analysis means to break apart. A critic's objective is to break apart the relevant matter to identify its strengths and weaknesses. The matter can be either mental thoughts or printed writings. An analysis of a chef's recipe and the resulting sample of food may indicate that the consistency is good but that it is a little too salty. The analysis identified a positive and a negative. In the review, the reviewer should provide a specific way that the weakness should be corrected, e.g., reduce the salt by a third. However, another critic, one who prefers salty foods, may argue that the salt content should be increased. The important point is that critical analyses often contain opinions that are beliefs that are less definitive than positive knowledge. Critical analyses may reflect the opinion of the critic rather than the truth. Thus, the connection between critiquing and skepticism is evident.

Criticism often carries the connotation of negativity. People who are often most critical in a negative way are generally avoided by others. A movie critic may pan the latest release of Hollywood's current star even though the picture in which he acted was very good. An anonymous reviewer for a professional journal returns a very negative review of a paper submitted for publication with the reviewer, strongly recommending that the paper be rejected. Such a decision is acceptable if the paper is really that poor. While biased reviews do occur, critical reviews of anything should provide knowledge that can be used to improve the work. If a weakness is identified, it is only fair to recommend the action that will correct the weakness. If the critiques are accurate and unbiased, then the review comments should be viewed as an attempt to help the author of the work. Negative critiques can actually be beneficial when they are accurate and accepted as an attempt to provide ideas for improvement.

10.7.3 ASSESSMENT OF THE CRITIQUING SKILL

Performing a critical analysis of a Picasso painting is quite different from critiquing a dining experience at a fast food restaurant. Yet, it would be of value to identify some general issues that may be relevant regardless of the object of a critique. Each critical statement must be accurate and stated in terms relative to the problem under investigation. Critical statements should be unbiased, unless the critic's bias is clearly stated. The following are some general qualities that can be used for assessment:

- *Quality*: A measure of excellence, usually evaluated using an ordinal scale measurement.
- *Quantity:* An interval scale of measurement of the amount of the characteristic.
- *Timeliness*: The relevance of the characteristic over periods of time.
- *Breadth*: The extent of applicability to a variety of measures.
- *Implications of decisions*: The future short-term and long-term consequences of every decision.
- *Sources of uncertainty*: The recognition that a value of each input and output is not precise.
- *Reliability*: The degree of trust that you have in all elements of the critique.
- *Dependency*: The necessity of external factors to function.

Interpretation of each of these would depend on the issue. For example, the quality of a Picasso artwork is different than the quality of a hamburger. Assessment criteria such as these can be the basis for arbitrating stakeholder conflicts. They are also useful for evaluating the overall value of an idea or a project. Critical analysis can be used in an introspective analysis following the completion of projects.

The nature of a critic of the works of others depends on the case. Critiquing a sports coach's leadership style is quite different than critiquing the latest women's fashion show at a Parisian salon. However, the following list tries to convey the general idea of critiquing, with the comments referring to a research effort:

- Correct or incorrect interpretation of the state of knowledge when establishing the goal and objectives of the problem.
- Thorough or inadequate assessment of the literature for establishing the state of knowledge.
- Good or poor experimental designs with their ability to ultimately test the research objectives.
- Appropriate or inappropriate methods of analysis.
- Correct or incorrect interpretations of the analyses.
- Accurate or inaccurate conclusions drawn from the results. Implications of the conclusions fully documented or inadequately identified.

Note that these critical assessment criteria cover all of the steps of the research process. In general, the criteria can also be used to assess the outcome of any project. Each one of the criteria is stated, so that the critic can evaluate the criterion from either a positive or a negative perspective, thus allowing assessment of the strengths or weaknesses of the issue.

10.8   PRIMARY SKILL: QUESTIONING

The critiquing step in the sequence identified a weakness in the knowledge base. New knowledge needs to replace the deficient bit of knowledge. Questioning is the link between the unknown and the known knowledge. Questioning is the use of interrogatory analysis in search of unknown knowledge. Critiquing is essentially the interrogation of the unknown knowledge in search of a direction to new knowledge. The analysis of the existing knowledge base should lead to a question that summarizes the weakness. Just as a police officer poses questions to a suspect in a criminal case, the person in pursuit of knowledge places the problem in the form of a question. The police officer expects the suspect will respond to the questions in a way that the response will shed light on the details of the crime. The critical thinker expects that a question will lead to advancement in the knowledge base.

The questioning skill may be better understood by treating it as the systems process: an input, the transfer function, and the output. The input is the statement of the weakness that was identified through a critical analysis. The transfer function is the uncanny insight that enables the person to discern the true nature of the weakness and its relation to a broader base of knowledge of the topic. The person needs to have an inexplicable ability to perceive the idea that allows the advancement of the knowledge base. The uncanny insight needed to expose the reason for the weakness is based on successful past experiences, but it requires the confidence to identify the link between the old knowledge and the new knowledge. This connection is imbedded in thoughts that lead to the question. The output is the question. Any failure to consider all relevant dimensions, such as the value dimension or the economic dimension, can lead to a biased question that ultimately will not produce the best outcome.

10.8.1   Developing the Questioning Skill

How does a person develop the questioning skill? Practice! It is important to recognize that questioning is the skill that serves as the link between critiquing and curiosity. It is an important link because it coerces the problem solver into concentrating on the important issue. Questions encourage the person to focus his or her thinking on the true nature of the weakness that was identified by a critical analysis. The questioning skill can be enhanced by posing questions about everyday activities. For example, at a local intersection, why does it seem that the traffic is most congested in the east-west direction? Someone who focuses on the question will be challenged to identify numerous reasons, with the concentration increasing the likelihood of generating many possible answers, one of which might be the best answer. A person should seek out newspaper and magazine Q-and-A columns, where a question is posed and then answered by a supposed expert on the subject of the question. Columns on health, medicine, eating, and fitness are common columns in general reading media. The best way of improving the questioning skill is to actually practice it.

To excel at critical problem-solving, questioning is one of the most important skills to possess and use. One benefit of questioning is that it forces the person to focus on the relevant issue, whether it is the overall goal of the project or a method of analysis. Some people may have the talent to critically question, but they may fail to apply it, as questioning can vary with a person's mood. For those who score low on the questioning spectrum, improvement is important and is possible with some effort. Growth in one's questioning skill can begin by focusing on one's everyday surroundings; it is not necessary to start with a complex technical issue.

10.8.2   QUESTIONING AND PROBLEM-SOLVING

Questioning is not a skill that comes naturally. It requires a sense of inquisitiveness and the discipline to transform a problem into the intermediate form of a question, i.e., problem-question-solution. Questioning and curiosity are often performed iteratively. A person may be curious, but lacking the knowledge to extend thinking to asking questions following the critical analysis will reduce the effectiveness and efficiency of the curiosity skill. Inquisitiveness can be viewed as the combination of questioning and curiosity. Inquisitiveness is one of the fundamental characteristics of a critical thinker, which is evident from the practice of asking questions.

Questioning is a primary tool of a critical thinker. Questioning aids problem-solving, encourages broader thinking, and helps the thinker gain a deeper perspective on the issue, i.e., it promotes both depth and breadth. Questioning creates a challenge for the critical thinker, as the person feels a responsibility to find an answer to the question, i.e., the challenge serves as motivation to find a solution. Questioning is often associated with the analysis phase of problem-solving, as this is the phase when problems are discovered or refined in scope.

10.8.3   THE PROBLEM-TO-QUESTION PRACTICE

The activity of questioning is often facilitated by the use of creative thinking methods, such as individual brainwriting or group synectics. It is the imaginativeness of the creative thought that helps to ensure that the question truly reflects the real problem. The easiest approach to questioning is to directly pose the problem as a question. The following are examples that illustrate a problem and a corresponding question:

PROBLEM	QUESTION
Excessive scatter in an x-y plot.	What new variable could explain the scatter?
Doing poorly on in-class testing.	What knowledge could improve test taking?
The need to develop a thesis topic.	What current hot issues interest me?
The need to develop a thesis topic.	How can I make unique changes to existing work?

Of course, more than one question could be developed for a given problem, as illustrated by the last two of the four examples. Once a good question has been identified, a person or a group can brainwrite or brainstorm on the question. Once a question has been developed to reflect a problem, an individual could write the question at the top of a blank piece of paper and then brainwrite (discussed in Chapter 3) potential answers. Brainstorming on the topic could also be done while outside for a leisurely walk, but this would not necessarily lead to a written list of ideas; however, while out for a walk, a person may be a little less distracted from the research environment and maybe a little more relaxed for creative idea generation. If a problem is not known, then the problem-question sequence may need to be iterative, with refinement of the problem statement that is the basis for the research. Introspective Idea Innovation (see Chapter 17) is useful for identifying research problems. The initial thought about the problem is the basis for the first question, which then serves as the foundation for a more refined problem statement. The problem-question sequence is repeated until the question seems like the basis for an analysis that will yield credible research results.

10.8.4 BENEFITS OF QUESTIONING

Why is it beneficial to pose a problem as a question rather than just a declarative statement? First, the question helps the person or group focus on the central issue of the problem. Just having a set of thoughts about the problem will generally not allow the person or group to focus on the fundamental issue that needs to be understood. Second, when the problem is posed as a question, finding a solution becomes a challenge. Most people like the uncertainty of a challenge, and complex problems encourage critical thinking. A question about a technical problem is similar to the challenge of a puzzle, but generally one that expresses a serious need. Third, since most problems have potentially more than one possible solution, the question encourages breadth of thinking to be sought, but does not immediately imply that one best solution can be found. Fourth, questions are often the product of inquisitiveness. Questions instigate inquisitiveness, which was identified as one of the fundamental characteristics of a creative thinker. Inquisitiveness reflects a mindset that is truly focused on the real problem. Fifth, posing a problem as a question may influence a person to modify his or her mindset to a mindset more conducive to problem-solving.

10.9 PRIMARY SKILL: CURIOSITY

Curiosity can be broadly defined as *an interest in knowing especially about a unique problem*. In a sense, the entire skill set as it will be defined herein could be interpreted as a means of satisfying curiosity, as we generally associate curiousness with a person who wants to gain new knowledge. Instead of using curiosity in this broad sense, it will be used herein as a narrow part of the skill sequence. Specifically, the term curious will be applied to the search for specific knowledge about the most effective means of improving knowledge of a model or a representation of a system.

While it may be possible to argue that all primary skills are of equal importance in research, curiosity is often considered to be the most important skill. It achieves special ranking because it is at the center of the skill set, with many of the other skills depending on its proper use. Curiosity is closely associated with the two preceding skills: critiquing and questioning.

10.9.1 DEFINITION

In this model of critical problem-solving, curiosity was previously defined as *an eagerness to identify ways of improving identified weaknesses*. More generally, curiosity is *an inquisitiveness for developing new knowledge*. In this case, the new knowledge would be the totality of the existing knowledge and the result of overcoming weaknesses in the existing knowledge. Curiosity is an important skill in this model of critical problem-solving.

10.9.2 Interconnectedness with Other Skills

Skepticism can arouse someone's sense of curiosity, while critiquing more closely directs the curiosity. In general, a person's curiosity arises when confronted by an unknown, such as an activity taking place in the immediate surroundings or a bit of knowledge that is not immediately understood. A curious person wants to understand the unknown, so a mental question arises, such as: how can the identified weakness be corrected? Of course, the initial question will be specific to the idea that is not understood. In most cases, the first question will shortly be followed by additional questions, possibly with some thinking taking place between each pair of questions. This sequence reflects the process that happens in a critical thinker's mind. The thinking is followed by another round of the mental activity. It is evident that a skeptical attitude is a precursor to a person's curiosity.

Curiosity is partly an attitude, and its application can be influenced by a person's state of mind. The person's attitude at the time influences the type of thinking that a person experiences. If a person is not in a curious thinking state, then an interrogatory analysis will likely not be conducted. A mood that reinforces a suppression of curiosity at any point in time might be due to a pessimistic feeling that exists. Progress toward solving problems can be inhibited by the mood at the time. Inhibitors (see Chapter 7) can suppress critical thinking just as they can suppress creative thinking.

Inquisitiveness, which is sometimes referred to as curiosity on steroids, was identified as one of the ten characteristics of a creative person. Everyone is curious about some issues, but the depth and topics about which a person is curious vary. An inquisitive person who employs an imaginative attitude will generally produce more creative outputs than those who rely on logical thinking. With respect to creative and critical thinking, inquisitiveness should be viewed from the perspective of gaining new knowledge. Of course, improving one's inquisitiveness is part of enhancing one's critical attitude.

10.9.3 Curiosity as an Identifier

If a person develops an attitude of curiosity in one's personal life, it should carry over to the problem-solving and research environments; however, the application of the skill is more difficult in a research environment, as broader knowledge will be required. For example, suppose that a student is enrolled in a fluid mechanics or hydraulics course, and Manning's equation is part of the lesson, where Manning's equation yields the velocity (V, m/s) of flow in a river:

$$V = R_h^{2/3} S^{0.5} / n \tag{10.1}$$

where R_h is the hydraulic radius (m), S is the channel slope (m/m), and n is the channel roughness, which is assumed to be dimensionless. If the student has some level of curiosity, he or she might ask one of the following questions:

- Would this equation be accurate for measuring the velocity of a mudflow?
- Why is n raised to a power of 1.0, while the slope has an exponent of 0.5?

- Would this equation provide more accurate estimates if the water temperature was a factor?
- Is the equation valid for both low flows and major flooding events?
- If the slope was represented by a logarithmic form rather than the power model form, would the model provide greater accuracy?

Note specifically the breadth of the questions. The first question seeks to know the extent to which the equation can be extended. The second question only deals with the values of the coefficients of the model. The third question addresses the benefits of including additional variables in the model. The fourth question is concerned with the applicability of the equation to extreme conditions. The fifth question suggests that the structure of the model might influence its accuracy. Thus, curiosity should be applied broadly. Other questions should follow immediately. Just training the mind to ask such questions will enhance one's curiosity. Mind training follows from the continual practice of questioning all observations.

As a second example, consider Goncharov's empirical equation for estimating sediment discharge rates (q_b, kg/s per meter of flow width) in rivers:

$$q_b = 2.08 \left(V/V_c \right)^3 \left(d / h \right)^{0.1} \left(V - V_c \right) \tag{10.2}$$

where V is the river flow velocity (m/s), V_c is the critical velocity (m/s), d is the mean sediment diameter (m), and h is the mean flow depth (m). A curious person may pose any of the following questions:

- Why do the terms have exponents of 3, 0.1, and 1?
- Should the river flow rate q (m³/s) be included as a variable, and if it were, would the prediction accuracy improve?
- Is the equation valid for both small brooks and large rivers?
- Does it matter if the particle diameters are nearly the same or quite variable?
- Would the importance of the particle diameter change if the exponent was 1.0 rather than 0.1?

Depending on the interests of the user, many other questions could be asked. Note the breadth that is inherent to these questions. The questions focus on the empirical coefficients that are dependent on the data used to fit the equation, as well as any theory used. The second equation seeks a critique of the theory or knowledge used to identify the variables that were selected to represent the physical processes. The third question shows that the breadth of use of the equation needs to be considered. If a solution to the fourth question is sought, then an assessment of the database would be needed. The fifth question focuses on the sensitivity of the sediment discharge rates to the inputs. Curiosity that leads to a set of questions is helpful in validating the applicability of the equation. Practicing and being curious will likely improve one's ability to solve problems, as curiosity is an important part of problem-solving. Each such question should lead to an experimental plan that will result in advancements in understanding that will extend the current state of knowledge.

10.9.4   IMPROVING THE CURIOSITY SKILL

How does a person improve his or her curiosity skill? Practicing being curious is the best approach to improving the skill. It is much like the development of the questioning skill; they are actually quite closely related. This educational growth is much like improving one's skepticism. A formal course on curiosity will not be available, so experience seems to be the only way for the development and improvement of the attitude. Being inquisitive about everyday activities is the first step. The type of questions identified will vary with the issue. As demonstrated with the Manning and Goncharov equations, any aspect of the variables, the empirical coefficients, the underlying theory, or the applicability of the equations are potential sources of questions. If the issue is not about an equation but instead about a problem, then questions about the origin or the applicability could reflect curiosity. Each question should reflect on any aspect of the problem where uncertainty is expected.

10.10   PRIMARY SKILL: EMERGENCE

General definitions of emergence can broadly apply to quite different occurrences; however, when it is applied in the domain of solving complex problems, the definition must be narrowed to a specific application. The word emergence could be interpreted as the materialization of an idea, new knowledge, or information, such as novel ideas that emerge from creative idea generation activities. This interpretation would make the word emergence applicable to the analysis stage of the research process; however, the true nature of the problem can rarely be immediately known for complex problems, so the initially accepted statement could be viewed as the emergence of the problem arising above the uncertainties in the current state of knowledge. Emergence could reflect the development of both performance criteria and decision criterion. Implications of results could emergence from the conclusions that were developed from the analyses. Thus, it is evident that emergence is a primary skill that is relevant to most stages of the research and decision processes.

Questioning demands answers! Answers need to emerge from the questions. The answers to the questions should be sufficient to create results from the analyses. The results should be sufficient to deduce conclusions. Thus, emergence is a skill with characteristics of both analysis and synthesis, and acts to connect the curiosity and consequence skills.

A concept is a general idea that is derived from experiences; this can be a set of ideas that form an understanding. Conceptualization, which is an element of the emergence skill, is a process that is used to mentally formulate ideas or theories. The need to understand is important throughout processes where problems need to be solved. A primary use of conceptualization occurs in the modeling process when a set of theories or experimental results are assembled and a new model of an actual system emerges. The modeler's understanding of the system is used to place the components in a realistic order and to model the interactions and relationships between the components. The resulting model is referred to as a conceptual model of the system under analysis.

The modeler's past experiences and understanding of the current state of knowledge are used to identify the most important components, the interactions between the components, the functional forms that are used to represent the components, and the selection of both the performance criteria and the decision criterion. Where the model is fitted with data, the experimental analyses will need to match the data with the estimates obtained from the model. Performance criteria, such as statistical metrics, are used to assess the quality of the fit.

10.11 PRIMARY SKILL: CONSEQUENCE

A consequence is *a logical result that follows from an action or condition*. Measured effects can be the consequence of an experimental analysis. The word *consequence* is often associated with a cause-and-effect activity. In experimental analyses, an effect needs to be significant for it to be important in solving a problem, especially a complex problem.

The consequence skill is usually associated with the latter part of the research process, especially in identifying the conclusions that are the consequences of the results. In modeling, the accuracy of model predictions or forecasts is the consequence of the quality of all aspects of the modeling process from the formulation of the model to the data used in the analyses. For example, if the only data that were available to calibrate a model were of suspect quality, then the model can only provide uncertain estimates of the system response. Similarly, if past empirical analyses were used as model components, then the results will depend on the relevance of the past analyses to the conditions for which the new model will be applied. A thorough assessment will be necessary to evaluate the consequences of the conceptualization process.

Consequences are important. We should want to know the accuracy of the consequences. The performance criteria and the decision criteria are useful indicators of the consequences of the appropriate problem-solving activities. Other measures can be used to characterize the consequences of the problem. These can be qualitative indicators that can be assessed in some way or quantitative criteria that can be objectively compared.

10.12 AUXILIARY SKILL: SELF-CONFIDENCE

Think about a friend whom you believe to be confident. What skills or attitudes contribute to his or her appearing to be confident? Of these attitudes, which one of them do you believe is your weakest characteristic? What actions could you take to overcome this weakness? What contributed to your being relatively weak in this characteristic? Has it kept you from achieving success in any way? These are important concerns, as a person's success is greatly influenced by his or her level of self-confidence.

Confidence can be defined as either *a trust in a person or thing* or *a feeling of assurance or certainty*. Self-confidence is *the conscientiousness of one's own powers and abilities*. Generally, those who lack self-confidence underperform. Those with self-confidence usually attain positions of leadership, have body language that suggests self-assurance, are good communicators, and often volunteer for the tougher assignments.

A number of actions can be taken to increase a person's self-confidence. He or she could mimic the conduct of those who are believed to be self-confident by volunteering for the tougher assignments. They could work to improve their communication skills, including written, oral, and listening. It is helpful to reflect on both past successes and past failures, as knowledge of oneself can be gained from both of them. What contributed to the successes? What could have prevented the failures? One could read one of the many self-help books on self-confidence. Receiving effective mentoring is an excellent way of gaining self-confidence. A mentor can identify the factors that contribute to success and actions that should be avoided to prevent results that stymie success. Just as remembering successes, introspective reflecting on past failures can be a source of changes in thinking that can lead to improvements in self-confidence. Active participation in creative thinking sessions will improve self-confidence because the person will recognize its benefits. Imaginative thinking has contributed to the success of individuals, which is widely documented in a variety of sources.

Being self-confident has many advantages. It is a primary input to the solution of complex problems. Confident individuals will approach problems with a more positive, I-can-do attitude. The confident person is more likely to produce novel outcomes, thus receiving greater recognition for the completed assignments, which should lead to more important assignments in the future. Financial benefits can accrue from the completed assignments that produce unique results.

Experience, which was discussed in Section 9.12, is a very important determinant in developing self-confidence. The greater the complexity of the problem, the more valuable the experience and the greater the improvement in self-confidence. Experience derives from active participation in a variety of activities, especially acting as leader of the efforts; breadth has positive implications to confidence, as it provides a greater set of experiences that can be transferred to solving future complex problems. Participation provides the opportunity to accumulate knowledge, especially as knowledge is added to all of the dimensions of critical problem-solving. Acting as part of a brainstorming group can improve a person's status. Active participation sends a signal to others that the person is confident. Of course, participation without significant contributions can be viewed negatively.

Confidence is an attitude that is spawned from values and experiences where success was achieved for solving problems of significance. Self-confidence refers to a feeling of assurance in one's own ability. Confidence is a primary attitude that is central to innovation. Self-confident individuals are better able to resist peer pressure that aims to violate values that are important to a decision. Confident individuals are more likely to be recommended for positions of leadership where they will have the responsibility to set organizational values that will be respected.

10.13   AUXILIARY SKILL: SELF-DISCIPLINE

Self-discipline, which is associated with industriousness, can be defined as *the ability to control one's own actions where a desired activity must be sacrificed in order to perform a less desirable activity.* Self-discipline, which is the result of training and possibly with some innate roots, is a type of behavior in which a person controls his or her actions to conform to expected conduct. If a person is truly inspired to work on an activity, then the effort does not represent an act of self-discipline. With self-discipline, a person will diligently work on an activity for which they lack interest but believe that it will fulfill a responsibility. It is important that the individual must knowingly control oneself by making a personal sacrifice in order for him or her actions to represent self-discipline. Self-discipline is usually viewed as a positive characteristic because it shows a willingness to sacrifice. However, if a person performs an activity and appears to be displaying self-discipline but the work is being done for selfish reasons, then the apparent self-discipline may not actually be evidence of good character.

Self-discipline is an act that is generally undertaken to fulfill a responsibility that is not of personal benefit. If a person believes that it is necessary to become self-disciplined toward a task that they really do not wish to do, some form of self-training is needed. Establishing a set of rules or guidelines is usually the first step in the training. For those who have trouble becoming self-disciplined because they violate the rules, it may be necessary to establish a set of punishments for any violation of a rule. While the use of punishments may seem to be adopting a negative approach to overcoming the problem, the person may establish a time frame for the punishment. Punishments could be dropped if the experience shows that a change of habit has eliminated the problem and the person is now self-disciplined. Establishing rewards for the timely completion of responsibilities is an alternative to punishment, so it is considered to be a positive approach.

10.14 AUXILIARY SKILL: PERSISTENCE

Persistence is another skill that is important in complex problem-solving and research. Persistence is defined as *a refusal to give up on the search for a solution.* Persistence is important in all phases of the problem-solving process. It is especially important when very difficult problems arise. For example, research problems are often stymied by a lack of sufficient data, both when the quantity of the data is inadequate and when the data fail to cover the range of experience that is needed. Once data are collected, persistence is usually needed in Step 4 of the problem-solving process, as it is here that deficiencies in the quality or quantity of data make it difficult to provide analyses that will be sufficient to provide accurate results. Missing data can confound the discovery of trends and such cases require persistence in prolonging the search for a solution. Outliers in data sets can cause unrealistic effects. Many researchers have debated about the ethicality of censoring outliers. Some argue that the measurement was made and should, therefore, remain in the data set regardless of its effect on the outcomes. Others counter-argue that outliers can be censored because if it is not, then it can introduce bias into the effects suggested by the analyses. Persistence is needed to ensure that missing data or outliers do not adversely distort or bias the conclusions. Even with perfect data, persistence is needed to ensure that the maximum amount of knowledge has been extracted from the data by way of analysis.

10.15   AUXILIARY SKILL: FLEXIBILITY

Some businesses use a flexible work schedule, which is often referred to as *flextime*; this business practice generally improves both organizational efficiency and the happiness of the employees. The flexible hours allow the employees some measure of freedom to set their own schedule in order to meet family responsibilities or just to avoid traffic problems.

Flexibility is another skill that is beneficial to critical thinkers. Flexibility can be defined as *an ability to accommodate change*. Flexibility can be a value in each step of the problem-solving process, from selecting a topic to study, Step 1, to application of the final decision in the last step. Flexibility is important in interactions with other participants who are involved in solving the problem. In a session for generating imaginative ideas, flexibility is important to the facilitator who is faced with a group that is having problems generating ideas. Similarly, the group can have problems with a facilitator who is not very competent, such as when the facilitator is not flexible in dealing with inhibitors within the group. If the stakeholders keep changing the problem statement because they are overly flexible, then the problem-solving group will have an array of problems. So flexibility can be a positive or a negative, but in either case, it can have a significant influence on the efficiency of the effort. Flexibility can increase or decrease the numerator of the efficiency equation, thus causing an appropriate change in efficiency. The numerator will greatly decrease in an inflexible problem-solving environment.

10.16   APPLICATION OF THE INTELLECTUAL DEVELOPMENT (ID) MODEL

The intellectual development (ID) model can be used to assess a person's status in critical problem-solving ability. It can be applied to each dimension individually with the thought that a collective analysis can be made collectively across all of the dimensions, which will provide an indication of the intellectual development for applying critical thinking to solve complex problems. It can also be used to suggest the personal growth needed to move to a higher stage of critical thinking. Table 10.1 provides the characteristics of the five stages of the ID model as applied to the skills dimension. The guidelines that are shown in Table 10.1 are just examples of the type of knowledge and abilities needed to function at the individual stages.

10.17   ASSESSING THE COMPETENCY OF THE SKILL DIMENSION

Individuals will vary in their ability to fulfill the responsibilities associated with each of the four dimensions. General criteria were provided for all dimensions in Chapter 9; these were intended to show the interconnectedness across the four dimensions. Experiences in solving complex problems represent a source of educational growth, which enables a person to enhance his or her abilities to complete responsibilities associated with each dimension. Specific criteria are needed to distinguish an individual's level of competency. If four levels of competency are sufficient to distinguish between the abilities for each dimension, then the model in Table 10.2 should

TABLE 10.1

The Growth of Knowledge of the Skills Dimension for Problem-Solving

I. Comprehend
- To identify and define skills that are commonly needed to solve complex problems.
- To identify skills that are relevant to critical problem-solving.
- To understand that advances in knowledge are greatly dependent on a set of skills.
- To recognize that skills are not independent of each other.

II. Experience
- To experience a variety of skills in cases that require conflict resolution.
- To experience skills and assess the positive effect that their sequential application has on problem-solving effectiveness.
- To experience case studies when inadequate skill development led to partial failures.

III. Analyze
- To analyze cases of professional decision-making to gain experience in applying the skills.
- To analyze the way that a person's mindset influences his or her application of skills.
- To analyze past cases to learn the effects of each skill in the success of problem-solving.

IV. Synthesize
- To synthesize outcomes to situations where the sequence of skills was not followed correctly.
- To synthesize the effects of various skills on the efficiency and effectiveness of decision-making strategies.

V. Create
- To create plans for using the skill set in solving a variety of complex problems.
- To create a plan to integrate skills with the values, thinking types, and mindsets for resolving complex problems.

TABLE 10.2

Assessment Criteria for the Skill Set Dimension at Different Stages

Stage I: Novice	Stage II: Specialist	Stage III: Master	Stage IV: Expert
1. Lacks courage and curiosity.	1. Lacks complex problem-solving experience.	1. Willing to question and critique.	1. Strong skill set use.
2. Procrastinator.	2. Only self-confident for simple activities.	2. Some curiosity.	2. Very experienced with complex problems.
3. No experience with critiquing.	3. Some self-discipline.	3. Stresses efficiency.	3. Uses iterative questioning and brainwriting.
4. Minimal with decision-making.	4. Hesitant to critique others.	4. Learns from some failures.	4. Learns from failures.
5. Difficulty in using mindset to apply skills.	5. Minimal leadership of stakeholders.	5. Some complex problem-solving experience.	5. Skilled at broad based decision-making.
		6. Difficulty in dealing with the media.	6. Excels at critiquing.

enable a person to have some idea of his or her strengths and weaknesses in terms of the skills dimension of critical problem-solving. Each of the items with a stage may have a different level of importance. The disparity between the different stages should provide a person with some measure of his or her critical problem-solving status and provide the knowledge needed to enhance their critical problem-solving skills in order to move to a higher stage.

10.18 SKILL SET EFFICIENCY AND EFFECTIVENESS

Problem-solving efficiency depends in part on the proper application of the skill-set process. Attempting to analyze the quality of the existing resources prior to fully understanding the problem is not efficient. The true problem needs to be recognized before collecting the resources, as the problem-solving attempts to identify advances to the state of knowledge from the collected resource base. This substantiates that Step 1 of the problem-solving process should precede acting on Step 2. While the skill set is often applied iteratively as new knowledge is gained, the over-all problem-solving efficiency will depend on the number of iterations required. It is difficult to effectively review the resources if the extent of the problem is not really understood.

Equation 2.2 provided the equation for problem-solving efficiency, with the general equation for computing the efficiency being the ratio of the output to the resource input. If the true nature of the problem is not fully understood, then we should expect that any recommended solution will not fully solve the real problem, i.e., knowledge will not be advanced as much as it would have been if the real problem had been identified. Using the skill-set process should lead to a better solution at a lower requirement of resources. Thus, the numerator of the efficiency equation will increase and the denominator will decrease, both of which will lead to an increase in overall efficiency.

10.19 ITERATIVE QUESTIONING

As one part of the skill dimension of critical problem-solving, questioning was briefly discussed in Section 10.8. However, questioning is a skill that deserves more attention. A Noble laureate (2004) in physics said, "Fundamental questions are guideposts; they stimulate people. One of the most creative qualities that a research scientist can have is the ability to ask the right questions." When answers are needed to find a solution to an observable problem, questioning is an important element of the research process.

Iterative questioning is a method that helps a problem solver to focus on the principal concern. Iterative questioning can be represented by the following six steps:

1. State the problem in the most accurate way.
2. Convert the problem statement to a question.
3. Provide an answer that best clarifies the question; this may require a short brainstorming list of ideas.
4. Based on the list of ideas, use divergent thinking to broaden the focus.
5. Additional questions may be needed to help understand the problem.
6. Use convergent thinking to identify an answer the question.

It may be necessary to treat Steps 3–6 in an iterative fashion until an acceptable solution is developed. Iterative questioning is preferable to just attempting to think of solution to the problem in an unorganized way. The continual use of questions helps to organize a person's thinking processes. Questions challenge the thinker to find a solution.

Questioning is an attitude of: (1) inquiring about a problem; (2) applying intellectual curiosity; (3) interrogating a problem; and (4) inspiring critical thinking. Questioning has the following benefits:

1. It helps to focus on the specific problem.
2. It leads to generalizations, i.e., divergent thinking.
3. It suppresses biases, as it eliminates focusing on one idea.
4. It is better for incorporating creative thought.
5. It ultimately requires convergent thinking.
6. It encourages the consideration of alternatives.
7. It should improve the overall efficiency of problem-solving.

With the exception of an additional requirement of time, questioning does not seem to have any negatives.

An important point about asking questions is that one question usually leads to a second question because the response to the first question indicates either that the original question was too narrow in scope or that it was too far off the mark. Answers to questions generally provide some knowledge but not enough to fully address the original problem. Answers to the additional questions provide responses that clarify the issue, which helps to approach a solution to the problem.

Questioning relates to creativity, with commonalities including the following:

1. They both depend on open-mindedness.
2. Both can use brainwriting.
3. They both require a curious attitude.
4. Both encourage identifying alternatives.
5. Both require an optimistic attitude.

Again, asking the right question is a key to success in uncovering new knowledge.

10.20 CONCLUDING COMMENTS

Let's talk about sports! Is the skill set for an offensive lineman in football the same as the skill set for a teammate who is a defensive safety? Is the skill set for a left fielder in baseball the same as the skill set for a starting pitcher? Even the skill set of a relief pitcher is different from that of a starting pitcher. Is the skill set for a basketball center the same as that as for a person in Canada's national pastime of curling? So the answers to these questions are obvious, specifically that the required skill set for any task depends on the demands for that task. Specialization has its positives and negatives. This observation is also true for problem-solving. The skill set that is required for a problem solver depends on the type of responsibilities that the task requires in order to effectively complete the task. Thus, the theme of this chapter has been the elements of the skill set required for someone who wishes to solve critical problems. The generality of the skills proposed herein enables the skill set to be the basis for solving a broad range of complex problems.

Nature sets the capacity for any skill, while nurture determines the extent to which a person reaches that capacity. This statement, while clearly debatable, may justify the belief that anyone can learn to be a critical thinker, but not necessarily to the same level as a friend or colleague. The accumulation of both knowledge and experience greatly influences the extent to which a person reaches his or her capacity in any endeavor. As stressed in this chapter, a certain number of skills are important if a person wants to be an effective critical thinker. They must first believe in the value of knowing the benefit of being a critical thinker. They also need to have the courage to accept the responsibility for directing the effort to solve a complex problem. Then they need to gain the knowledge and experiences needed to advance their skill set in the skills that are important to think critically. Everyone can reach their capacity to learn the skills through education and experience. This chapter has identified some of the most important skills needed to be a critical thinker and identified ways that knowledge of the skills can be obtained.

10.21   EXERCISES

10.1. Identify criteria that can be used to characterize the quality of teaching. For each of these, indicate why it is relevant and important. If possible, rank the items in terms of importance.

10.2. Assume that you eat at different restaurants on a regular basis. What criteria would you use to rate the dining-out experiences? Can you develop quantitative scales to quantify the worthiness of each criterion at any one restaurant?

10.3. Where would the skill model fit into the problem-solving process?

10.4. Identify ways that a person can gain experience in evaluating your beliefs about politics, religion, and your choice of a technical specialty.

10.5. Courage is a primary skill, which relates to the fear of failure. Discuss the similarities and differences when the concept of courage is applied to a soldier in a war zone and a person who is involved in critical thinking.

10.6. Define the word *curiosity* and discuss ways that curiosity is important to a person wishing to improve his or her own curiosity ability.

10.7. Identify criteria that could be used to critique a fraternity party.

10.8. What thoughts might a curious person have about any empirical equation?

10.9. What thoughts might a curious person have about an equation that represents an unproven theory?

10.10. Identify ways of developing questions for a given problem statement.

10.11. What factors can limit a person from developing their self-confidence?

10.12. What visual factors (kinesics) suggest that a person lacks self-confidence? Has self-confidence?

10.13. Identify three factors that contribute to a lack of self-confidence. For each of the factors, identify the best way that the problem can be overcome.

10.14. Discuss the importance of having an *I-can-do* attitude in conducting research.

10.15. Identify past accomplishments that have contributed to your "I can do!" attitude.

10.16. Propose a process of developing a person's self-confidence. Associate the steps with those of the concept of critical thinking.

10.17. Discuss the value conflict between self-discipline and a lack of commitment.

10.18. Discuss values that are relevant to persistence.

10.19. In what way is flexibility needed when the facilitator is not properly prepared for the brainstorming session?

10.20. Discuss the ways that biasedness and honesty are in conflict.

10.21. Discuss why biasedness can be an inhibitor to progress in problem-solving.

10.22. What actions should a person take if a colleague who is working on the same project is procrastinating to a point that the person is being hindered in completing his or her part of the work.

10.23. Consider the words optimism and negativism to be two endpoints of a spectrum. Discuss the attitudes of the people who fall into the intermediate area between the two ends of the spectrum.

10.24. Based on the concepts of critical thinking, what criteria would you use to judge a science fair?

10.25. If you saw a dog but did not know people called it a dog, what would you call it? Is your name of the dog based on logical thinking or is it an emotional response? Is your name an instance of creativity?

10.26. Define curiosity, inquisitiveness, and questioning. Discuss how they may relate to problem-solving.

10.27. Propose a metric for assessing a person's inquisitiveness. It may be informative to begin by identifying the steps in developing any metric.

10.22   ACTIVITIES

10.22.1   ACTIVITY 10A: EDUCATIONAL DEVELOPMENT

Select one of the critical thinking skills identified in this chapter and find a definition for it. Also make a list of synonyms that reflect elements of the definition; this should help appreciate the breadth of the definition. For each academic level (elementary, middle school, high school, and college) develop an educational activity that could be used to enhance a student's ability at the skill.

10.22.2   ACTIVITY 10B: A NEW CHAPEAU

Add a new hat to de Bono's model of thinking; identify the color and location of the hat in the sequence of de Bono's model in Chapter 2. The intent of the new hat is to ensure that skills are properly accounted for in the problem-solving process. Thus, this is a problem in innovation. Provide a few example ideas that illustrate the application of de Bono's model when skills are important.

10.22.3   Activity 10C: Sticky Notes

Adhesives vary in the force that holds the tape to a surface. Post-it notes are char-acterized by a low adhesive force. Transparent tape is able to withstand a greater separation force, while forms of epoxy are intended to permanently bond the two materials. One model for computing the failure rate of an adhesive (F_R) is:

$$F_R = g \, k \, F \, / \, w \qquad\qquad (10.3)$$

where F = applied load (kg) needed for separation, w = tape width (m), g = gravity (m/s^2), θ = angle relative to the surface to which the tape is attached and which the tape is pulled, and k = a constant that depends on θ. Consider the experiment where a piece of tape is attached to a smooth horizontal surface. A spring scale should be attached to one end of the tape. The end of the tape to which the scale is attached is raised to an angle θ. The scale is pulled at the same angle as the raised tape. Failure is assumed when the tape begins to separate from the surface. During an experiment, measurements of F, w, and θ would be made for various types of adhesives. If the empirical results for the same values of the variables show considerable uncertainty, i.e., estimates of F_R are not well reproduced, what questions would you ask in order to search for an improvement to the model? For each option, develop an experimental analysis to improve the accuracies of the model.

10.22.4   Activity 10D: A Scale for Questions

The objective of this activity is to create a ranking system of the usefulness of the questioning skill. We have scales for ranking hurricanes, earthquakes, and music. First, identify factors that should influence the quality of a question; this might include factors such as knowledge content or the level of thinking. Then develop a quantitative rating system that could be used to indicate the way that a well-posed question will assist in finding a solution to a problem.

10.22.5   Activity 10E: The Game of Ten Questions

The intent of this activity is to innovate a player's inquisitiveness and improve the player's ability to focus on questioning. The game is for two players: the *host* who knows the answer and the *player* who asks the questions. The objective is to identify the name of the equation before accumulating ten NO answers. The equations should come from a high school physics book and could be an equation that relates to veloc-ity, forces, work, energy, conservation of energy or momentum, heat, sound, light, machines, and direct current circuits. Other general topics could be used if the host and player agree. The player wins if he or she identifies the equation or concept prior to getting ten or fewer NO responses. Hint: it might be better to ask questions such as "can I eliminate the topic of heat?" rather than "Does it relate to heat?" YES and NO are the only allowable responses of the host. The player is allowed 30 seconds to pose a question. If the player uses more than 30 seconds to state another question, a NO penalty is given.

10.22.6   Activity 10F: Turn Up the Heat

The wall of a building has several layers of material from the warm inside wall to the cold outside wall. The layers are interior wall board, insulation in the air pocket, cinderblock, brick, and aluminum siding. You want to determine the heat transfer coefficient for the wall as a whole. What questions would you ask in order to get the information needed to make an analysis of the coefficient?

10.22.7   Activity 10G: The Nosy Skill

Curiosity is said to have killed the cat, but a lack of curiosity can kill the opportunity to find the best solution to a problem. Curiosity is one of the primary skills of the skill dimension. Because of its importance, create a model process of curiosity by proposing a set of steps that illustrate the important elements of the curiosity skill.

10.22.8   Activity 10H: The Dirty Stream

A local stream is causing concerns among the local residents. The stream and its surroundings are trash laden. Excessive weeds and brush also detract from its appearance. On occasion, drug paraphernalia has been found by the stream. The local community wants to clean up the area. Use the sequence of eight skills to summarize a solution to the problem.

11 The Value Dimension of Critical Problem-Solving

CHAPTER GOAL

To show the importance of the value dimension in solving complex problems.

CHAPTER OBJECTIVES

1. To identify and define values that are important in critical problem-solving.
2. To discuss the measurement and balancing of values.
3. To explain the way that values interact with the other dimensions of solving complex problems.

11.1 INTRODUCTION

Values have directed the actions of many important people. In many cases, people have made significant sacrifices because of value conflicts. Galileo (1564–1642) was a believer in the truth of the Copernican theory such that he was willing to spend the last years of his life in confinement rather than give into a belief system with which he disagreed. Thomas Jefferson (1743–1826) stated: "Science can never be retrograde; what is once acquired of real knowledge can never be lost." Once J. Robert Oppenheimer (1904–1967) realized the potential use of nuclear bombs, he developed personal value concerns that caused him to oppose further development of the technology. He suffered in his career because of his value concerns. Socrates (469–399 BC) paid the ultimate price for his gadfly actions of being disrespecting the gods and his promotion of these values to the youth whom he taught. He chose to drink the poison cup of hemlock rather than renounce his convictions. These cases show the extent to which people will sacrifice because of values: Galileo's freedom, Oppenheimer's reputation, and Socrates life.

The decisions that people make and the actions that they take are often influenced by the values that are important to them; however, other factors like economics and personal safety can influence decisions and actions. The set of values and the importance that a person places on each one are referred to as the person's value system. A person who has a poorly developed value system will allow values to minimally influence his or her decisions. Selfishness will greatly influence his or her decisions and actions. A person with a strong value system will use values like honesty and fairness in making decisions more than economic considerations. A person's value

DOI: 10.1201/9781003380443-11

system includes both the values and the weights that the individual places on each value to reflect the importance of the value. The criterion used to set the values of the weights will depend on the problem and reflect the quality of the person's value system. Similarly, a person's personal economic system is the set of economic-based decision criteria that a person uses in making financial decisions. Some people believe that saving for a rainy day is more important than spending the money to be the first to have the latest cell phone option. Such people place greater weight on the saving criterion than on the convenience criterion. The point is: a person needs to know the relative importance of the values in his or her value system. A critical thinker must be willing to balance his or her value system to other personal systems, such as the economic, entertainment, and health systems.

What is the origin of our personal value systems that each of us use to make decisions? Do organizations have value systems? Does society have a value system? If so, how can the value system of a society be defined? Can life experiences really change our beliefs? Value systems influence a person's inclination to adopt attitudes that reflect critical problem-solving and these attitudes influence decisions and actions. For this reason, the value dimension of critical problem-solving is central to decision-making. It is worthwhile reviewing both the values and the associated attitudes that are important to a person who wishes to better understand the way that he or she makes decisions and enhances his or her critical thinking abilities.

This chapter focuses on values and value systems, both personal and organizational values. An important component of critical problem-solving is the roles that value systems play in decision-making.

11.2   CRITICAL THINKING: THE VALUE DIMENSION

Values form the basis for one of the four dimensions of the critical thinking model. While it may be the easiest dimension to discuss, it may be the most difficult dimension to gain the experiences that are necessary to be a critical thinker.

11.2.1   DEFINITIONS

A value is defined as *a principle, standard, or quality considered desirable in attending to all responsibilities.* A person's value system is *the collection of values that he or she considers important and the decision criteria that are applied to weight the values.* Inherently, the values may be weighted according to importance, but the magnitudes of the weights may be somewhat flexible depending on the complexity and nature of the problem. The value system influences, maybe controls, decisions and the actions that a person takes.

How does a person's value system develop? Is a person's value system hereditary-based or does it depend on the person's environment and experiences? The origin of a person's value system is just one aspect of the nature-versus-nurture controversy, but it is an important aspect. Recent advances in the field of genetics suggest that genetics (i.e., nature) sets a person's capacity for an issue, but environmental determinants (i.e., the nurturing) have important influences on the shaping of a person's value system and the extent to which the person reaches the capacity set genetically.

Experiences seem to be the factor that determines the extent to which a person reaches his or her capacity.

Beliefs are deep-seated feelings that a person considers important. Attitudes are applied reflections of a person's beliefs and are sensitive to the conditions at the time when the person acts, i.e., short-term feelings. Beliefs reflect the integration of a person's values and experiences. It is the values and attitudes that are evident from a person's overt actions. Values and attitudes should overlap. A person's beliefs will influence the decisions that the person makes, the ways that he or she approaches problem-solving, and the degree to which creative thinking influences his or her actions. However, the values used for each of these tasks will differ.

Some additional values that are important to critical problem-solving include:

- *Integrity*: Rigid adherence to a code of behavior that stresses honesty and responsibility.
- *Diligence*: A belief in the steady application of effort toward the specific fulfillment of specific goals and the advancement of related knowledge.
- *Dependability*: Reliably completes all responsibilities efficiently for all stakeholders.
- *Open-mindedness*: Using brain processes that are receptive to new ideas and different ways of thinking.
- *Variety*: The ability and propensity to adapt one's thought processes based on the nature and complexity of a problem.
- *Liberty*: The condition of not having any self-imposed restrictions to act with unusual thinking processes.
- *Boldness*: Having the resolution to adopt imaginative, uncommon thinking processes, especially in the face of external inhibitors.

These values are incorporated into one's value system through experiences. They can be enhanced by new experiences that result in successes in part because the person acts on the values and recognizes his or her connections to the success. Less successful experiences may weaken the person's adherence to the values and the associated beliefs.

The following are definitions of values that could be important in many complex-problem cases:

- *Knowledge*: Understanding accumulated through education and experience.
- *Fairness*: Justice to all stakeholders and in compliance with established rules.
- *Honesty:* The practice of acting without deception or fraud.
- *Wisdom*: An understanding of what is true and right and the practice of using good judgment in applying that understanding.
- *Prudency*: The exercise of good judgment.
- *Unbiasedness*: Impartial and without preconceived judgments.

It is evident that these values are correlated. Honesty and unbiasedness are interdependent by way of truth. Fairness and wisdom are also correlated through integrity.

Since values are so important, they are often an important decision criterion in most cases of decision-making. However, the interdependency of the values should be considered in the weighting of the individual values.

Other values will be important in specific cases, and such values should be identified and considered in the decision. Confidentiality, credibility, perseverance, industriousness, tolerance, and equity are a few examples of values that may be important in specific situations. Public health, safety, and welfare are very often principal values in decision-making.

11.2.2 IMPORTANCE OF VALUES

Just as it is important for decisions to be technically correct, the moral correctness of decisions is just as important. Technical aspects of a problem are generally easier to weight and assign values of relative importance than are the decisions that depend on value issues. Very often, the technical aspects of a problem can be judged quantitatively, such as with a benefit-cost ratio or using an expected value criterion. Compared to the technical aspects, value issues are more qualitative in nature, which makes it more difficult to assign a numerical score to them. How would the scenic beauty of a lake in the wilds be quantified or even rated? Thus, balancing quantitative and qualitative issues complicates decision-making especially for complex problems. Given the importance of value issues, such as public safety or environmental quality, it is generally not possible to exclude them from the decision-making. Thus, where both quantitative and qualitative decision criteria are involved, identifying an acceptable means of evaluating cases is necessary to make a decision. This task would require a performance criterion that can enable value-based objects or experiences to be assigned a value.

Solving complex problems is different from solving simple problems in part because of the greater number of performance criteria that need to be evaluated. If all of the criteria could be reduced to a quantitative metric, such as a benefit-cost ratio, then decision-making would be relatively easy. Unfortunately, too many problems involve decision criteria that cannot be quantified in a form such as benefits or costs. The inclusion of performance criteria that reflect values is one such constraint on decision-making with some semblance of creativity. How much value can be placed on public safety? Can the aesthetical qualities of a site be quantified? For example, people who own shoreline property are very much against wind farms located in the coastal waters where they can see the blades from their shoreline. Yet, many other people recognize the energy and CO_2 reduction benefits of these farms. Thus, the values of the involved stakeholders conflict, with aesthetics challenged by energy concerns. While the wind farm issue is certainly important and complex, the point of the dilemma is that not all performance criteria can be placed into a quantitative form. How would the aesthetical issue be judged? If decision makers wanted to transform these value aspects of the wind farm to dollars, it is doubtful that the stakeholders who are involved in the case could agree on a value-to-dollar conversion rate. Even if they were willing to try to make the conversion, their inherent, but legitimate, biases would yield quite different quantities. Such differences create conflicts. Thus, since the value issues are often the basis for major performance criteria, the value dimension of the problem becomes a major decision issue, especially for complex problems.

11.2.3   ASSESSMENT OF VALUE IMPORTANCE

It is difficult, likely impossible, to quantitatively rate value issues such as public safety or aesthetics, but some way of ordering the values is necessary when making a decision. If values are not weighted in some way, then they will likely not adequately and accurately influence the final decision, which is counter to the belief that value issues are important. Furthermore, stakeholders who have value-sensitive but conflicting interests will greatly complicate decision-making. The stakeholders will pressure decision makers to be more sensitive to the values that are relevant to the stakeholder's interests.

Ordering the values would be the first step in incorporating values into the decision process. If values are considered important criteria in the evaluation of alternative decisions, identifying the relative importance of values becomes necessary. Obviously, the importance of values will vary with the project. Public safety may be the most important value in a public works project while confidentiality may be the most important value issue in an organizational decision. The proposed model for incorporating values that are important in critical thinking may need to involve a ranking of levels of importance.

Statisticians classify statistics on three scales, which can be used with this model of value balancing:

- *Nominal*: Values are placed into mutually exclusive categories that are not distinguishable in relative terms; these categories are discrete and qualitative.
- *Ordinal*: This class also assigns values to categories, but unlike the nominal scale, the categories can be ranked in accordance with importance, but the difference in ranks cannot be specified, i.e., the magnitude of the interval between two categories cannot be specified.
- *Interval*: Values on this scale can be assigned a quantitative value according to some criterion, with the interval between two items known and important.

The problem with weighting values on each of the scales is quite obvious and would vary with the problem. The interval scale allows for quantitative analyses, but concerns with the uncertainties of the numerical values assigned could hinder decision-making, especially when the stakeholders have very distinct and different objectives. The uncertainty is less of an issue with the ordinal scale, but the obvious disparities in ordinal ranks could cause over-valuing and under-valuing of many human values. These general concerns have and will continue to complicate the inclusion of values into decision-making.

11.2.4   IMPORTANCE OF VALUES: STEPS OF THE RESEARCH PROCESS

So far, the discussion of values has centered about their use in making decisions, which is the last step of the six-step decision-making process. The focus on decision-making does not imply that values are not relevant to the other five steps of the process. The primary values previously identified are relevant to Step 6 (i.e., decision-making).

Final decisions should reflect values throughout the decision process. The following is a partial list of values that are separated to show their relevance to the individual steps of the problem-solving process:

1. Problem identification
   - *Knowledge*: To identify the fundamental problem that needs investigation, the critical thinker who is responsible for placing the problem in the proper context must have knowledge that is based on both education and experience.
   - *Variety*: To fully appreciate the potential implications of the problem statement, a critical thinker will need to have a variety of experiences.
2. Resource collection
   - *Fairness*: To provide unbiased assessment of the state of knowledge for any discipline requires acknowledgment of all past efforts on the topic, which implies fair treatment.
3. Experimental design
   - *Knowledge*: To use the knowledge obtained in Step 2 to develop testable hypotheses.
   - *Wisdom*: To have the vision to recognize whether or not the experimental analyses will adequately verify the hypotheses.
   - *Industriousness*: Due to the exponential growth of research, the volume of work currently available is considerable. Thus, searching for all sources of knowledge and data will require an industrious attitude, which will be necessary to provide the data that would be necessary to complete the experimental analyses.
4. Analysis
   - *Diligence*: To provide the effort that is needed to ensure that the analyses provide the most knowledge that could be extracted from the data.
   - *Honesty*: To conduct all analyses honestly is vital to the accuracy of the decision.
   - *Competence*: To apply oneself to the maximum extent possible.
5. Synthesis of outcomes
   - *Truth*: To ensure that the results of Step 4 are properly interpreted in developing the research outcomes.
   - *Credibility*: To present the outcomes in a way that all future uses of them are ethical and cannot be misinterpreted.
6. Decision
   - *Wisdom*: To communicate the results such that implications ensure safety in the decision.
   - *Honesty*: To be honest in all communications related to the research.

Diligence and perseverance are always values to which a critical thinker must adhere. The critical thinker must be diligent in pursuing all knowledge relevant to the decision about which they are responsible. Any deficiency in knowledge, either technical or value related, can lead to an incorrect decision. For this reason, a criterion for being identified as a critical thinker is to make every effort to obtain knowledge of all aspects of a case. Other values relevant to critical thinking include the following:

- *Equity*: The quality of being impartial.
- *Fidelity*: Faithful to duties.
- *Prudency*: Use of good judgment.
- *Confidentiality*: Assurance that the person will improperly disclose information.

Other values will be important for other problem types. A variety of values are important to critical thinkers. Values are central to the subject of critical thinking, which is the reason that values are the basis for one of the four dimensions of critical thinking. Values influence all aspects of decision-making including the effort expended, the accuracy of their results, the promptness of the work, the knowledge gained to advance the state of knowledge, and the reputations of those involved. More effective decisions will be made if a critical thinker truly knows the beliefs and skills that interact with his or her value system.

11.3   THE ORIGIN AND CHANGING OF VALUES

As a general rule, parents and the extended family try to instill in children a basic set of values, ones that they themselves have found to provide a good basis for daily decision-making. Primary values such as honesty, self-discipline, kindness, prudency, and diligence are often central to early childhood development. These characteristics are values in the sense that they are the principles that adults believe are important for children to develop, to form the basis for their value systems, and to control their actions. From values such as these, the children develop both positive and negative beliefs and attitudes; for example, the value of honesty translates to the attitude that cheating on tests is generally not acceptable. Also, diligence is a positive value; since laziness is counter to diligence, then laziness is considered to be irresponsible. The values that children learn at an early age remain in force until the events and experiences of their lives place doubt in the child's mind about the never-changing truth of the value. These experiences may result in a change in the child's value system.

11.3.1   EFFECT OF EXPERIENCE ON VALUE SYSTEMS

Experiences in life can change a person's basic beliefs. For example, a student who becomes aware that classroom cheating goes unpunished may irrationally think that cheating is now considered acceptable conduct; then the student's fundamental belief about honesty may change. When the student enters the business world, the altered belief that cheating is an acceptable practice may cause a corresponding change in his or her values and attitudes such that he or she now considers cheating in the business world to be acceptable. This temporally dependent experience of classroom-to-workplace cheating suggests that beliefs can change with both time and experience. The issue can be extended even to the point when the student believes that honesty is not a core value and that dishonesty is acceptable and, in some circumstances, can rightly influence behavior. Experiences can change a person's existing value system over periods of time. The extent of change will be influenced by the strength of the experience.

Values and beliefs influence a person's inclination toward or away from seeking experiences that involve critical problem-solving. If a person's experiences have led to the belief that only other people can be successful at innovation, then he or she will avoid learning about critical thinking. The person will pursue non-innovative solutions to problems and thus not accumulate experiences that are needed to become a critical thinker. Ultimately, the person will accomplish less than he or she might have been capable. Conversely, experiences that reinforce an I-can-do attitude will lead to greater success. For example, a successful researcher will likely act based on the values that led him or her to adopt an "I can do!" attitude. If the person has a positive-oriented value system, he or she will likely act as an innovator and will likely achieve beyond his or her expectations. The question is: "What experiences lead to changes in one's value system and ultimately heighten or suppress an innovative/creative attitude?" Successful experiences with creative activities will likely influence one's attitudes and willingness to become more open to participate in critical problem-solving.

Life experiences can create new beliefs and values, as well as cause change to existing beliefs and values. To become a critical thinker a person should seek out experiences that will cause them to believe strongly in both a selfless value system and the value of imaginative idea-generating methods. While individuals have had informal experiences with imaginative idea generation techniques, such as being involved in a brainstorming session, they often lack sufficient personal experiences to take full advantage of the merits of being a critical thinker on a regular basis. In a sense, they lack the full benefits of having a critical thinking attitude because they have not had the quality experiences that are needed to produce critical thinking skills and attitudes. Also, they may not have had the ability to recognize the knowledge that could be gained from the experience. They likely have not thought about their experiences in ways that would improve their critical thinking ability. A lack of knowledge about introspective analysis can prevent the learning that is needed to become a critical thinker. The lack of experiences that promote critical thinking can prevent a person from developing the mindset and values that are needed to identify novel solutions to problems. This illustrates that experiences, values, and beliefs are interdependent.

11.3.2 TEMPORAL EFFECTS ON VALUE SYSTEMS

Time can be a factor in value assessments. Climate change may be the most obvious issue where time is important. Some people argue that, since they will not be alive in the year 2100, they are not overly concerned with climate change. Other people argue that each of us has a moral obligation to leave a healthy planet for future generations and that poor decision-making at this time can have serious consequences in the future. Obviously, value conflicts can be time dependent. How can time-dependent (future) values be weighted so that they can be balanced with the worth of current value issues? This temporal-sensitive nature of values is another aspect of value decision-making that complicates problem-solving. As value issues become more important to any conflict resolution, the cases are considered to be more complex. Thus, problem complexity can be time dependent.

In summary, it appears that innate factors combine with life experiences to create a person's value system. Growth of a person's value system is a direct contributor to the growth of a person's critical thinking abilities.

11.4   ORGANIZATIONAL VALUE SYSTEMS

Just as individuals have value systems, organizations have value systems. An organizational value system is the collection of values that are embedded in the organization's policies and practices. The value system may be explicitly stated in the form of a code of conduct or just a set of policies that implicitly indicate expected conduct. The value system will be reflected by the decisions that the organization makes and the actions that the organization takes. Organizational decisions must be monitored by those in positions of leadership. Often, the organizational value system is not explicitly stated or changed until after a problem has surfaced. Then the policy statement is developed, but such practice can distort the value system, as it may place too much emphasis on the problem that recently occurred. A quality value system should reflect the breadth of values for all of the organization's responsibilities. It should be a broad statement that can be applied to a range of problems. Leaders need to establish the organizational value system well before value conflicts challenge the organization, as existing organizational policies are usually insufficient to cover almost all issues that can arise.

An organization's value system is much like a personal value system. The experiences of an organization can cause changes just as personal experiences change a person's value system. A change in leadership is often accompanied by changes to the organization's value system. New leaders generally want to change the organization's policies and practices; however, the changes in policies may or may not be sensitive to values. Someone new to the leadership position may wish to have a company value system that is associated with the organization's economic system, which may place less emphasis on values and more on profits. Employees of the organization will need to recognize the breadth of implications of such changes.

An organization's value system becomes part of the personal value system of each employee. If the employee does not believe that the organization's value system supports his or her own personal value system, then conflicts may arise and the seriousness of the conflict may be detrimental to the employee's success within the organization. Since values influence critical problem-solving practices, the organization's value system will influence the way that decisions are made.

11.5   BALANCING VALUES IN AN UNCERTAIN WORLD

The incorporation of qualitative human values into decision-making is problematic. It is difficult to convince stakeholders who have quite conflicting viewpoints to agree on a way of rating values so that the values can be integrated and weighted with the other decision-making factors. Ideally, the importance of the value issues that are relevant to a complex problem could be transformed into quantitative decision criteria that can be balanced with the other decision factors, such as the technical and economic dimensions. Unfortunately, this ideal situation has not been achieved to everyone's satisfaction.

Stakeholders can be legitimately biased toward the obligations that are important to their cause. While some bias is expected and is reasonable, in some cases stakeholders may distort their assessments of the importance of the values relevant to their own cause. This distortion will make decision-making more difficult. It introduces a significant uncertainty into the decision-making. Mathematical analyses are available for accounting of uncertainties. Computing a value for the decision criterion with and without the uncertainty can provide some measure of the importance of the uncertainty in making the final decision. If the effect is significant, then the values and their weights must be reconsidered, with greater care taken in assessing the importance of each weight. The objective of this sensitivity exercise would be to assess the importance of the decision to the uncertain weights assigned to the value issues. The outcome would be a decision as to whether or not it would be necessary to expend additional resources to develop better assessments of the accuracy of each input related to each value issue.

11.6   APPLICATION OF THE INTELLECTUAL
DEVELOPMENT (ID) MODEL

The intellectual development model can be used to assess a person's status in critical problem-solving ability. It can be applied to each dimension individually with the thought that a collective analysis can be made across the dimensions, which will provide an indication of the intellectual development for applying critical thinking to the solution of complex problems. It can also be used to show the personal growth needed to move to a higher stage of critical thinking. Table 11.1 provides the characteristics of the five stages of the ID model as applied to the value dimension. These guidelines are just examples of the type of knowledge and abilities needed to function at the individual stage.

11.7   ASSESSING THE COMPETENCY OF THE VALUE DIMENSION

The ID model can be used to measure the intellectual growth within a dimension; however, a person can function at different levels across the five abilities. Individuals will vary in their ability to fulfill the responsibilities associated with each of the four dimensions. Experiences in solving complex problems represent a source of educational growth. Of course, the quality of experience will influence the level of growth. Criteria are needed to distinguish an individual's level of competency. If four levels of competency are sufficient to distinguish between the abilities for each dimension, then the model in Table 11.2 should enable a person to have some idea of their strengths and weaknesses in terms of the value dimension of critical problem-solving. The disparity between the different stages should provide a person with some measure of their critical problem-solving status and provide the knowledge needed to enhance their critical problem-solving skills in order to move to a higher level.

TABLE 11.1
The Growth of Knowledge of the Value Dimension for Problem-Solving

I. Comprehend
- To identify fundamental values and provide definitions of each that are relevant to critical thinking.
- To understand the value basis of personal attitudes.
- To understand that a set of personal values constitutes a value system.
- To understand that values can take on different levels of importance depending on the issue.
- To understand that a person can have a set of professional values that are needed to make decisions in professional life.

II. Experience
- To use case studies to gain experience in assessing the effects of values in making decisions.
- To experience decision-making that involves actual value conflicts.

III. Analyze
- To analyze the value basis of important decision criteria.
- To analyze the way to critique the value decision criteria of stakeholders.
- To analyze organizational codes of conduct on the basis of values.
- To analyze professional problems to recognize the ways that values influence decisions.

IV. Synthesize
- To synthesize value-based codes of conduct for application to different organizations.
- To synthesize the outcomes of complex problems of value-based decisions.

V. Create
- To create solutions to complex problems where it is necessary to balance values of conflicting stakeholders.
- To create the elements of selfless attitudes for use in solving complex problems.
- To create a procedure for integrating value criteria with criteria for other dimensions, such as the technical and economics dimensions.

TABLE 11.2
Assessment Criteria for the Value Dimension in Problem-Solving

Stage I: Novice	Stage II: Specialist	Stage III: Master	Stage IV: Expert
1. Selfishness influences decisions.	1. Values confined to self and employer.	1. Willing to compromise values.	1. Knows the way to balance values.
2. Actions are either right or wrong.	2. Lacks experience in using values for making decisions.	2. Hesitant to weight values.	2. Beliefs reflect values.
3. Little breadth in understanding values.	3. Does not recognize balancing of values.	3. Some sensitivity to stakeholder conflicts.	3. Unbiased decision maker.
4. Fails to recognize uncertainty in decisions.	4. No experience with multiple stakeholders.		4. Firm in supporting high value standards.
			5. Tends to be selfless.

11.8 THE EXPERIENCES-VALUES-BELIEFS MODEL

Experiences can influence a person's value system, which influences the person's beliefs. A person's beliefs influence the way that a person reacts to experiences, i.e., values influence experiences and experiences influence beliefs. Thus, the relationship between experiences, values, and beliefs is circular. A starting point for the triad is not evident, if one exists at all. It seems reasonable that relations exist between beliefs, values, and experiences. All of these influence the way that each person approaches problem-solving, whether they use logical thinking or imaginative thinking methods. Yet, the values that a person holds influence the way that he or she will act. Understanding the connections between experiences, values, and beliefs is important.

Success at problem-solving depends on a number of factors, most notably the person's attitudes. Seemingly gifted individuals may not be able to perform novel research because of the attitudes that influence their thinking and actions. For example, a pessimist may not have the confidence to even make an effort to solve a problem; this reluctance to act implies a lack of problem-solving courage. The likelihood of success at problem-solving will improve following experiences that increase one's optimistic viewpoint, self-confidence, and self-discipline. This improved attitude then encourages the person to seek knowledge on ways to improve one's creative thinking skills and attitudes. Ultimately, success at quality experiences should lead to greater success, with success breeding self-confidence. A person who has responsibilities to solve problems needs to recognize the effects of values and attitudes on success and that advancing one's critical problem-solving skills and mindset will make future success more likely.

In order to better understand the important synergisms between problem-solving and critical thinking, a model that reflects on their interactions will be proposed. Efforts at critical problem-solving are influenced by both values and attitudes, which direct a person's actions and decisions. Experiences influence both the beliefs and attitudes that a person adopts and can influence a person's value system. Experiences in life can lead to changes in both the values and beliefs. At any point in time, a person's values and beliefs will influence the person's responses to future experiences and these experiences can lead to changes of attitudes. It should be evident that beliefs, values, and experiences are a triad that influences a person's thinking and decision-making and a change in any one of them can lead to a revision of the other two.

11.9 CONCLUDING COMMENTS

Values are generally an important component in making decisions for complex problems. The primary difficulty with values is the wide range of beliefs about the importance of individual values. People have different life experiences, and it is the nature of these experiences that influences the importance that a person places on different values. The unwillingness of egoistic people to consider the value priorities of others can greatly skew a person's value system and his or her decisions. While biases in the weighting of values are to be expected and are even legitimate, extreme biases can distort decisions. Decision-making is more accurate when a decision maker recognizes the priorities of the individual stakeholders who are involved in a problem; in such cases, the decision process is also less complicated.

Why are values part of critical thinking? Critical problem-solving must be viewed as more than just the generation of wild-and-crazy ideas of creative problem-solving. In fact, it is possible that some complex problems could be solved without imaginative thinking being part of the decision process; however, it is likely that value issues will be a major part of the decision-making of every complex problem. Value issues usually have a significant influence on decisions for complex problems. The uncertainty of weighting of values also contributes to the difficulty in assessing the importance of each value and even including value issues in the decision process. Yet, the decision makers must contend with the many difficulties associated with incorporating values into the evaluation of alternative solutions to complex problems.

It is important to differentiate between creative attitude and critical problem-solving. The former is a disposition, i.e., tendency at a specific time, while the latter reflects action that is based on a collection of attitudes and depends on the issue being assessed. Creative attitudes are often based on one's ability to be imaginative, to be curious by nature, and to question the existing state of an issue, which are issues related to all four dimensions. Having these abilities improves one's chances of trying to solve complex problems such that the outcome is more unusual and superior to outcomes based solely on logical thinking.

A general premise to critical problem-solving is that the different dimensions are not independent of each other. Knowledge of values supports the educational growth of skills such as critiquing and questioning. Truth, unbiasedness, and fairness have obvious connections to critical problem-solving skills. Not appreciating such values may distort a person's perspective on the skills. Thus, the educational effort should simultaneously provide advancements in knowledge of skills and values, which reflect two of the four dimensions discussed herein.

Values are not universal truths, even within a culture, even within a family, and even for a person over time. This arguable truth proposes the thought that critical problem-solving is not a one-size-fits-all process. Yet, we may be agreeable that some values should be held universally high. Values, such as public safety, knowledge, diligence, perseverance, and fairness, should be considered important to everyone. Other values may be considered as principal values by an individual, but not by everyone. In spite of the concern that it may not be possible to declare a universal set of values, one can argue that the value dimension of critical problem-solving is a necessary dimension.

11.10   EXERCISES

11.1.   Is cheating in a classroom environment as a student correlated with cheating in the business environment later in life? Discuss.

11.2.   How do experiences in life influence changes in a person's value system? Provide an example and discuss the relation.

11.3.   What types of experiences have a positive influence on the development of an I-can-do attitude? Can experiences discourage the development of an I-can-do attitude? Explain.

11.4. What are examples of life experiences that influence the development of a person's sense of honesty? Compassion of the plights of others? Work ethic?

11.5. Life experiences can change a person's belief system. What life experiences could change a person who is selfish to a person with a selfless attitude?

11.6. When trying to balance values, what general factors influence the weights that are assigned to values?

11.7. What does it say about a person who distorts the weights applied to values? Consider both assigning a weight that exceeds the expected value and a less-than-expected weight.

11.8. Given the definition of diligence in Section 11.3, what factors contribute to a person generally being less diligent than society would expect of the person?

11.9. Why is variety considered to be a value?

11.10. A Bible verse is: "Wisdom is better than rubies" [Proverbs 8:11]. What is meant by the adage?

11.11. A Bible verse is: "Many shall run to and fro, and knowledge shall be increased" [Daniel 12:4]. Interpret this adage in terms of education.

11.12. Why are values difficult to quantify?

11.13. Three scales are discussed in Section 11.3.3. Provide three examples of each scale.

11.14. Two values are indicated as being important in problem identification (Section 11.3.4). Identify a third value that could be important and explain the way that it is important to problem identification.

11.15. Obtain a copy of the Hippocratic Oath. Identify the primary values on which it is based. Use the values to synthesize a code of conduct for people involved in animal welfare.

11.16. How could values be involved in the application of the efficiency equation (e = output/input) to the solution of problems?

11.17. From the standpoint of Eq. 2.4, discuss the ways that biased thinking causes reduced creative efficiency.

11.18. What biases might stakeholders have if they were involved in a case of environmental degradation?

11.11 ACTIVITIES

11.11.1 ACTIVITY 11A: THE RATING GAME

Part I: This is an activity for a group of four people. The objective is to demonstrate the difficulties in weighting values. Initially, the group should not have any discussion of values. Each person gets a sheet of paper with the following value issues listed at the top: public safety, environmental health, wildlife preservation, climate change, and food for the needy. On the sheet of paper, each person should rank these in importance from the most important (rank = 1) to the least important (rank = 5). The person should write a brief statement of justification of their ranking of each value.

Part II: Following the individual analysis, a group discussion should follow; the group should *not* provide a group ranking of the five value issues—discussion only.

Part III: Following the group discussion, each individual should provide a second ranking. Then the group should discuss the reasons that weighting values is so difficult.

11.11.2 ACTIVITY 11B: AESTHETICS VERSUS FUNCTIONALITY

Consider the case of the design of a new bridge for a moderate sized city. The objective of this activity is to critically assess the importance of values in critical decision-making. First, identify who might be the stakeholders in the bridge design; this list would include anyone who might have an interest in the bridge project before, during, or after completion of the construction. This list of stakeholders could include society and various interest groups. Environmentalists who recognize environmental issues with the bridge construction can be considered stakeholders. Even those who are concerned with the aesthetics of the bridge can be considered a stakeholder group. Second, for each stakeholder on the list, identify the value issues that would be individually relevant to their position. Third, within each stakeholder list, assign weights to each value. The weights should indicate the relative importance of the value to the stakeholder. Each person should develop a system of assigning weights unique to them—no discussion. Fourth, consider the public official who would be making the final decision. Discuss how he or she would balance the values and the weights in deciding whether or not to build the bridge and if the decision is to build, which bridge option. Also, discuss the accuracy and uncertainty of the weights.

11.11.3 ACTIVITY 11C: WIND FARMS ARE AN EYESORE

Obtain information on wind farms in coastal areas, with special attention given to the positives and negatives. (*a*) Based on your beliefs and attitudes, list the positives and negatives in order of importance. Do the positives outweigh the negatives? Justify your response. (*b*) Now place a numerical value on each item according to your assessment of their importance. A 0-to-10 or 0-to-100 scale can be used. Find the total score for the positives and the total score for the negatives. Discuss the result. (*c*) Repeat parts *a* and *b* but now assume that you are a property owner along the coast and the wind farm would be seen from your property. How does the rating change from you as a person in the city to a property holder? Compare the totals with those of part *b*. (*d*) Repeat part *b* but assume that you are a representative of the energy company. Discuss the result. (*e*) Repeat part *b* but now assume that you are an environmental activist. (*f*) Create a table of the ratings for part *b* through *e*. Discuss the results. What do the results imply about biases and the problem of making decisions that are value sensitive?

11.11.4 ACTIVITY 11D: HOW DO I RATE?

Take a moment to rate yourself on each of the following values and attitudes: honesty, kindness, and perseverance. Assume that each value or attitude is on a continuum.

For those characteristics that you do not have a perfect rating, identify reasons that your rating is toward the negative end of the continuum. Then develop a plan that will lead to progress toward the more positive end of the continuum. Discuss the attitudinal problem with a mentor or read a self-help book on the issue. Such an assessment and attitude change will increase your problem-solving-efficiency and lead to greater success.

11.11.5 Activity 11E: Little White Lies

We can probably agree that it is wrong to tell your mother that her special dinner was great when you really thought that it was no better than a dinner at the local fast food restaurant. But in comparison to the dinner issue we could also agree that it is more wrong to lie about your name as an author of a report that you only copy edited. Maybe lying, or more broadly being honest, is a spectrum of values rather than a good-bad dichotomous scale. Develop a scale for the value honesty that covers the severity of the honest-dishonest spectrum using four intermediate points. Provide examples to show the intermediate points between the ends of the spectrum.

If this is a group activity, use four intermediate points and have the individuals separately identify, i.e., name, the intermediate points. Use the variation in the ratings for any one value as an indication of its uncertainty. Discuss reasons for the uncertainties.

11.11.6 Activity 11F: Covering Up Wrong Doings

Loyalty is believed by some companies to be an important value, but it can be a failing if it requires lying or doing something unlawful. Identify organizational and colleague pressures that may be used to prevent proper professional conduct when an unlawful act is done within an organization. Also, identify rationalizations that are used to succumb to the pressures to cover up wrong doings.

11.11.7 Activity 11G: Balancing to Make a Decision

In urban areas, sound barriers are installed along the sides of roadways where traffic volumes are high. While the barriers control the noise, they are criticized on the basis of aesthetics and their constraint on the movement of wildlife. This imbalance in values determines where these barriers are placed. Develop a value-based metric that can be used to reflect the different effects of barriers. Discuss the ways that the metric could be used to balance the positive effects and the negative effects.

11.11.8 Activity 11H: Drone Bombing

Drones are now being used to bomb military targets that are far removed from the location of the person who controls the drone. Some people have moral concerns with this form of warfare. Identify the values that are relevant to both sides of the decision to use this tactic. Discuss and balance the values and the decision.

12 The Mindset Dimension of Critical Problem-Solving

CHAPTER GOAL

To show the ways that a person's mindset influences success in critical problem-solving.

CHAPTER OBJECTIVES

1. To provide an overview of mindset characteristics.
2. To introduce mindset inhibitors that can have negative effects on success and discuss ways to overcome the inhibitors.
3. To acknowledge the interconnectedness of the mindset dimension with the other dimensions of critical thinking.
4. To discuss the effects of inhibitors such as pessimism, procrastination, and a lack of self-discipline on the quality of creative output.
5. To identify ways that a person can consciously make positive changes to his or her mindset.

12.1 INTRODUCTION

The advancements of Scientific Methods in the late eighteenth and early nineteenth centuries created a mindset that promoted change. Many sections of society were not ready for change, which spawned a pessimistic frame of mind with many people. The pessimism was evident in some of the popular writings of the day. In 1798, Thomas Malthus (1766–1834) published a book that suggested the trends in society were leading to severe poverty and considerable levels of starvation. Malthus based his pessimistic views on the apparent disparities in population growth rates and the supply of resources. He suggested that an exponential growth in population would overwhelm the linear growth in food and resources. The Industrial Revolution experienced increases in factory production rates and related declining rates in home manufacturing, which led to the formation of groups like the Luddites who would raid factories and destroy the machinery. The writings of authors like Charles Dickens (1812–1870) portrayed a society that was poverty-stricken. At that time, his description of city life was quite depressing to the public.

DOI: 10.1201/9781003380443-12

For a noticeable number of years, this pessimistic mindset limited advancements within many sectors of society.

The content of this chapter will prepare a reader with basic knowledge of the mindset dimension, with special emphasis placed on defining terms and discussing the roles that the mindset plays in enabling a person to develop the abilities of a critical thinker. Unless a person has good control of his or her mindset, his or her mindset can act as an inhibitor to critical problem-solving; therefore, overcoming inhibitors such as pessimism should be a goal of anyone who wishes to be a critical thinker.

12.2 INTRODUCTION TO THE MINDSET DIMENSION

Of the four dimensions of critical problem-solving discussed herein, the mindset dimension is likely the most difficult to characterize. It has many components and most of the components of this dimension are subject to considerable uncertainty, even in providing definitive definitions. The components of a mindset also vary from person to person and for any one person from day to day. For example, some people are very moody, with the mood varying over short periods of time; other people are much less moody and subject to less frequent changes in mood. A moody disposition may be evident in the decisions and actions of the person. Even a person's beliefs can change, but generally over a longer time frame than changes in the person's mood. Beliefs, attitudes, dispositions, and moods are all components of a person's mindset dimension of critical problem-solving, but it is difficult to provide a clear distinction between the components.

A person's mindset largely controls his or her decisions and actions. For example, a person who is in a procrastinating mood is difficult to motivate to act on his or her assigned responsibilities. Conversely, when a curious person confronts a puzzling problem with his or her research, it is difficult to entice him or her from the lab where he or she is acting on the unexplained event. Mindset states range from the positive, such as an optimistic attitude, to the negative, such as making decisions when in a pessimistic mood. Some individuals have considerable control over their moods and can adjust the mood to fit the conditions and the demands on their time. Other individuals do not seem to have control over their moods, which can adversely affect their productivity and happiness. Not having the ability to control one's mood can limit a person's ability to solve problems. A critical thinker will generally have excellent control of his or her moods. As a person advances to become a critical thinker, he or she improves in the control of his or her mindset.

The mindset dimension is important to a critical thinker because it has a significant influence on the application and effectiveness of the other dimensions of critical problem-solving. If technical knowledge is considered to be the fifth dimension, the mindset certainly influences one's ability to solve problems that have a technical basis. When a person is in a pessimistic mood, he or she will lack the confidence to believe that he or she can resolve a technical difficulty. A person's mood is very influential in his or her willingness to meet all of the value responsibilities. The mindset influences the person's application of the skills in the skill set, such as courage and questioning. For example, a pessimist may not have the courage to accept responsibility for a challenging job. Thus, the connectedness between the mindset dimension and the other dimensions is quite evident and therefore influential in meeting responsibilities.

A person's mindset reflects his or her collective memory based on both life experiences and education, and the innate ability to understand and assimilate the experiences and education into a set of mindset components, which include principles, beliefs, attitudes, dispositions, and moods. Of interest here are the mindset components as they relate to the solution of complex problems. A person's state of mind is the subset of mindset components that govern his or her thinking at a particular time and about an issue of interest. A mood could dominate thinking and be the controlling state of mind. At other times, a principle could control the state of mind. The component that controls the thinking, which influences decisions and actions, depends on the relative strengths of the components at the specific time.

12.3 DEFINITIONS

Mindset is a difficult word to define, as it can be applied to a variety of concepts, so the definition should be coordinated with its specific use. Thus, before providing a definition for mindset, it would make sense to define the framework to which it will be applied, which specifically is the mindset dimension of critical problem-solving.

Definitions of the word *mind* mention words like thought, feeling, memory, behavior, opinion, and purpose. The definitions for these words imply that the word *mind* refers to conscious or unconscious processes of the brain; however, for our introductory purposes herein, the specific mental processes and lobes of the brain are not of immediate importance. Some definitions of the mind also include the words imaginative thinking, as imagination refers to the direct lack of reality and the ability of the person to create mental images or concepts. For our purposes herein, mind can be defined as:

> A **mind** reflects the conscious and unconscious processes of the brain that generates ideas or concepts, real or imaginary, for the purpose of making decisions and directing actions.

This definition, when combined with the word set, yields a useable definition of an important dimension of critical problem-solving:

> A **mindset** is the sum of conscious and unconscious processes of the brain, including the generation of imaginative ideas or concepts for use in making decisions and directing actions with the objective of solving problems, in a way that enables these processes to be integrated with other important dimensions.

While this definition does not require a problem to be complex, the heart of the discussion of the application of the mindsets of critical thinkers will focus on the mental activities used in the solution of complex problems. For the purpose of problem-solving, this definition recognizes the interrelationships with the other three dimensions, i.e., values, the skill set, and thinking types, as well as other dimensions specifically relevant to the problem of interest.

One problem in an attempt to define mindset is the inability to provide precise, objectively defined terms for mindset components. Mindset components are mental constructs, but they contribute to the uncertainties in decision-making because of their changing nature. Some definitions of beliefs suggest that they are little more than opinions, while principles are often viewed as universal truths. One of these definitions seems too subjective while the other seems too restrictive. Beliefs can change rapidly, i.e., over the short term, while principles are assumed to be stable over a longer period of time. The following definitions characterize the mindset components as the definitions apply to critical problem-solving:

- *Belief*: A sympathetic conviction that is generally accepted as true, but often reflects societal conventions declared by a person's culture.
- *Principle*: A relatively fixed attitude that serves as a mode of action or thought that remains constant over long periods of time but can be changed following significant experiences.
- *Disposition*: A person's usual manner of emotional response providing an inclination toward an action, with the inclination being somewhat ingrained. A person's disposition can influence his or her behavior, i.e., actions.
- *Mood*: A temporary state of mind usually held for a short period of time.
- *Attitude*: A prevailing state of mind toward an idea, person, or group—a state with a tendency toward a belief. A state of mind with cognitive and affective components that are based on an accumulation of knowledge.

When confronted with a problem, a person's thinking pattern greatly influences the way that he or she responds. Thinking is a process that involves formulating an idea or concept in one's own mind and reflecting on a sequence of mental actions that extends from identifying the problem to the development and verification of a conclusion to the problem. Some people associate thinking with the invention of an idea or concept. This is just part of the thinking process. Thinking is much more as it reflects the action of the person's mind. Thinking is a process. The thought process begins with a problem or a concern and ends with a decision or conclusion. The definition of thinking is also important relative to the ways that individuals solve complex problems. Thinking involves mental action usually dependent on an array of mindset components.

Before the issue of critical analysis as a problem-solving method can be addressed, the word *attitude* needs to be more fully understood. An attitude is a state of mind with a cognitive component that is based on facts and an affective component that is based on emotions and feelings. Attitudes are influenced by knowledge and produce inclinations toward certain decisions. A person's attitude at any point in time is influenced by the disposition on the issue and can greatly influence the person's conduct when assessing alternative responses, when solving problems, and when making decisions. Facts that differ from those that led to a specific attitude can cause a change of behavior at any point in time that is counter to the inclination of the attitude. An attitude is an indicator of the person's likely thoughts and actions in the immediate future. A biased attitude creates an emotional response that may override a belief or disposition and prevent finding the best decision. Attitudes can change quickly. For example, a moody person's attitude can swing from optimism to pessimism in a relatively short period of time. A person's attitude may not be properly interpreted by other people. For example, a person who is believed to hold contrary thoughts about an issue is often accused by those who have an opposing opinion on the issue of having a bad attitude. Attitudes do influence actions, but an attitude is *not* an action.

12.4 MINDSET AND CONTROL ISSUES

It seems that the time frame over which each of the mindset components is in force is difficult to know. The difficulty in identifying the component that dominated a particular decision makes it difficult to attribute success or failure. Moods influence actions over relatively short periods of time, whereas principles are more deep-rooted attitudes that direct the general trend of thinking. A mood that causes thought that conflicts with a principle may lessen the influence of the person's principles on actions taken and decisions made. A disposition is based on principles, but its strength varies more than that of principles. A fundamental issue that is relative to decision-making is that each of these mindset components exerts an emotional force. The extent to which a critical thinker has control over the different mindset components determines the decision and subsequent action. It is also indicative of the person's ability as a critical thinker.

At the time when a decision must be made, a person's attitude can influence the quality of thinking and, therefore, the decision. For example, a person who is currently experiencing a negative mood may not have the proper mindset to access from his or her mind the knowledge and experiences that would usually influence his or her normal thought processes. For a person who is usually optimistic, a state of pessimistic mood may distort functioning of the mental processes that would be applied under normal circumstances. The mood might cause the person to make a pessimistic decision in spite of a situation characterized by positive conditions and a set of principles that would normally lead to an optimistic decision. A facilitator of a brainstorming session needs to judge the moods of the participants, as their moods can influence the participants' reactions to the problem and to the actions of the facilitator. If the participants were recently informed that end-of-year bonuses would not be given, the moods may contribute to an ineffective brainstorming session. As another example, consider a researcher who at the time when his or her creative abilities are needed is suffering through a pessimistic mood; at that time and because of the mood, he or she will likely not have the needed creative mindset to be an efficient or effective problem solver.

A person's problem-solving productivity depends in part on his or her state of mind during the time that is required to solve the problem. If a person is aware of his or her mood at the time when a decision must be made or some responsibility must be performed, the state of mind may have less negative consequences or the person may understand the way to modify his or her mood. A critical thinker can generally control his or her current state of mind such that both poor thinking and ineffective decision-making are avoided; however, if the strength of a negative mood happens to be strong, the person may not be capable of making the decision that would be best over the long term. To develop into a critical thinker, a person must develop the ability to assess his or her mood and then control it so that it does not adversely influence decision-making. If a person frequently makes a conscientious recognition of his or he mood, then he or she will be better able to adjust the mood when an important decision needs to be made. Continual awareness of mood is characteristic of critical thinkers, as this practice improves decision-making.

The definition of the term mindset is broad because of the roles that it must fulfill as a factor in critical problem-solving. First, since the mindset dimension encourages imaginative idea generation, it must have the capacity for both affective and cognitive thinking. Generally, imaginative thinking depends on a well-developed emotional system. Second, mental processes that influence decisions and actions are not constant over time; therefore, when a person is making a decision, the mindset dimension will need to address the duration that a person's attitudes and beliefs apply. The short term reflects the state of mind and depends on the person's attitudes and moods at the time when a decision is being made. A person's long-term mindset is largely controlled by the principles that have evolved from experiences and education. Therefore, any decision will likely depend on the relative strengths of the moods and principles. It will also depend on the ability of the decision maker to recognize and control any mood that would prevent him or her from allowing a positive principle to override any negative mood. The durations of beliefs and moods vary over time, so the roles that the temporal variations of a mindset play in decision-making must be understood and recognized, with the degree of variation an indication of the uncertainty that could have adverse consequences to decision-making. Third, the mindset dimension must acknowledge the formation and change of mindset states, again for both the short term and the long term. A person's beliefs and even principles change, which will obviously influence decisions and actions. Some aspects of the mindset are short-lived while others are assumed to be held for a relatively long period of time. A principle is generally considered as a long-term conviction, while beliefs are associated more with moderate periods of time. A person's beliefs can change in response to significant experiences. Conversely, a mood reflects a temporary state of mind for a relatively short-term state of mind and usually dependent on the relative involvement of emotions. Thus, the time frame relative to a problem that needs a solution must be considered.

The ability to control the mindset components to produce an effective and efficient state of mind at the appropriate time is a positive attribute. Being able to consistently control one's state of mind is usually necessary to make unbiased decisions. If a person has not had sufficient experience with complex problems to adequately understand and develop his or her mindset, then regardless of the person's state of mind at the time when a decision needed to be made, the person may not make a good decision. As an example, honesty is usually a preferred value-system component, but a person who has had some poor experiences with respect to honesty may have principles that undervalue honesty. Thus, he or she may make decisions that fail to emphasize honesty regardless of his or her frame of mind at that time. Only after having experiences that stress the merits of honesty will the person begin to make a corresponding change to his or her mindset; however, under certain mindset conditions, the person may still opt to make decisions that reflect dishonesty because the strength of the honesty component is not sufficiently ingrained. This is an example of the importance of having control of the mindset components for good decision-making. Societal and organizational principles may to some degree influence a person's mindset.

12.5 MINDSET AND DECISION-MAKING

A person's mindset will influence the functioning of the processes that are associated with the other dimensions, specifically his or her value system, skill set, and the selection of thinking types. As an example, if a person has not had the experiences that are necessary to develop the curiosity skill, then regardless of the person's mindset at the time, the person will not make decisions and take actions that reflect a well-developed sense of curiosity. If the person has developed the curiosity skill, then the mindset will influence the functioning of the curiosity skill at that time. The same idea is true for the other skills in the critical thinking skill set. A person's mood or disposition can either prevent or encourage the effective use of each skill. The general point is that a lack of control of one's mindset can lead to poor decision-making that fails to assign the proper weights to competing principles. Every decision maker needs to have good control of his or her affective and cognitive mental processes.

The concept of a mindset is not limited to individuals; organizations have mindsets. Organizations, as well as individuals, have value systems that influence a person's mindset and are used in making decisions, although it is the leaders of the organization who collectively make the organizational decisions. Collectively made decisions generally reflect a weighted average of the mindsets of the leaders of the organization. Just as societal values influence individuals, organizational decisions can be influenced by cultural norms. This collection of principles and beliefs forms the company's policies and practices, which represent the basis for the organization's mindset. A consistent, collective set of principles should be used to make decisions. The leaders of an organization must continually control the organization's mindset by maintaining the policies and ensuring that the employees understand the policies. Any lack of mindset control can hinder good decision-making. Conflicts can arise if the elements of a person's mindset are in conflict with the apparent beliefs of the employing organization. These internal conflicts can hinder decision-making if they are not controlled. Decision-making will be less subject to the effects of conflicts when the organization hires employees who have commonalities with the organization's mindset.

An individual who has little experience at decision-making is not expected to have a well-developed mindset, which can lead to inconsistent decision-making. One purpose of a mentor is to assist the mentee in developing a professional mindset. Mentoring is most effective when the mentor places emphasis on developing a mentee's mindset. A mentor should informally evaluate the mentee's mindset to understand his or her principles and any tendency to have unproductive moods. The mentee's status on the optimism-pessimism continuum should be part of the assessment. Mentors need to act in ways that ensure that their mentees are assigned experiences that lead to the development of good mindset components. For example, the optimism-to-pessimism spectrum is one of the more important components of the mindset dimension; therefore, a mentor needs to direct the mentee to experiences that promote optimism. The mentor should encourage the mentee to avoid pessimistic people. Over time, the mentor needs to observe the mentee's actions and use the actions to develop a better understanding of his or her mindset.

Successes and failures can have a significant influence on an individual's mindset. Failures can initially encourage the development of a pessimistic mindset. A mentor should make the mentee aware of the potential effects of failures. To enhance a person's ability to place failure in the proper context, it may be necessary for the mentor to help the mentee assess the effects of past experiences that involved any level of failure. Introspective analysis is one way of placing each failure into a more positive perspective. Bad experiences can be good teachers. The analysis should specifically consider the influence of a failure on beliefs and principles. Failures can be learning experiences, but failures should not be allowed to distort any of the mindset components. The best time to assess a failure is immediately following the failure. An immediate analysis of a failure can ensure that future failures do not occur because a previous failure was not adequately analyzed. Just as experiences influence the development of a skill, past experiences also influence a person's ability to control the components of the mindset.

12.6 APPLICATION OF THE INTELLECTUAL DEVELOPMENT (ID) MODEL

Improving one's own mindset depends on experiences. Possibly the most important experiences are the ones that make a person aware of the potential influence of the mindset dimension. This awareness may result from a significantly positive experience or an introspective analysis following a failure. The acceptance of advice from a mentor can also cause mindset maturation. Recognizing that other people benefit from the control of their mental processes is another alternative. But recognizing its importance is the first step in progressing through the development of the mindset dimension.

The Intellectual Development (ID) model can be used to assess a person's status in critical problem-solving ability. It can be applied to each dimension individually with the thought that a collective analysis can be made across the dimensions, which will provide an indication of the intellectual development for applying critical thinking in the solution of complex problems. It can also be used to show the personal growth that would be needed to move to a higher stage of intellectual development relative to critical thinking. Table 12.1 provides the characteristics of the five stages of the ID model as applied to the mindset dimension. These guidelines are not a complete list but only examples of the types of knowledge and abilities needed to function at the individual stage.

Experiences influence a person's state of mind and have the potential to alter his or her mindset. The ability of a person to react positively to experiences that influence his or her state of mind depends on a variety of factors. Certainly, any negative stress, whether personal or related to the work environment, can suppress any interest in improving one's thinking. We certainly know that a person's personal life can influence his or her thinking. Positive feelings are associated with nice weather, so

TABLE 12.1

Growth of Knowledge: Mindset Dimension of Critical Problem-Solving

I. Comprehend
- To understand the potential influence of one's mindset on actions and decisions.
- To identify the vocabulary of mindset components (e.g., moods, beliefs).
- To recognize the ways that optimism and pessimism influence the growth of knowledge.
- To acknowledge that having control of one's state of mind is important to advance learning.

II. Experience
- To experience the influence of one's mindset on decision-making.
- To experience by way of applications the changes in mindset components and the effects of the change on decisions and actions.

III. Analyze
- To analyze complex case studies to appreciate the effects of mindset conditions on actions.
- To analyze past decisions to detect the effects of controlling state-of-mind thinking to decision-making.
- To analyze the effects of constraints demanded by alternative stakeholders on the solution of complex cases.

IV. Synthesize
- To synthesize the effects of alternative mindsets on actions in solving complex problems.
- To synthesize changes in decisions expected for changes in a state of mind.

V. Create
- To create scenarios via simulation to show the effects of mindset uncertainty on decision-making.
- To create complex problems where mindset components influence leadership actions.
- To create situations that demonstrate the effects of alternative states of mind on the solutions of complex problems.

weather and other environmental factors, even a clean work place, are experiences that influence a person's state of mind. Physical activity is usually a positive factor unless it involves a failure, such as a loss of a tennis match or a did-not-finish result of a 10-K run. Being around people who have positive mindsets can also influence a person's state of mind. A person's state of mind can influence a friend's state of mind. It is also likely that internal issues, such as the response to a dopamine high, can cause a person to temporarily have an optimistic state of mind. Of course, the negative side of each of these factors can suppress the development of a positive mindset.

12.7 ASSESSING THE COMPETENCY OF THE MINDSET DIMENSION

Individuals will vary in their ability to fulfill the responsibilities associated with each of the four dimensions. Experiences in solving complex problems represent a source of educational growth, which enables a person to enhance his or her abilities to complete responsibilities associated with each dimension. Criteria are needed for each dimension to distinguish an individual's level of competency. Table 12.1 provides a potential critical thinker with some measure of the knowledge that he or she has about the mindset dimension. It does not indicate the expertise with which the

TABLE 12.2
Assessment Criteria for Critical Problem Solver Ability: Mindset Dimension

Stage I: Novice	Stage II: Specialist	Stage III: Master	Stage IV: Expert
1. No tolerance for ambiguity.	1. Sensitive to inhibiting factors.	1. Sensitive to organizational inhibitors, but not to peer inhibitors.	1. Ignores inhibiting pressures.
2. Pessimist.	2. Accepts decisions of others without question.	2. Good self-discipline.	2. Full control of mood and disposition.
3. Lacks self-confidence.	3. Not able to address uncertainty.	3. Questions the influence of other people's state of mind.	3. Optimist.
4. Narrow interests.		4. Minimal experience with uncertainty.	4. Has sought breadth.
5. Frequent mood changes.			5. Deals well with uncertainty.

person applies the knowledge. A person may have good knowledge of an issue, but he or she may still be poor at applying the issue. If four stages of competency are sufficient to distinguish between the abilities to apply mindset knowledge, then the model in Table 12.2 should enable a person to have some idea of their strengths and weaknesses in terms of the ability to apply the knowledge of each of the five levels of knowledge of the mindset dimension of critical problem-solving. The disparity between the four different stages should provide a person with some measure of their current ability to apply the knowledge.

Recall that the intent of the Intellectual Development (ID) model is to characterize the knowledge of a person. Of course, a person may have better knowledge of one dimension than knowledge of a different dimension. Table 12.1 provides criteria for identifying the knowledge of the mindset dimension. A person can be at a high level of knowledge for a dimension but be constrained in its application because of personal characteristics. Similarly, two people can be at the same knowledge level but have quite different capacities for applying the knowledge. Thus, a second criterion can be used to characterize a person's ability to apply the knowledge for a specific critical problem-solving dimension. Table 12.2 provides criteria relative to the mindset dimension that can be used to distinguish between the application levels. Four levels should be sufficient for allowing a person to recognize a limitation on his or her ability to apply the knowledge for the dimension.

12.8 SELF-DISCIPLINE: A MINDSET CHARACTERISTIC

Self-discipline is *the ability to control one's own actions where a desired activity must be sacrificed in order to perform a less desirable responsibility.* Self-discipline was previously mentioned as an auxiliary skill (see Section 10.13). It is also a mindset characteristic. A self-disciplined person is more likely to approach his or her problem-solving responsibilities in a timely, respectful manner. A person who is willing to forego a desirable activity in order to fulfill a less desirable activity must have good control of his or her state of mind. Therefore, the person is likely less sensitive to mood changes and, therefore, more successful at fulfilling responsibilities. Conversely, a person who lacks self-discipline will appear to lack commitment to the endeavor, which will suppress the likelihood of high-quality effort and output, which decreases his or her chances of success at completing the responsibilities. Any lack of self-discipline reflects a lack of mindset control. A person who has to force himself or herself to complete a responsibility lacks self-discipline even though they may complete the responsibility on time. It is the need to compel oneself to act as opposed to enthusiastically acting that reflects the mindset deficiency.

An association between self-discipline and procrastination is real. People who lack self-discipline are often procrastinators. Generally, procrastination reflects a combination of a lack of self-discipline and greater than normal selfishness. Those who lack self-discipline are often viewed as being immature, as selfishness is often viewed as a sign of immaturity. To avoid being considered immature or irresponsible, the person must overcome the procrastination fault and adopt attitudes that encourage self-discipline. Correcting the procrastination problem is generally accompanied by an improvement in self-discipline.

12.9   PROCRASTINATION: A MINDSET CONSTRAINT

Whenever I asked a class to raise their hands if they frequently procrastinate, most of the students would raise their hands; however, many would comment that their tendency to procrastinate was because they had to prioritize many responsibilities. While I rarely procrastinate, I do procrastinate when it comes to the housework. Thus, it appears that both the nature of the responsibility and the number of competing responsibilities can influence the likelihood of a person procrastinating.

12.9.1   DEFINITION

Procrastination can be defined as the *needless postponement or an unnecessary delay of a responsibility*. It could also mean *to delay the transmission of knowledge*. Procrastination is essentially a stealing of time, i.e., a time water. It is the misuse of time, as responsibilities are delayed with the time spent performing unnecessary tasks. Procrastinators seem to have the mantra: "Don't do today what you can put off until tomorrow," with the tomorrow indicative of a time in the future following the point in time when the responsibility should have been completed. It is important to stress that the definition indicates that the delay is unnecessary. Thus, a very busy person who has many responsibilities and delays some of them to attend to other more important responsibilities is not procrastinating.

It may be helpful to view the definition of procrastination in terms of the four dimensions of critical thinking. Courage is the first skill in the skill set, and any needless delay at the beginning of a task may yield a poorly defined problem statement, which will be stealing time away from effective completion of the tasks associated with the skill sequence. The definition of procrastination has obvious connections to values such as accountability, dependability, diligence, fairness, industriousness, self-discipline, and trust. Thus, the mindset constraint of procrastination limits the proper application of the value dimension. A procrastinator will likely delay the time before considering and selecting the best type of thinking.

12.9.2 EFFECT OF PROCRASTINATION ON EFFICIENCY

A common myth about procrastination is that the delay is acceptable because the procrastinator works better under pressure, and so the delay will not result in a lowering of the quality of the output. Unfortunately, for those who adhere to the basis of this myth and needlessly delay in completing responsibilities, evidence clearly indicates that the myth is untrue, i.e., studies have shown that any delay in fulfilling a responsibility is accompanied by a reduction in the quality of the output. In terms of efficiency, the lower quality of the output (i.e., a decrease in the numerator of Eq. 2.2) and any time delay in starting the work (i.e., an increase in the denominator) mean a reduction in efficiency. Furthermore, when one member of a team procrastinates, it often delays and hinders the completion of the responsibilities by other members of the team. The sections of their responsibilities that depend on the procrastinator's work will be delayed until the procrastinator acts and completes his or her work. This can adversely influence the attitude of colleagues toward completing their responsibilities as well as their opinion of the procrastinator. Thus, procrastination typically causes both a less-than-best output and a reduction in efficiency, which is the reason that procrastination is considered to be a mindset constraint. Both personal and organizational efficiency are negatively influenced by procrastination.

12.9.3 PREVENTING PROCRASTINATION

How can a leader discourage team members from procrastinating? Because procrastination is so common, many books have been written on the topic, with each of these identifying multiple ways of overcoming the practice. The most commonly proposed method of reducing or eliminating procrastination is the use of a TO-DO list. In the case of complex problem-solving, two general alternatives should be considered: (1) the project should be separated into a long list of tasks each of which would require a short period of time to complete and (2) a list of just a few major tasks, but ones that would take a relatively long time to complete. Both lists should be a series of brief but descriptive phrases for each element of the problem for which the procrastinator is responsible. So a single large task that may take several days to complete is divided into a number of sub-tasks, each of which will take only a few hours to complete. Thus, when it is separated into a numerous set of smaller tasks, the complex problem does not seem as overwhelming as the short list of complex tasks. The completion of each small task will act as a reward to the procrastinator. The option of a long list of simple tasks is generally preferable as it is less stressful on the procrastinator. However, the TO-DO-list option must be approached with maturity, which is something that procrastinators usually lack. Some habitual procrastinators have found that self-imposed punishments for not completing their work on time can serve as motivation to be more punctual. Also, self-designed rewards for completing each item on the TO-DO list on time and of good quality are possibly a better option, as they are more positive in nature than the negativity that underlies the punishment options.

Many other methods for controlling procrastination have been proposed. The trick is for the procrastinator to identify the conditions that encourage him or her to procrastinate and then select the method that is most effective for suppressing the bad habit. The bottom line is that the habit of procrastination can be overcome.

12.9.4 PROCRASTINATION AND RATIONALIZATION

Rationalization is very often a controlling factor to a procrastinator. Rationalization is *the making of excuses to avoid responsibilities and to justify the irresponsibility of the act*. Rationalizers falsify the reasoning in part to avoid making changes to their behavior; thus, rationalization allows the individuals to avoid changing their value systems. Before the practice of procrastination can be overcome, the procrastinator must acknowledge that his or her procrastination is a significant problem and that it truly does lead to lower quality output. He or she must renounce the myth that both productivity and output quality are not influenced by procrastination, and that responsibilities should be completed on or before the deadline.

Critical and creative problem-solving can be time consuming, as the problems to which they are applied are usually complex. Thus, procrastination prevents the devotion of the necessary time needed for effective imaginative thinking, while rationalization provides the false justification. Additionally, the last-minute efforts that result from procrastination prevent allowing adequate time for either idea generation or effective idea evaluation, both of which are important to all types of complex problem-solving. Almost all of the important characteristics of critical thinkers identified in Chapter 9 are compromised by any tendency to procrastinate.

12.10   PESSIMISM: A MINDSET CONSTRAINT

Creativity is grounded in the belief that imaginative thinking is a positive act. Generally, those who use methods of critical problem-solving have an optimistic outlook on the ability to find novel, effective solutions to complex problems. Those who largely harbor doubts about the benefits of imaginative problem-solving tend to be pessimistic, as they believe that logical thinking is sufficient to solve complex problems and just as effective as the use of imaginative thinking. Pessimism is one end of an attitude spectrum, with optimism being at the other end. The optimism-pessimism continuum is an important decision element of the mindset dimension.

12.10.1   DEFINITION

Pessimism is *an inclination to adopt a negative view of a situation, i.e., a belief that the bad will generally exceed the good*. A pessimistic attitude can result from one of many factors, including some innate causes. Of course, with respect to the merits of critical problem-solving, the cause of a person's pessimism is likely a set of diverse experiences that resulted in some level of disappointment or failure from which the person learned to transfer the feelings of despair to other situations, including a negative feeling toward imaginative problem-solving. Failures that are not followed immediately by a self-introspective analysis appear to contribute to the development and growth of a pessimistic attitude.

12.10.2 Dimensional Aspects of Pessimism

While pessimism is directly associated with the mindset dimension of critical thinking, it is actually relevant to all of the dimensions. Pessimism is relevant to values because a pessimistic attitude can distort the application of values. Values are generally presented as positive attitudes, so a pessimistic attitude is counter to the effective balancing of values. Similarly, pessimism has a negative influence on the effectiveness of skills, such as curiosity and courage. For example, a lack of courage to approach complex problems can be traced in many cases to a pessimistic attitude. Pessimism represses the likelihood of attempting to generate creative ideas to solve complex problems. Any reduction in problem-solving efficiency due to an individual's pessimistic attitude will lessen the chance of finding the optimum solution and simultaneously continue to instill negative attitudes within both the person and the organization. Pessimism can control the mindset of a person and even the organization itself. A pessimistic moodiness by leaders can be a significant contributor to an organization's general state of mind. Organizational pessimism does certainly not contribute to organizational pride or happiness among the employees. Pessimism could influence the functioning of other dimensions, such as those related to safety, economics, and the cultural characteristics of an organization.

12.10.3 Pessimism: A Constraint on Problem-Solving

Pessimism would have a negative effect on the successful completion of each step of the problem-solving process. A pessimistic person will likely lack the courage to enthusiastically pursue tasks related to overcoming weaknesses in the state of knowledge. A pessimistic assessment of the problem will lead to a less well-defined statement of a weakness in knowledge, as the pessimist will adopt an incomplete view of the problem, which will lead to less useful statements of the goal and objectives. If the individual does not understand the real problem, then he or she will not be able to develop performance criteria or a decision criterion that would be good evaluator of alternative solutions. Since the pessimist will not truly appreciate or understand the problem, he or she will likely develop less beneficial experimental plans. The plans developed by a pessimist will likely reflect negative thinking, which means that the plans will undervalue the chance for success. Given the pessimist's limited view of creativity, he or she will inherently limit thinking to logical ways and will be less inclined to use creative thinking methods, especially when imaginative thinking methods could lead to a more novel outcome for the complex problem. Pessimists may likely have had a greater failure rate in past problem-solving experiences, which likely results in the continuing use of the failure attitude. Past failures may have reduced the quality of his or her ability to conduct critical analyses and critical syntheses. For the final step of the problem-solving process, the pessimist will be less inclined to recognize the wide spread implications of the findings; thus, the work will have less impact on society, which represents a lower-than-possible output.

12.10.4   Pessimism as an Inhibitor

In defense of their lack of success in the past, pessimists generally downplay the usefulness of creativity. A pessimistic attitude is likely to have other negative consequences. Pessimistic people tend to act as inhibitors to colleagues. As a result, colleagues generally avoid associating with pessimists. Pessimism may cause a person to procrastinate, as he or she will not likely want to start working on an activity about which they believe there are few prospects for success. Pessimists generally do not seek new knowledge due to the inner feeling of failure. With respect to contributing to the advancement of knowledge, a pessimist who recognizes a weakness in a published document or a report will argue with himself or herself that this will not lead to a better solution, so the pessimist makes hasty judgments that discourage further study. The pessimist does not provide the necessary time for ideas to properly develop into productive output. In summary, pessimism seems to be counterproductive to the advancement of knowledge.

This introduction to pessimism might seem to be overly negative in tone; however, it is important to understand that pessimism is one end of a spectrum with optimism being the other end. A pessimistic person likely does not always act at the extreme end of the spectrum. At times, the person's actions may reflect a sensitivity to optimistic thinking. The person should make every effort to overcome a dominant tendency toward pessimism and move more toward the optimistic end of the spectrum. The person should only use a pessimistic attitude when it is beneficial.

The general concern is that the attitudes and actions of pessimists will reduce the efficiency of problem-solving. Pessimism can create stress that acts as a constraint on progress because it influences the person's mindset. Stress also influences the quality of the output, so it has a two-fold negative influence on the overall efficiency.

12.10.5   Overcoming Pessimism

A person who wishes to become a true critical problem solver cannot continue to be a pessimist. While pessimists are often very critical, which in some cases can be a positive ability, they often fail to take the steps beyond identifying the problem; thus, they fail to recommend a solution to a problem. Curiosity and questioning are the two skills that follow the critiquing skill, and pessimism really prevents worthwhile action when a person tries to apply these two skills. Pessimism suppresses the person's curiosity and limits the flexibility in questioning. Thus, a pessimistic mindset constrains the application of the skill set such that critical thinking skills are less efficient than possible. The pessimistic attitude prevents the individual from believing that the most novel solution can be found. Therefore, to become a critical thinker, the pessimist first needs to recognize the inhibiting nature of pessimism. To overcome the constraint of pessimism, it is best to start by reviewing past personal experiences of pessimism and for each instance of failure identify an alternative optimistic action and the possible changes in outcomes that would have occurred if an optimistic perspective had been followed. This action reflects counterfactual thinking. Recognizing both the improvement in the outcomes and the effects on efficiency that would have resulted from having an optimistic perspective is a productive step toward overcoming a pessimistic nature.

Critiquing was identified as one of the primary skills of a critical problem solver. In one sense, critiquing is a pessimistic action as it is used in the search for a deficiency; however, it is also an optimistic skill. Optimism encourages the search for a way to advance knowledge, and this is the reason that critiquing needs a basis in optimism. For critiquing to be effective a person needs to have an optimistic critical attitude. Pessimism can help to identify a problem by identifying weaknesses, but courage, confidence, and optimism are needed to transform the idea into a positive outcome. The critiquing of existing work is of little value if a person maintains the pessimistic attitude throughout the search for a solution.

12.11 ASSESSMENT OF COUNTER-SKILL ACTIVITIES

A counter-skill is a skill that retards progress. Counter-skills, such as procrastination and pessimism, can hinder a person's progress toward becoming a critical problem solver. A person should recognize the strength of his or her ability to apply a skill, whether or not it has a positive or a negative influence on his or her thinking. It is important for a person to have some measure of the potential effects of counter-skills. Therefore, a person should assess whether or not counter-skills are contributors, or suppressors, of his or her overall critical thinking ability. An unbiased qualitative indicator of a counter-skill may be useful to a person to understand his or her strengths and weaknesses. While an objective criterion for evaluating the influence of a counter-skill on the person's decisions is not currently in use, some form of assessment of the influence of counter-skills on a person's overall ability to solve problems should be made. A self-introspective assessment of counter-skill influence will aid in the person's efforts to become a critical thinker.

For procrastination, a person could apply an ordinal-scale measurement index, such as the following:

Rating	Criterion
1	Frequently procrastinates
2	Procrastinates more than ideal; has started to assign penalties for procrastination events
3	Occasionally procrastinates, but only for short periods of time
4	Never procrastinates

Any score less than four would suggest that some effort needs to be made to overcome the practice of procrastinating. If the score is less than four, a diary should be started with all past instances of procrastination cited along with any rationalizations that were used at that time to justify the procrastination. Any conditions that encouraged procrastination, such as stress, should be noted. The rating should be used to encourage improvement toward overcoming the fault. This diary should be maintained until a rating of four is achieved.

Pessimism is another counter-skill that should be part of a self-evaluation. Pessimism is likely more difficult to assess than procrastination, but a person might start by establishing an ordinal-scale measurement rating, such as the following:

Rating	Criterion
1	Frequently pessimistic
2	More often pessimistic than optimistic
3	More often optimistic than pessimistic
4	Almost always optimistic

Again, such a rating scale may be used at least to begin a self-evaluation that would indicate the extent of the problem and serve as motivation to undergo change.

The important point is: the reasons for a low rating on either of these scales needs to be identified. A plan should be developed to overcome any weakness. Without improvement, a low rating on a counter-skill may detract from good ratings on primary skills.

12.12 CONCLUDING COMMENTS

The negative mindsets of significant parts of the society during the Malthus-Luddite-Dickens period were overcome during the latter part of the nineteenth century by a number of factors. Better knowledge of medicine and nutrition reduced death rates. The opening of the virgin lands like the Western U.S. expanded the resource base. Technological changes in agriculture reduced the Malthusian concern over food shortages. Urban planning and government regulations improved life in the cities. Factors such as these created a more optimistic societal mindset, and thus these factors contributed to greater advances in knowledge.

Is the mindset dimension the dominant of the four dimensions for critical problem-solving? A bad mood can suppress potential positive application of the mental skills like curiosity and questioning. An unproductive mood can discourage using multiple thinking types. A bad disposition at a specific time can promote the use of poor values. Thus, it appears that a person's mindset has some control over the beneficial application of the values, skills, and thinking type dimensions. As much as each of the four dimensions is individually important to success, the interdependency of the four dimensions must be respected as a controlling force on success.

A person's state of mind can act as an inhibitor to success. A period of pessimistic thinking, procrastination, or any mood disorder acts as a deterrent to focusing on task completion. A person who recognizes that he or she suffers from any counter-skill must learn to recognize when he or she is experiencing an event and determine a way of deviating from the inhibiting state of mind. While these periods may seem to occur at random times, the occurrences are usually accompanied by indicators, which if recognized can serve as the basis for a change in a mindset component. As it can be difficult for an individual to recognize the indicator events, having a mentor or friend who is able to recognize such indicators is important, as he or she can alert the person of an indicator event.

Some procrastinators and pessimists would deny suffering from either of these two maladies. They would likely argue that they are realists, specifically that they recognize life in a more realistic way. They would believe that having a pessimistic attitude allows them to confront problems with a more realistic outlook—specifically that the novel outcome from imaginative thinking is not worth the greater effort that is needed for the use of imaginative thinking methods. The procrastinator would provide rationalizations that support the fallacious myths about procrastination and success. Both pessimism and procrastination limit problem-solving efficiency as they are mindset counter-skills that generate negative mindset components.

Attitudes are time dependent. Mood swings and pessimistic feelings can control day-to-day actions. Negative attitudes are tempered over time, as the time from the most recent application of the attitude lengthens. Factors that contribute to the negativity become less influential with the time between incidents. Similarly, positive attitudes can become less influential with time, again because of the difficulty in recalling the sources of the positive feelings that established the attitude. This time dependency is sometimes referred to as the sleeper effect. As time passes, the memories that influenced the development of an attitude, positive or negative, become less memorable, which leads to a softening of the attitude itself. Thus, changes to the mindset components influence decisions and actions, with the actual effect of change depending on the mental significance of the experiences.

12.13   EXERCISES

12.1. How might a person's mood influence his or her decisions on everyday activities such as the choice of lunch, the type of TV program to watch, or the number of beers to drink?
12.2. Identify the elements of the mindset of an organization.
12.3. What factors could influence a change in a person's mood?
12.4. Find definitions for emotions and mindset. Discuss their relationship?
12.5. Do values cause a person's beliefs or do the person's beliefs cause the person's values?
12.6. How might a person's pessimistic attitude influence a person's honesty?
12.7. How are moods and dispositions similar? Different?
12.8. What types of experiences are relevant to the Analyze and Synthesize phases of the ID model for the mindset dimension?
12.9. What effects could failures have on a person's future mindset applications?
12.10. What are characteristics of someone who is self-disciplined? Who lacks self-discipline?
12.11. Why does procrastination suggest a lack of self-discipline?
12.12. Why do procrastinators believe the myth that they do better work when they wait until the last possible moment to complete their responsibility?
12.13. Why do you believe that a reward system to discourage procrastination is more effective than a penalty situation?
12.14. Under what circumstances do you believe that a penalty system is effective for controlling procrastination?
12.15. What is rationalization? How is it used by procrastinators?
12.16. What are the outward signs of a pessimistic person?

12.17. Is there a connection between pessimism and self-confidence? Justify your response.

12.18. Are there advantages to being a pessimist?

12.19. What effect would you expect a person's pessimistic attitude to have on his or her decision-making?

12.20. Why is being an optimistic thinker relevant to the use of critical thinking methods?

12.21. Why is it important for a creative person to have an affective nature?

12.22. How can an optimistic person mentor a person who tends to be a pessimist?

12.23. How can a mentor help a mentee become more self-confident?

12.24. Define the words *responsibility* and *faithfulness*. Discuss ways that the two values are interdependent.

12.25. List the six steps of the decision process. For each step, discuss the ways that a pessimistic attitude acts as an inhibitor.

12.14 ACTIVITIES

12.14.1 ACTIVITY 12A: PICK UP THE TRASH!

To illustrate the effect of mindset on actions, consider the case of a community that has to start the recovery from a level-four hurricane. Assume that the community has an optimistic mindset. Outline the approach that they would use to clean up the community. Then assume the adjacent community has a pessimistic mindset; outline their approach to clean up. Discuss the results of differences in mindset.

12.14.2 ACTIVITY 12B. GO AT IT!

This activity is a two-person debate. One person takes the position as a procrastinator and the other person as a non-procrastinator. The procrastinator states one of the myths about the benefits of being a procrastinator and provides a defense of the myth; rationalizations are permitted. Then the non-procrastinator is given equal time to support the premise that the myth is false. All myths should be debated. Time limits may be placed on the speaking times. In addition to the benefit of communication, a purpose of the activity is to use the elements of the four dimensions in supporting the two positions.

12.14.3 ACTIVITY 12C: DO NOT DO TODAY WHAT YOU CAN PUT OFF UNTIL TOMORROW!

To illustrate the effect of mindset on actions, consider the case of an aide to a politician; the election is only one month away. The politician's aide is in charge of getting out the vote. Outline the approach that the aide would take if he or she is self-disciplined. Then assume that the rival politician's aide who has the same responsibilities as the other aide is a procrastinator. Outline the approach that the aide would take in meting his or her responsibilities. Discuss the effect of self-discipline on success.

12.14.4   ACTIVITY 12D: THE MOODY DUDE

Provide a definition of the word mood. Identify all possible types of moods. For each one, identify the outward and inward signs of the mood. Discuss ways that knowing signs of moods can help a person to control his or her moods when solving complex problems.

12.14.5   ACTIVITY 12E: DOES DENNIS THE MENACE HAVE SELF-DISCIPLINE!

(1) Define self-discipline. (2) Identify five activities where self-discipline is a positive characteristic. This might be writing a report, organizing your desk and workspace, or getting out of bed on Sunday morning. (3) Develop a metric that can be used to rate self-discipline. (4) Use your metric to rate your own self-discipline for each of the examples. (5) Use the five ratings to assess your self-discipline and discuss whether or not you need to improve your skill. While this activity may focus on self-discipline, its real purpose is to show the value of self-assessment and the development of performance metrics to use in the assessment.

12.14.6   ACTIVITY 12F: THE RATING GAME

To improve our knowledge of the optimism-pessimism spectrum, identify factors that reflect a person's tendency toward pessimism or optimism. To identify a person's ranking on the spectrum, develop a more realistic metric than the metric provided in Section 12.9. Develop the metric with consideration of the factors identified. This activity illustrates the creation of a new metric.

12.14.7   ACTIVITY 12G: COUNTER-SKILL BALANCING

Create an assessment model that measures the combined effects of pessimism and procrastination. The assessment criteria in Section 12.9 for the two counter-skills may only be a starting point for revision. The new model should yield a single-valued metric that accounts for pessimism, procrastination, and the combined effect.

12.14.8   ACTIVITY 12H: BEAUTY IS IN THE EYES OF THE BEHOLDER!

Part I: Obtain pictures of small lakes, with some of the pictures showing well-maintained areas and others that are not properly maintained. Create a list of the characteristics that contribute to or detract from the overall aesthetical quality of each lake area. COMPLETE PART I BEFORE READING THE PROBLEM STATEMENT OF PART II.

Part II: Using the list of characteristics in Part I, develop a quantitative metric that could be used to rate the aesthetical quality of the sites.

Part III: Obtain a few more pictures of lake areas and use the metric to rate the aesthetical quality of the sites. If the metric is not effective for these new sites, assume that the metric has not been verified and recommend changes to it. This activity illustrates the need to create and verify a performance criterion.

12.14.9 ACTIVITY 12I: PLUTO THE PESSIMIST

Part I: How would two dogs, Pluto the pessimist and Otto the optimist, react when confronting a burglar? Develop a list of the ways. To the dogs, the burglar is a complex problem that they must solve.

Part II: Select a complex human problem and use the list for the burglar problem to provide potential solutions to the human problem. This activity applies synectics to demonstrate that it can be used as part of an introspective analysis while helping to solve a mindset problem.

12.14.10 ACTIVITY 12J: THE APPLE CORE

Identify three personal core beliefs. Provide definitions for each one. Identify three values that are relevant to each belief. Use these examples to discuss the origin of a person's beliefs. Then discuss the way that this relates to the mindset.

13 The Thinking Dimension of Critical Problem-Solving

CHAPTER GOAL

To show that successful problem-solving partly depends on selecting the types of thinking that are best for the characteristics of the problem.

CHAPTER OBJECTIVE

1. To summarize important characteristics of creative thinkers.
2. To contrast logical and imaginative thinking.
3. To show the use and importance of divergent-convergent thinking.
4. To define various types of thinking and show the ways that they are important in problem-solving.

13.1 INTRODUCTION

Antoine Lavoisier (1743–1794) proposed a theory that combustion involved the combination of the burning substance with oxygen in the air with water being the product. His work was based on experiments made by Henry Cavendish (1731–1810). Cavendish did not properly interpret the results of his own experiments. What thought processes led Lavoisier to make the proper interpretation and Cavendish to misinterpret the experimental results? What mindset made Cavendish and others fail to recognize and acknowledge the new knowledge that Lavoisier was proposing? They did not want to abandon old knowledge and modify their thinking based on the results of new experimental analyses. This was poor critical thinking on the parts of Cavendish and others. Alexander Fleming (1885–1955), who is credited with the discovery of penicillin, based his findings on observations of the experimental effects of a mold on a bacterial culture. What thinking led Fleming to the discovery that was not recognized by others? Finding solutions to complex problems requires flexible thinking, which is evidenced by Lavoisier and Fleming. Evidently Lavoisier and Fleming used thinking characteristics of the mindset and the thinking types necessary to apply their critical thinking skills.

Children can think up creative excuses. Many school teachers have heard the excuse from students who failed to turn in an assignment: My dog ate the homework!

DOI: 10.1201/9781003380443-13

Many mothers who have had one of their children who broke a glass have heard an excuse blaming a younger sibling. How many kids have stated: "I missed the school bus because..."? What are some of the endings to this statement that teachers have heard? Young people can think very quickly when they are in trouble. Many of their excuses are quite creative. As they get older, kids often lose their ability to create excuses. How many adults have heard other adults provide wild-and-crazy finishes to the remark, "I forgot my COVID mask because..."? People think creatively to get out of jams, but maybe less so to solve complex problems. Creative ability can actually be very valuable for those who need to solve complex problems. If a person has lost his or her youthful imagination, the ability to solve complex problems may suffer.

Creative thinking is sometimes viewed as a two-step activity, namely the generation of ideas and then their evaluation. This narrow perspective may limit the effectiveness of the process of creative decision-making, and especially learning about the decision process. Engineering design could not realistically be presented as a process of two steps, such as (1) considering alternative designs and (2) implementing the design that is evaluated to be the best. This oversimplification under-values a number of design tasks. By considering the other specific steps of the design process, better designs will result as other factors relevant to the design process will receive proper consideration. The same is true of creative problem-solving. It is important to understand all of the mental processes that are used to creatively solve complex problems.

Everyone thinks, but we often question the quality of some people's thinking. The process of thinking can be defined as *the use of mental processes to reason or reflect on an issue usually for the generation of an idea and using the reflection in decision-making.* While we continually think, thinking is a process that begins with a premise and ends with a conclusion. It is a coordinated, sequential mental activity and it is important to recognize thinking as a process. A person who knows about the different types of thinking and can properly apply them will likely make more creative and useful contributions to the solution of complex problems. While critical thinking is a central focus herein, other types of thinking are also important. We all know of logical thinking, but maybe not upward-downward thinking. Knowing the various types of thinking should lead to more effective thinking and problem-solving. The need to know alternative types of thinking is an important dimension of critical problem-solving as it provides the opportunity for flexibility in problem-solving and leads to a greater variety in alternative solutions. Being adept at matching a problem with the most appropriate type of thinking is an important indicator of a critical thinker.

13.2   TYPES OF THINKING

Reasoning, which is a word often used when referring to the decision-making process, is *the application of mental processes, especially the drawing of conclusions or inferences from observations, facts, or hypotheses.* This definition also suggests an association with decision-making, but the beginning and ending points of reasoning are far less clear. Reasoning, it seems, is a general term that is part of the decision process or problem-solving process. The ideas generated from reasoning are based on logical and/or imaginative thinking and are integrated to select the best solution. The term will be treated as such in the following sections.

Alternative methods of thinking are available to assist in generating ideas. The following is a list of methods of thinking, with some based on the generation of divergent ideas and others based on facts and observations:

- *Abstract thinking*: A form of thinking associated with mysteriousness; practical considerations are avoided in the generation of ideas.
- *Additive-subtractive thinking*: A form of counterfactual thinking where outcomes that did not occur are assumed to have occurred or outcomes that did occur were assumed to be different; thus, facts are added or subtracted.
- *Analogical thinking*: Idea development is based on logical inference and the recognition of similarities to other ideas; ideation based on some similarity, but having unique differences.
- *Cognitive thinking*: Ideas and possible solutions are generated by having the capacity to discern the true nature of a problem and the ability to innovate solutions that depend primarily on facts and observation.
- *Computational thinking*: The generation of ideas by an electronic machine that uses logical reasoning and correlations with associated knowledge.
- *Counterfactual thinking*: Generating ideas that seem counter to ideas that would be expected from the available data, facts, and experiences.
- *Creative thinking*: Focusing on the use of imaginative idea development during the divergent phase of idea generation. Ideas are developed from mental imaging, often from a disordered illusion of reality.
- *Integrative thinking*: Developing ideas and solving them by combing reason and imagination.
- *Lateral thinking*: Providing responses or ideas that are similar to existing ideas.
- *Linear thinking*: Another term for logical thinking. The solution results from a direct, i.e., linear, outcome of facts and existing knowledge.
- *Logical thinking*: Standard thinking based on logical analysis, which uses facts and observations as the motivators for reasoning.
- *Stochastic thinking*: The application of random thoughts to spur idea development. The thoughts could result from external sources, such as words taken from a dictionary or magazine, with the randomly accessed words being connected to the issue and leading to possibly relevant ideas.
- *Upward-downward thinking*: A form of counterfactual thinking based on hypothetical cases of better or poorer outcomes. Upward thinking would take a negative experience and imagine a positive outcome, while downward thinking would generate a negative outcome from a positive experience. A method of thinking usually associated with a need to reverse the direction of thinking or acting.
- *Vertical thinking*: Idea development based on fundamental characteristics of the issue and with minimal divergence, which is similar to logical thinking in some respects.
- *Visual thinking*: Creating ideas and possible solutions through mental imaging from experiences and from visual sensing of the immediate environs. Constructed images used in place of words. Pictostorming is an application of visual thinking.

Thinking types that are based only on fact and observation are collectively considered as logical thinking methods. Each of these 15 methods, except for those related to logical thinking, intend to encourage the person to deviate in some way from facts and existing knowledge. In a sense, the thinker can temporally deviate from reality in the pursuit of seemingly far-fetched imaginative ideas, with the hope that these ideas will lead to a solution that is superior to a decision otherwise found. The extent of divergence from expected solutions would partly depend on the imaginative powers of those participating in the group. Creativity inhibitors (see Chapter 7) can limit thinking and constrain the novelty of any outcome.

The important point is that complex problems have unique characteristics and by knowing different types of thinking, the decision maker will be able to match the problem and the most appropriate type of thinking.

13.3   THINKING TYPE VERSUS CHARACTERISTICS

Each of these thinking types depends on the thinker having and employing certain characteristics. These characteristics are necessary in order to attain the knowledge needed to apply the skills. The following primary characteristics are important to various types of thinking, including creative idea development:

1. *Realistic*: Based on objectively verified inputs.
2. *Emotive*: Based primarily on feelings or emotions.
3. *Incongruous*: Based on seemingly disparate facts or observations, but having some measure of commonality.
4. *Reversionistic*: Based on the ability to mentally process a contrary thought.
5. *Correlative*: Created through reciprocally related inputs.
6. *Imaginative*: Based on make-believe thoughts.
7. *Conjectural*: Based on inferences or incomplete evidence, possibly even guesswork.

Table 13.1 shows the characteristics that are relevant to each of the types of thinking. Except for the first characteristic, the other characteristics reflect some level of imaginativeness. An important observation from Table 13.1 is that different thinking types should be applied for the conditions to problems that have the characteristics to which they are best suited.

13.4   LOGICAL THINKING

In general, the word logical means *reasonable on the basis of existing knowledge*. In reference to the activity of thinking, the term logical thinking can be interpreted as *a process of reasoning that is based on facts, existing knowledge, and previously proposed ideas, events, and statements*. Logical thinking is considered the basic type of thinking. It does not involve imaginative thoughts. Observations from past experiences or existing knowledge are the principal sources of information that are used in making a logical assessment. General knowledge can play a central role in discussions based on logical thinking. This sole dependency on past knowledge

TABLE 13.1

Fourteen Types of Thinking versus Seven Primary Characteristics That Are Important to Employ the Method

Thinking Type	1	2	3	4	5	6	7
Abstract	X						
Additive-subtractive	X			X			
Analogical	X		X		X		
Cognitive	X			X	X		
Computational	X				X		
Counterfactual	X			X		X	
Creative		X				X	X
Integrative		X				X	
Linear	X						
Logical	X						
Stochastic						X	X
Upward-downward	X			X	X		
Vertical	x						
Visual						X	X

1 is Realistic; 2 is Emotive; 3 is Incongruous; 4 is Reversionistic; 5 is Correlative; 6 is Imaginative; 7 is Conjectural.

and/or data suggests that logical thinking by itself is of limited use in solving complex problems, such as when the problem is related to research that is on the cutting edge of the state of knowledge. Logical thinking can obviously be part of problem-solving, but the more difficult problems often require more than one type of thinking. Creative thinking is believed to be contrary to the definition of logical. The logical thinking requirement of rationality that is based on existing knowledge does not mean that some logical thinking cannot be combined with imaginative thinking. The best alternatives are often the result of using multiple thinking types. In fact, in most cases decisions that have been the result of creative thoughts also involved logical thinking. At some point in problem-solving, it may be necessary to combine ideas that were developed with imaginative thinking with ideas based on logical thinking.

Logical thinking is given special notice, as it is often referred to in oral arguments. For example, one of the participants accuses the other participant as basing his or her statement on illogical thinking. The accuser believes that an idea is not truthful and can, therefore, not be a logical thought. The truthfulness of facts or ideas is necessary for a thought to be logical. It then follows that logical reasoning depends on the amassing of logical ideas in the development of a conclusion.

13.5   A PERSPECTIVE ON LATERAL THINKING

Next to logical thinking, the most commonly used name of a type of thinking is lateral thinking. In some cases, the two terms are used interchangeably even though they should not be.

13.5.1   ANALOGOUS INTERPRETATIONS

Before providing an applied definition of lateral thinking, it may be instructive to provide a conceptual interpretation. The term *lateral* is commonly used by announcers of football games. A team uses a lateral pass when the quarterback goes back into the pocket and cannot find an open receiver. If the quarterback believes that the pocket is collapsing and that he may get sacked, he often tosses the football sideways to a running back who is at about the same yard line as he is. This is referred to as a lateral pass. The pass is made to have the ball in the hands of someone who has more options than that of the under-siege quarterback and to buy time until either a receiver gets clear of the defensive players or a hole opens up in the defense where the running back believes that he can advance the ball. The lateral pass is one option that teams believe can increase their overall yardage efficiency. At the time when the quarterback makes the lateral pass, an immediate improvement in yardage is not made; instead, the lateral pass is an attempt to increase the likelihood of later improvement in yardage.

The term lateral is also used in reference to water supply pipelines. A lateral pipe is a smaller pipe that takes water from the water main and distributes it to sites that do not have direct access to the main pipe system. The availability of the lateral pipes makes water distribution more convenient to community residents and greater overall efficiency of the water distribution system. The use of the word *lateral* does not suggest a major change to the water supply system. Like the lateral pass in football, a lateral waterline increases efficiency; however, in this case, it is in the supply of water to the public.

13.5.2   DEFINITION

Lateral thinking is a term that is not commonly associated with creativity. Disagreements exist over the application of the word *lateral* to the process of thinking. Unfortunately, a couple of definitions of lateral thinking are currently in vogue, with the major problem being that the alternative definitions suggest quite different philosophies of thinking. A common interpretation suggests that lateral thinking corresponds to piggybacking and hitchhiking (see Chapter 3), which are idea generation terms associated with minimal immediate advancement, such as the lateral pass in football. Another definition suggests that lateral thinking relates directly to the development of novel ideas such as divergent thinking and is analogous to the downfield pass in football. In reading any literature on the subject, it is important to know the author's underlying interpretation of the term. Does the use of the term correspond to the lateral pass or the downfield pass? In order to show a difference between logical and imaginative thinking types, I prefer to use lateral thinking more in line with logical thinking, not imaginative thinking.

13.5.3 NOVELTY VS. PIGGYBACKING

General definitions given herein may involve a unique interpretation. My definition of lateral thinking falls more in line with the definition of a lateral pass in football. Returning to the football analogy, a divergent pass would be one that is thrown downfield into defensive coverage. It is a riskier play than the lateral pass, but the downfield pass has a greater potential for an immediate significant gain in yardage. The divergent pass in football is much like the development of a divergently generated novel idea in brainstorming. Lateral ideas should be viewed as ones that have less potential for providing a novel solution than would a divergent idea, but are likely to be ideas that require less effort to generate. That is, a lateral idea is one that is not too different from the ideas already proposed. Assuming that lateral thinking produces lateral ideas, it then follows that lateral thinking is likely of less value than thinking methods that require greater imagination. In the piggybacking view of idea generation, piggybacking is much like the lateral pass and, therefore, should be encouraged as part of a brainstorming or synectics activity; piggybacking does have some benefits just as the lateral pass does have the opportunity for benefits in a football game.

The level of novelty of an idea for a research problem is an important criterion to use in distinguishing between divergent and lateral thinking. In terms of the generation of an idea, the word novel would suggest that the idea is both new and unusual. This thought suggests that novelty is associated with divergent thinking, but not with lateral thinking. The connection of novelty to the words innovation (i.e., a change unexpectedly different) and creative (i.e., developing new ideas through imaginative thinking) should be evident. Unfortunately, we cannot always distinguish whether or not an idea is divergent or lateral. These two terms should *not* be viewed as dichotomous concepts; instead, they should be viewed as ends of a spectrum.

Lateral and divergent thinking can also be compared, as they are the ends of a spectrum rather than two opposing methods. Both methods of thinking are used to generate ideas. The position within the spectrum reflects the degree of association with ideas already expressed, i.e., the extent to which a generated idea is novel versus an idea that is similar to an existing idea. While a lateral idea is very similar to a previously generated idea, divergent ideas reflect a significant departure from ideas previously generated. When involved in creative idea generation, ideas based on divergent thinking require the thinker to have a curious, questioning attitude. Ideas based on logical thinking depend more on extrapolations of currently known knowledge. An individual who lacks the imagination skill set may have difficulty applying thinking types that depend on the curious, questioning attitude.

13.6 DIVERGENT-CONVERGENT THINKING

Divergent thinking is a central part of critical problem-solving. Unlike logical thinking, divergent thinking encourages deviation from the existing facts and knowledge. The imaginative element of divergent thinking generally leads to a set of ideas that initially appear to lack an association with the problem under investigation. Very often, the divergent condition is based on a situation that is more familiar than the experiences that depended on imaginative thinking. This familiarity makes it easier for the person to propose novel solutions to the situation. As an example, synectics is a creative thinking method that applies divergent-convergent thinking. In the divergent portion of the exercise, ideas are creatively generated based on the synectics topic. In the convergence phase of divergent-convergent thinking, imaginative ideas from the divergent phase are transformed into potentially useful solutions to the original problem. Most of the previous discussion has suggested that divergent-convergent thinking is the primary approach to the development of novel ideas. Lateral thinking is the junior partner to the divergent-convergent thinking model, as ideas already proposed are modified to reflect an added element of knowledge. The outcome of a divergent thinking session should be a variety of ideas that potentially can produce novel solutions.

In terms of conducting novel research, divergent-convergent thinking is believed to be an effective tool for identifying novel solutions. Divergent thinking is usually thought of as the thinking that leads to ideas that seem to deviate significantly from previously identified potential solutions to the original problem. The belief is that a novel solution to a problem will more likely be generated from an idea that itself suggests novelty. Divergent ideas may not have an immediately obvious connection to the original problem, and supposedly, it is this disconnect that illuminates a better solution; however, every disconnected idea requires refinement, which is the convergent thinking phase that leads to a solution. In practice, all methods of thinking should be applied, especially those with characteristics that best match the characteristics of the problem.

13.7   DIVERGENT THINKING

Is divergent thinking automatically critical thinking? Assume that the problem at hand requires the use of the conservation of mass for water flow in pipes, specifically in the form of the continuity equation, $Q = V * A$, where Q is the discharge (cubic meters per second), V is the velocity (m/s) of the water flowing in the pipe, and A is the cross-sectional area (m^2) of the pipe. If the idea generation session focused primarily on the velocity and area, then this would reflect logical thinking as the discussion is based on facts and existing knowledge, i.e., the continuity equation. If a group member proposes trying to solve the problem using an alternative form of the conservation of mass, then the discussion is still based on facts, but the ideas are then based on divergent thinking. For example, the idea might be to use the form that $Q = U/t$ where U is the volume (m^3) and t is the time (seconds). This form is still based on facts, but the thinking is somewhat divergent from the $Q = V * A$ form. Thus, it could be considered as divergent thinking, but we might consider the idea to be in the middle portion of the logical-creative spectrum. As an alternative divergent thought, the water pipe problem may be approached using the flow of electrical current in a copper pipe, specifically identifying ideas that relate to elements of Ohm's law. How about the movement of ants through small passageways in the ground? How about cars passing through toll booths on the turnpike? Note the degree of divergence in each of these possible alternatives. Which of these alternatives would lead to the best solution would depend on the original problem, but all alternatives should be evaluated.

Critical thinking can be illustrated using the pipe flow example. Assume that a participant in a brainstorming session proposes using the flow of people through the doors of a tech store on Black Friday, i.e., the day after Thanksgiving day. This idea is still based on the continuity equation, but uses masses of people rather than masses of water, which is in some way imaginative thinking. The flow-of-people situation would be more familiar to the participants of a synectics group than would be the water mass in a pipe, as they have likely experienced the flow of people, such as when passing into or out of an athletic stadium. Most individuals have actually experienced the flow of people through a narrow opening. Such experiences prepare the individuals for discussing the pipe flow problem. A personal experience that involves the flow of people would serve as an inspirational force to solve the complex problem. Therefore, the idea of a mass of people is both creative and divergent. The generality of the divergent idea will encourage imaginative thinking. The generality will also relieve the problem solver from concentrating on the technical details of the actual problem. The use of a situation that is not directly related to the real problem allows consideration of imaginative and divergent ideas. These examples suggest that ideas based on divergent thinking can have an association with seemingly unrelated but more familiar situations.

The selection of a divergent idea can be an important factor in the solution of a problem. The best divergent ideas have the following characteristics:

1. They should be based on simple life experiences, ones about which most people are familiar.
2. The danger or risk level in the divergent idea should be on par with that of the real problem.
3. The connection of the divergent idea to the real problem should be evident.
4. The divergent idea should avoid a dependence on any technical issue about which the participants may not be familiar.
5. The divergent idea should motivate emotional involvement.

The benefits of divergent idea generation are given as follows:

1. They enable a broader perspective to be taken.
2. They allow emotional thinking, not just cognitive thinking.
3. Preconceived ideas about the solution to the real problem are minimized.
4. Stress related to finding a technical solution is less because the technical complexities are not as important as ideas from divergent thinking.

The benefits of using a simple idea as the divergent topic are given as follows:

1. Wild-and-crazy ideas can be used.
2. Emotional thinking can complement rational thinking.
3. The participants will be more open to deal with silly issues than with complex technical issues.

An argument could be made that the flow of people is still logical thinking because the velocity and area are still relevant. The two cases, the flow of water versus the flow of people, have some differences such as that people are discrete while the masses of water are continuous beyond the molecular level. This distinction would be a judgment call by the facilitator, but brainstorming on the people-mass situation may provide some unique solutions to the water flow problem because the participants of a brainstorming session can better visualize the people-mass situation.

Logical thinking can be contrasted against divergent thinking, as they differ in several ways. Logical thinking is based solely on the analysis of existing facts and knowledge with the thought that an acceptable solution to problems can be developed through a direct analysis of such information. Some people view a direct analysis of facts as a constraint, yet it is the common mode of thinking. In my interpretation, those who use divergent thinking have the attitude that incorporates methods of imaginative thinking in problem-solving with facts and knowledge. Therefore, the thinking allows a broader base of experience, including imaginative thinking. Convergent thinking involves the evaluation of imaginative solutions from the list of divergent ideas, which may be combined with the existing knowledge to develop the final solution to a problem.

13.8   CONVERGENT THINKING

Convergence can be interpreted as a movement toward a common point, but from different directions. In transportation, regional roads converge toward the center of the city. Similarly, rivers converge, such as the Allegheny and Monongahela Rivers merging to form the Ohio River at Pittsburgh. In terms of the development of solutions to research problems, the hope is that novel ideas obtained from divergent analysis converge to a feasible solution to the original problem that initiated the research. The convergence phase of the divergent-convergent process intends to transform a novel idea into a useful solution by way of an imaginative connecting idea. The convergence phase of the process involves the assessment and refinement of the feasibility of the generated ideas.

The convergent thinking phase begins with the long list of ideas that were generated during the divergent activity, with each of the ideas having the potential to provide a solution to the original problem. Each idea must be thoroughly evaluated, with both positives and negatives of each idea identified. Having a critical attitude is important here. One task that is part of the convergent phase is to identify connections between the divergent idea and the original problem. A connection between the ideas and the problem will converge to solutions; however, each alternative solution may or may not be a realistic solution. The initial step is to evaluate each idea from Phase I and determine which of them are feasible and which are not. The ideas from this short list of feasible solutions must then be evaluated in detail to identify the best of the feasible alternatives. Performance criteria are metrics that can influence a decision. Ideas that are not part of the best solution should not be discarded at this point, as the group may need the alternatives at a later date to re-evaluate the initial solution. The list of feasible alternatives may also be used to extend the research or provide the basis for other novel research topics.

Once a list of divergent ideas has been developed, convergent thinking is used to transform each seemingly disconnected divergent idea into a realistic solution.

The convergence-to-solution transform allows for emotional thinking and reduces stress associated with trying to solve a complex problem using only logical thinking. However, the convergence-to-solution transform has the following complicating factors:

1. The need to transform a non-technical idea into a technical framework.
2. The intellectual effort to transform a simple idea into the solution of a complex problem.
3. The necessity to eliminate extraneous aspects that can materialize from the generality of the divergent idea.
4. The necessity to eliminate any emotional aspect of the convergent idea to a practical solution.

These issues are not usually troublesome for a person who had had experience in critical problem-solving.

What characteristics enable a person or group to develop divergent ideas and then transform them into a novel, feasible solution to a problem? Examining the convergent thinking problem from the perspective of converging traffic entering a city, it should be evident that efficiency depends on planning, tolerance, optimism, and confidence. These same characteristics apply to success with a convergent thinking activity related to research. Just as planning can reduce bottlenecked traffic, a facilitator who is involved in a research activity can plan the idea evaluation process so that it is effective and efficient. Tolerance is required to manage delays due to traffic accidents, and similarly, tolerance is necessary to avoid inhibiting practices in seeking a feasible solution to the research problem. Optimism that drivers should act rationally in their decision-making is important, just as the rational evaluation of generated ideas will lead to an effective solution. Table 13.2 provides definitions of four important characteristics and their applications to both the transportation problem and creative problem-solving.

Convergence requires both belief and thinking, which was a duality identified as being fundamental to creative thinking. Convergent analysis requires the same critical thinking characteristics as used during the divergent thinking phase; however, the focus is on solutions rather than on idea generation. Individuals who are involved in divergent-convergent problem-solving must be curious, as well as being confident that imaginative ideas can be transformed into a workable solution. Convergent thinking requires an optimistic attitude and the expectation that an indefatigable effort may be required. Perseverance is necessary as research in search of a novel solution is not simple; problems almost always arise when trying to solve complex problems.

13.9 STOCHASTIC THINKING

With respect to problem-solving, the word *stochastic* can be interpreted as guessing or drawing a conclusion from incomplete information. In statistical analyses, the word *stochastic* implies randomness. The extension of these definitions to the realm of imaginative idea generation would suggest that stochastic thinking could be

TABLE 13.2

Example of Convergent Thinking

Characteristic	Definition Relative to Convergent Thinking	Application to Transportation Example	Application to Creative Problem-Solving
Planning	A procedure identified beforehand that outlines the steps needed to achieve a stated objective.	Develop plans that emphasize reductions in bottlenecks.	Requires an experienced facilitator to transform imaginative ideas into realistic solutions.
Tolerance	A productive deviation from the most obvious solution.	Tolerance for driver errors; for accidents; for delays due to heavy traffic.	Creative thinking encourages deviations of imaginative ideas from commonly accepted ideas in order to better understand the solution to the problem.
Optimism	A belief that the best possible outcome will result.	Expects plans to yield the optimum flow of traffic by avoiding activities that cause bottlenecks (accidents, adverse weather conditions).	To avoid inhibiting factors such as negative criticisms of ideas.
Confidence	A state of trust that the expected will occur, often reflected in the courage to approach the problem using imaginative thinking.	Trust that the plans will be effective in reducing traffic and overcome inefficient travel experiences.	Trust that critical thinking methods will direct thinking in a way that a unique solution is developed.

interpreted as the development of ideas using an uncontrolled source of information. One example of stochastic thinking for the purpose of idea generation would be to randomly open a dictionary and identify the first noun on the left-hand page; then use this word to create an idea that would be a relevant entry to a brainstorming list. The selection of a page at random is the stochastic element of the activity. Connecting the word and the problem is the imaginative thinking part. Alternatively, any article in the room could be selected as the stimulating object. For example, a desk or light fixture could be the object on which to generate the imaginative ideas. The effectiveness of stochastic idea generation depends on the ability of the participants to ideate the connections. This task of association is best achieved by selecting participants who tend toward imaginative thinking. It is also important that inhibiting factors are kept from limiting the flow of ideas.

The concept of stochastic thinking refers to the reflection on or drawing of an association between the problem under consideration and a totally unrelated experience or event. For example, the random experience might be any one of the following:

- Espying a stray mutt on the ride to work in the morning and using a distinguishing characteristic of the mutt to generate a creative idea.
- Use the first word in the title of the movie that you watched the night before.
- Identify the basis of an argument that you recently had with a friend.
- Randomly identify an equation, such as $E = mc^2$, that is commonly recognized and use it as a basis for a new idea.
- Use a keyword that you heard during an advertisement on the radio.

If we consider de Bono's six-hat thinking process (see Section 2.5), it seems reasonable to suggest that stochastic thinking would most likely be involved with the red, black, and green hats. We may even creatively imagine a new hat that specifically inserts stochastic thinking into the process. For example, a violet hat could be used to represent stochastic thinking and would be inserted as a step between the red and black hats. The stochastic thinking could lead to novel ideas or extend ideas developed from the emotive thinking step, but the stochastically generated ideas would be available for the counterfactual assessment, i.e., the black-hat step. Stochastic thinking is also a useful tool for a facilitator of a brainstorming session to use when the group is having trouble contributing ideas.

13.10 COUNTERFACTUAL THINKING

Heat transfers from cold bodies to warm bodies! Water freezes at 100 deg C and boils at 0 deg C! These are facts in the world of counterfactual thinking. The premise of counterfactual thinking is that new ideas might follow if the facts of reality are suspended, i.e., if we live in a world where events follow both laws and principles that do not apply in reality. The suspension of actual laws should encourage freer thinking, such that fantasy becomes the reality of a counterfactual world.

If the problem of interest was to design a revolutionary new submarine, we could start by identifying laws that apply to the oceans. In our world, pressure increases with depth, sound travels faster in water (1450 m/s in the ocean, but only 334 m/s through a dry atmosphere), and wave speed decreases slightly with increasing depth. We might allow our thinking to believe that pressure does not increase with depth in the ocean. In our counterfactual world, we might entertain the idea that sound is transmitted less efficiently through the ocean water than though a dry air. We might also counterfactually think that wave speed does not change or increase with depth. The suspension of facts associated with the oceans may enable novel ideas to be generated and thus help in the design of this innovative submarine. By thinking in our counterfactual world, maybe the sub could dive deeper, maneuver more easily, or exceed the speeds of currently designed subs.

The facts to be suspended would depend on the problem. Ohm's law might be ignored for brainstorming on the design of new electrical equipment. Water runs uphill, right! It actually does when the pressure gradient is amenable. Where a problem assumes that Archimedes' principle applies, the assumption that the buoyant force is not equal to the weight of the displaced fluid might lead to a novel idea. What facts would be relevant to use counterfactual thinking in the design of a more fuel-efficient airplane?

One of the steps of the problem-solving process is to identify all relevant knowledge. We see here that one application of the existing knowledge might be to ignore it for the purpose of generating a counterfactual knowledge base. Without being constrained by the existing knowledge, novel ideas may materialize when the counterfactual knowledge is applied. If counterfactual idea generation is to be used in a brainstorming session, the facilitator will need to identify, assemble, and in some way distribute the relevant knowledge to the participants.

13.11 UPWARD-DOWNWARD THINKING

Upward-downward thinking is similar to counterfactual thinking in that the generation of ideas, at least initially, is in the direction opposite that of the direction one would rationally think. The method would be most useful for problems that focus on directionality, which is where something is increasing when you want it to decrease or it is decreasing when you want it to increase. The generation of ideas is oriented toward the direction in which the issue is currently moving, which is the wrong direction, but the idea evaluation is in the desired direction, but using the ideas generated for movement in the wrong direction. Two examples (see Table 13.3) can be used to illustrate the directional issue. For the sales case, the fact that sales and likely profits are declining is a concern because increasing sales is the desired trend. Thus, ideas related to continuing the decline are generated, such as skiing down a slope, but the idea will then be used to find ways of creating an increase. At first, this opposite direction may seem counterproductive, but it does increase the motivation to find a solution. It also encourages imaginative thinking because reality has been suspended. Ideas about slowing the rate of moving down the slope may lead to ideas for moving up the slope, i.e., increasing sales.

A directional flip-flop may be easier to appreciate from the outcome of another example. Assume that the problem is a decline in productivity in the machine shop of a company. Obviously, management wants productivity to increase, so they are alarmed about the decline. For the brainstorming session on the problem, the idea generation will be on ways of making the decline continue, with the following being a few of the ideas generated:

TABLE 13.3

Effect of Directions in Upward-Downward Analyses

Problem	Current Direction	Desired Direction	Idea Generation Direction	Idea Evaluation Direction
Sales	Declining	Increasing	Decline	Increase
Crime	Increasing	Decreasing	Increase	Decrease

Give a bonus for anyone who can decrease their productivity.
Allow longer coffee breaks.
Reduce the work hours without a reduction in pay.
Sabotage the equipment so that it breaks down more frequently.
Put the machine on a platform so that it is more difficult for the workers to use.

Note that none of these ideas is overly imaginative, but wild-and-crazy ideas would be possible, such as:

Cut out part of a giraffe's neck so it is less tall.
Make the clock speed up so that it appears that the work is being done in a shorter time period.

The next step is to flip each of the downward ideas into an upward trend framework, which yields the following ideas:

Give a bonus for increased productivity.
Shorten coffee breaks.
Cut hours without loss of pay for increased productivity.
Buy more efficient equipment.
Make working conditions more comfortable.

Note that each of the ideas that indicated a decline in productivity is transformed into an idea that suggests the correct direction. The task is to use the ideas, including the two wild-and-crazy ideas, to develop a reasonable solution. The central point here is that it may be easier to develop ideas for the current condition of the problem, but those ideas can be transformed into ideas that reflect the correct direction.

13.12   APPLICATION OF THE INTELLECTUAL DEVELOPMENT (ID) MODEL

The Intellectual Development model can be used to assess a person's status in critical problem-solving ability. It can be applied to each dimension individually with the thought that a collective analysis can be made across the dimensions, which will provide an indication of the intellectual development for applying critical thinking to solve complex problems. It can also be used to show the personal growth needed to move to a higher stage of critical thinking. Table 13.4 provides the characteristics of the five stages of the ID model as applied to the thinking type dimension. These guidelines are just examples of the type of knowledge and abilities needed to function at the individual stage.

13.13   ASSESSING THE COMPETENCY OF THE FOUR DIMENSIONS

Individuals will vary in their ability to fulfill the responsibilities associated with each of the four dimensions. Experiences in solving complex problems represent a source of educational growth, which enables a person to enhance his or her abilities

TABLE 13.4

The Growth of Knowledge of the Thinking Type Dimension for Problem-Solving

I. Comprehend
- To identify the fundamental characteristics of different types of thinking.
- To understand the role of thinking types in making decisions.
- To recognize that selection of the best thinking type to use depends on the characteristics of the problem.

II. Experience
- To experience problem-solving that involves interactions between thinking types and the thinking skills.
- To experience complex problems where thinking type influences decisions.
- To experience the use of cognitive and affective thinking in actual problems.

III. Analyze
- To analyze solutions to past cases to identify the most effective thinking types for certain general categories of problems.
- To analyze the interactions between the thinking types and the other dimensions of critical problem-solving to understand the influence of dimensional interactions on decisions.

IV. Synthesize
- To synthesize complex problems and show the effects, good or bad, of various thinking types on outcomes.
- To synthesize the effects of variations in stakeholder decision criteria on the effectiveness of a solution to problems that use different types of thinking.
- To synthesize the interacting effects of different levels of uncertainty in the use of types of thinking.

V. Create
- To create guidelines for mentoring novices about thinking types.
- To create outcomes for complex problems subject to different thinking types.
- To create guidelines for selecting thinking type to resolve specific types of problems.

to complete responsibilities associated with each dimension. Criteria are needed to distinguish an individual's level of competency. If four levels of competency are sufficient to distinguish between the abilities for each dimension, then the model in Table 13.5 should enable a person to have some idea of their strengths and weaknesses in terms of the four dimensions of critical problem-solving. The disparity between the different stages should provide a person with some measure of their critical problem-solving status and provide the knowledge needed to enhance their critical problem-solving skills in order to move to a higher level.

13.14   CONCLUDING COMMENTS

Assume that most people have considerable experience using logical thinking, but very limited experience with other types of thinking. Additional experiences with logical thinking will prepare the problem solver only to solve problems that lend themselves to solutions that are based on basic facts and existing knowledge. Such

TABLE 13.5
Assessment Criteria for the Thinking Type Dimension at Different Stages

Stage I: Novice	Stage II: Specialist	Stage III: Master	Stage IV: Expert
1. Logical thinker only.	1. Piggybacking is the only imaginative thinking used.	1. Some imaginative thinking experience.	1. Uses affective and cognitive.
2. Biased thinker.	2. Problems recognizing the need to coordinate thinking type and problem characteristics.	2. Broad use of thinking types.	2. Correctly matches thinking type to problem characteristics.
3. Avoids imaginative idea generation.		3. Beginning to connect problem type and thinking.	3. Easily applies multiple thinking types.
4. Does not understand that complex problems have unique characteristics.	3. Limited breadth in thinking types.		4. Develops novel ideas.
			5. Extensive experience in complex problem-solving.

individuals are pursuing depth in experience, but complex problem-solving needs multiple types of thinking that reflect breadth. Without knowledge based on broad experience in thinking types, the best solution to a problem may be difficult to find. Problems themselves have specific characteristics, so the best decisions will be found only if the characteristics of the thinking type correspond directly to the characteristics of the problem statement. As society's problems become more complex, the thinking abilities of the problem solvers must grow at least at the same rate.

Trying to solve a complex problem with a thinking type that does not match the characteristics of the problem is inefficient. Using an inefficient thinking type means that the transformation from input to output will yield an output that is less than the level that would be possible if the correct thinking type was used. The thinking type is like a transformation function in mathematical modeling. When a function is not ideal, the level of the output is less than the level that could be achieved with a proper model.

13.15   EXERCISES

13.1.   Define the word *attitude* and discuss its application to the ethical conduct of professional engineers. Define the word *thinking* and discuss the way that it applies to resolving an ethical conflict.

13.2.   Develop three educational activities that could be used to develop the creative attitudes of engineering undergraduates.

13.3.   What are general characteristics and weaknesses of knowledge of a subject?

13.4.   Can "clear headed" thinking involve the use of emotions? Discuss.

13.5.   Define the word *logical* and explain its use in the term *logical thinking*.

13.6.   Contrast logical thinking and creative thinking.

13.7.   Contrast logical thinking and counterfactual thinking in their potential to solve unique problems.

13.8.   What factors contribute to illogical thinking?

13.9.   Is creative thinking illogical?

13.10. Synonyms for the word *logical* include coherent, valid, cogent, and relevant. Define each of these words and discuss whether or not they could adequately be used in place of the word *logical* in the term *logical thinking*.

13.11. Develop an analogy for lateral thinking and explain its correspondence to the steps of the research process.

13.12. Provide reasons that divergent-convergent thinking is considered to be the basis for creative thinking.

13.13. Where does divergent-convergent thinking fit into the steps of the research process?

13.14. Problem statement: birds are being killed because they fly into the blades of the turbines on the wind farm. Develop three imaginative ideas for this problem statement.

13.15. For each of the following problems, provide three divergent ideas: (a) failure of sensors in a self-driving car; (b) a bridge collapse due to scour around the supporting bridge pier; (c) water leakage into a house due to the poor design of its green (vegetation covered) roof. Wild-and-crazy ideas are encouraged.

13.16. For each of the following problems, provide three divergent ideas: (a) being turned down for a job; (b) damage at a construction site due to high winds; (c) an accident due to unexpected air bag inflation. Wild-and-crazy ideas are encouraged.

13.17. The problem is the increasing crime rates in the local college town. The following three divergent ideas are proposed to reduce the crime rate: (a) cutting back on drinking at fraternity parties; (b) reducing text messaging by drivers; (c) requiring the mayor of the college town to lose weight. For each of these divergent ideas provide a convergent idea that could lead to a solution to the crime problem.

13.18. The following three divergent ideas are proposed to reduce the crime rate: (a) have the college football team score more points in their games; (b) reduce the number of stray cats that are roaming the campus; (c) reduce the number of U.S. Senators in Congress from 100 to 50. Develop convergent ideas to correspond to each divergent idea.

13.19. Why is stochastic thinking considered to be a creative thinking method?

13.20. Identify and explain three new ways of generating ideas using stochastic thinking.

13.21. Why would critical thinking important in research?

13.22. How can counterfactual thinking be used in the assessment of failures in research?

13.23. Assume that you are thinking about going to a movie. What are the steps of the thinking process that would lead you to a decision?

13.24. Explain why thinking applies to all steps of the research process.

13.25. What factors contribute to the thinking ability of a person?

13.16 ACTIVITIES

13.16.1 ACTIVITY 13A: CREATIVE BIOGRAPHY

The following is an activity that could be used to enhance critical thinking and develop a creative attitude. Start by assembling a set of disparate items, such as a copy of a famous painting, an empirical equation that you have used, a copy of a famous speech, the words of a popular song, a short biography of an infamous person, and a brief summary of a philosophical position such as existentialism or logical positivism. Evaluate each of the items and find a concern about each; then recommend ways that the concerns could be rectified. This practice will emphasize your ability to critique the works of others, i.e., perform an aspect of critical thinking. Also, identify connections between the different items; this practice encourages creative thinking as it forces you to use your imagination. Being critical of a wide variety of things is the first step to developing a critical attitude, as being critical adds breadth to any natural inquisitiveness that a person has. Positive criticism can be as helpful as negative criticism. Now rewrite the biography of the infamous person in a way that makes him appear to be a better person than he or she really was; try to incorporate the disparate items into the false biography. This part of the activity is intended to demonstrate biasedness and the impact that it can have on honesty.

13.16.2 Activity 13B: Selflessness Thinking

The objective of this activity is to develop a new thinking type. The central focus of the new type should be selflessness. State the overall objective of your proposal, the steps of its application, and discuss the decision criteria. Provide an example application of the type. Provide a creative name, yet one that is appropriate for the thinking type.

13.16.3 Activity 13C: Illogical Thinking

Logical thinking is the most commonly used method of thinking. Logical thinking is based on existing knowledge and facts. Develop a new brainstorming method called illogical thinking. State its objective, procedure, advantages, and disadvantages.

13.16.4 Activity 13D: Congestion

The problem is severe traffic congestion at an intersection with traffic control using four STOP signs. Develop three or more imaginative ideas that illustrate divergence; none of the ideas should specifically relate to rust or bridges. For each idea, generate a convergent idea and then indicate a solution to the traffic problem for each divergent-convergent pair.

13.16.5 Activity 13E: Rust

The problem is rust development on a steel bridge, such that the safety of the bridge is being questioned. Develop three or more very imaginative ideas that illustrate divergence. For each idea, generate a convergent idea and then indicate a solution for each divergent-convergent pair that will limit the development of rust on the bridge.

13.16.6 Activity 13F: Polar Bears

You are required to make an oral presentation on the plight of polar bears due to the adverse effect of climate change on the polar ice cap. You want a catchy opening. Using divergent thinking combined with imaginative thinking methods, develop a brainwritten (or brainstormed) list of possible openings. These could be humorous, frightening, and any other genre. This activity is intended to improve the ability to communicate effectively.

13.16.7   ACTIVITY 13G: TAME THE BULL

This activity requires three half-sheets of paper, one piece for each person in the group. Each sheet is then separated into four rows and two columns, with the left column only wide enough to include the following: (1) problem, (2) wild-and-crazy, (3) wild-but-not crazy, and (4) neither wild nor crazy (the solution). Each one of the pieces of paper is used for a different one of the three problems: (1) removing a crooked politician from office; (2) prevention of forest fires; (3) the need to improve STEM education in schools. Write the problem statement in the box for Column 2, Row 1. Each player is given one of the sheets; he or she enters a response in the second box in Column 2 (i.e., wild and crazy) and passes the sheet counterclockwise to the next person who uses the idea in box (2) to generate an idea, which is entered into the box of Row 3, Column 2. The sheet is passed counterclockwise to the third person who provides a solution based on the thinking of the previous response. After the three sheets have been completed, the group members discuss the way that each box idea was transformed into a solution by way of divergent-convergent thinking. The objective of this activity is to show that a wild idea can converge to a potentially useful idea.

13.16.8   ACTIVITY 13H: ARCHIE WANTS TO DATE VERONICA

Part I: A friend is interested in getting a date with a new neighbor, but your friend lacks the courage to approach the new neighbor. Brainstorm a list of ideas on ways of increasing your friend's courage to approach the neighbor. Items in the list can be practical, but creative ideas are strongly encouraged. PART I SHOULD BE COMPLETED BEFORE THE DIRECTIONS FOR PART II ARE READ.

Part II: Courage is important in approaching problems in a research environment. Use the list of Part I to identify ways that a professor could help his or her new research assistant to have greater courage in addressing problems that arise in research.

14 Leadership of Critical Problem-Solving

CHAPTER GOAL

To identify ways that knowledge of critical problem-solving can advance the goals of an organization.

CHAPTER OBJECTIVES

1. To summarize the benefits of having a leader who has the knowledge of the four dimensions of critical problem-solving.
2. To identify the use of critical problem-solving to promote positive characteristics in employees, including self-confidence and an optimistic outlook.
3. To identify a leader's responsibility to direct an inhibitor-free organization.

14.1   INTRODUCTION

A leader is the person who establishes and monitors the policies and practices of an organization or group. Leadership is the art of planning, organizing, and conducting the activities of a group. Leadership is critical to success. This perspective is true now, it was true in the past, and we can expect it to be true in the future. Joan of Arc was very successful as a young soldier because of her leadership skills during battles. She even acted as a leader after she was burnt at the stake by her adversaries. Her fellow soldiers continued to fight to achieve the goals that Joan had established before she was killed. One tenet of leadership is that subordinates need to have someone or a principle around which they can rally. Joan of Arc provided the ideals around which the troops rallied.

The opposite experience was true of Ferdinand de Lesseps, a French engineer who started to build the Panama Canal in the late nineteenth century. He lacked the necessary technical competency, which represents both the technical dimension of critical problem-solving and the skill dimension. His failure was also due to his lack of other leadership skills. He lacked knowledge of the value dimension as evidenced by the financial mismanagement that occurred during his tenure as the chief engineer. He lacked the mindset dimension as he believed that the project would fail from the start; his pessimistic view was not helpful. He lacked the thinking skill dimension as evidenced by his unwillingness to try new ideas when confronted by problems on the site; he wanted to use the knowledge that he accumulated from his work on the Suez Canal,

DOI: 10.1201/9781003380443-14

which was quite different because of both the topography and the climate of Panama. deLessups was a national hero in France when he accepted the job on the Panama Canal but because of his failures and lack of leadership on the project, he died in disgrace.

Leaders are problem solvers. A leader's effectiveness depends, in part, on his or her ability to motivate employees to solve complex problems. This responsibility requires many abilities and attitudes. First, the leader needs to provide a work environment that controls organizational inhibitors to creativity. Second, the leader needs to provide creative work responsibilities that are sufficient to meet the employees' needs for workplace satisfaction. Third, the leader needs to have the values, skills, and mindset characteristics of a critical problem solver. Fourth, a leader needs to have strong communications skills, which can be enhanced using creative problem-solving methods. These needs are minimal requirements for success.

The content of this chapter focuses on the critical thinking responsibilities that are needed for a leader to achieve organizational success. It is not a broad discussion of all of the leader's responsibilities. A leader must have demonstrated exceptional proficiency in all of the dimensions of critical thinking, including the project-specific dimension such as technical proficiency. He or she must adhere to and promote high value standards. A leader must be able to adapt his or her thinking style to the characteristics of the problem and use all of the components of the skill-set model. A leader must know his or her belief system and be willing to make changes when experiences dictate that changes are needed to improve. These factors are central to one's growth in critical thinking ability.

14.2   ORGANIZATIONAL CREATIVITY

An organization that promotes critical thinking is one that provides an environment that adopts policies and practices in support of imaginative problem-solving. Such an atmosphere can contribute to the success of both the individual employee and the organization as a whole. Clients generally recognize when a creative culture is a central part of the foundation of an organization. They will be more likely to entrust their complex problems that need to be solved with a company that specializes in innovative solutions, especially when the problems require critical problem-solving skills. Just as creative individuals are characterized by certain skills and attitudes, creative organizations can be recognized for its creativity promoting qualities. The management of an organization promotes critical thinking by:

- Rewarding creative ideas of employees.
- Encouraging employee creativity.
- Providing the resources that are needed to facilitate creative collaborations.
- Removing all policies and practices that inhibit creative idea development.
- Instituting policies to promote critical thinking.
- Not limiting employee's future opportunities when past creative efforts did not fully succeed.
- Initiating policies that eliminate routine work.
- Providing training and educational opportunities to learn about critical problem-solving.

If such policies and practices are not part of an organization's culture, success may be constrained, and potential clients may take their work to more progressive companies. Many of these characteristics will require minimal resources to adopt for the workplace; however, if they are part of the company's official plan of organization, they will likely increase problem-solving efficiency because of the increase in morale among creative employees. Establishing some of the characteristics may require reorganization of company personnel. Critical thinking characteristics should be used as decision criteria in the recruitment of new employees.

Each of these eight guidelines is presented as a positive statement. It can be instructive to analyze each of these eight positive acts from a negative point of view. Companies that do the reverse of these eight actions are essentially suppressing critical thinking. Companies can be unwilling to reward new ideas as they recognize that creative ideas usually take more time to develop, analyze, and evaluate; it is a matter of being willing to allocate resources. Pessimists will generally criticize ideas in a way that suppresses creative solutions. Companies that are reluctant to provide resources for innovative activities argue that they lack the extra funding or resources, including any support staff that would be necessary to provide a creative environment. Even verbal encouragement may be withheld, which also means that advancements that may have resulted from critical problem-solving are lost. In many businesses, direct awards or rewards are rare and often a superior is acknowledged for the creative work of a subordinate: leaders should ensure that the creator is rewarded, not the supervisor of the creator. Even when efforts at creativity are not successful and may have used extra resources, it should not be assumed that creative thinking will always lead to disappointing results. With any failure, the organization should perform a self-evaluation of the failure and take the steps needed to ensure that the same problems do not reoccur in future efforts when critical problem-solving skills are applied. While critical problem-solving will not always provide a novel solution, company policies and practices can enhance the likelihood that methods of critical thinking will at least be cost-effective over the long term.

While many training and educational resources are available and should be used, mentoring programs within an organization are often the most productive approaches to problem-solving. Mentors are more familiar with organizational policies, practices, and resources, as well as the types of jobs that the organization pursues. As leaders, mentoring is especially valuable to the success of young employees. Mentors can encourage creative problem-solving and protect the mentee by discouraging peer-inhibiting practices that occur within the organization. Mentors can provide advice that will help the mentee become more professionally mature. Leaders should encourage mentoring, possibly even establishing a mentoring program within the organization. Even providing group mentoring has many benefits to the organization.

14.3 ORGANIZATIONAL INHIBITORS

14.3.1 PEER-INDUCED INHIBITORS

Subordinates who are new to an organization can be very intimidated by personnel who have more experience than they do and have been employed by the organization for a longer period of time. If the novice makes recommendations for policy changes that are related to increasing creative related practices, he or she may have a fear of getting poor performance reviews submitted by the more experienced employees. Thus, a novice may feel intimidated to the point that he or she would not want to propose creative solutions to problems. This feeling would represent an inhibiting office climate, which may be real or only imagined. A leader has the responsibility to discuss the issue with the novice and provide assurance that creativity is encouraged and rewarded. Employees of an organization should report inhibiting actions to organizational administrators, preferably the immediate supervisor unless it is he or she who is acting as the inhibitor. The leadership should then discuss the issue with the offending person and report back to the novice that the issue has been addressed.

14.3.2 Self-Inhibitors

A leader should have sufficient experience to recognize a novice's self-inhibiting attitudes and alert the novice to the effect of such feelings or beliefs. Procrastination, pessimism, and a lack of self-discipline are a novice's most detrimental inhibitors. Procrastination is probably the greatest inhibitor to productivity for a critical thinker as it causes unnecessary delay and contributes to a negative attitude within the organization. A pessimistic attitude is too often a self-inhibiting problem. A lack of self-discipline is another inhibitor, especially when a person is involved in team work, as others on the team may depend on the work of a person who lacks self-discipline. Brainwriting can be a useful way of overcoming self-inhibiting habits. Specifically, state the inhibitive habit and then write out the potential effects of the habit and ways of overcoming it.

14.3.3 Routine Work

Studies have shown that on-the-job boredom decreases productivity and efficiency. Routine work creates boredom. Thus, routine work can be a major inhibitor of critical thinking, and leaders can improve organizational productivity by minimizing routine work. The word *routine* is defined as *lacking in originality* or *a set of mechanically performed activities*. Henry Ford recognized the boredom that assembly lines caused, but at that time computerized robots were not available to replace humans. Thus, those on assembly lines found the work boring, which created a need for so-called efficiency experts. These experts were needed to identify ways of offsetting the decline in productivity that resulted from the boring existence on an assembly line. The definition of efficiency of Eq. 2.2 can be applied to workplace efficiency.

Compared to the time of Henry Ford, workers have experienced considerably more education and are much less willing to be involved in routine work. Thus, the leader of a group must institute programs that eliminate routine work, as it is a primary inhibitor to productivity. This responsibility is especially true of those who are capable of performing creative work. Critical thinkers are especially sensitive to the burden of routine work, so it is a leader's responsibility to unburden those who need to do critical problem-solving from the scourge of routine work. Meeting this responsibility can be accomplished by minimizing the need to do unnecessary paperwork, eliminating unnecessary meetings or at least hold well planned meetings, and ensuring that support staff are available to reduce the burden on critical thinkers' routine work.

14.4   LEADERSHIP AND VALUES

The value dimension of critical problem-solving is very relevant to many of a leader's roles. The importance of some values, such as honesty and truth, is obvious. But other values are also very important to the long-term success of a company. Leaders must be fair to the employees and clients. The leader who assigns problem-solving assignments to employees must be fair to each employee and not play favoritism in assigning the better opportunities to favored employees. Employees are often very sensitive to the quality and challenge of their assignments. Correspondingly, the leader must be fair to both the company and the client by ensuring that the job will be completed successfully, as a client may not respect the leader's responsibility to be fair in the assignment of opportunities to employees. It is sometimes difficult to balance this aspect of fairness, as the clients and employee may have competing goals.

The leader must serve as a role model with other values. All work should be conducted in an unbiasedness way. Accuracy in all work is important. A leader needs to ensure that the organization and its employees are always accountable to the public, to the profession, and to the client. As indicated in Section 10.15, flexibility is important and interacts directly with accountability and tolerance, both of which are important organizational values. A leader's failure at value responsibilities can have negative consequences to the reputation and the economic well-being of an organization, as well as to the careers of the employees.

14.5   LEADERSHIP AND THINKING TYPE

As indicated previously, problems are labeled as being complex because of the significant breadth in responsibilities, conflicts between stakeholders, and the uniqueness of underlying weaknesses in the current state of knowledge. Complexity is not just because of technical complexity. In fact, if technical issues are the only problem, then the problem is likely not complex. Complex problems often have multiple solutions, each of which may be reasonable based on some set of decision criteria. The selection of a decision criterion is an important decision for a leader. Any uncertainty in the selection of decision criteria can influence the accuracy and quality of the decision. A leader must recognize that each possible decision is subject to uncertainties to a varying degree and develop plans to assess the effects of the uncertainties on decisions. One solution procedure is generally not sufficient to solve a complex problem because of the broad, unique characteristics associated with any complex problems. This uncertainty is the reason that solutions should be based on critical problem-solving. Accounting for values, following the proper skill sequence, and considering alternative thinking types are central to dealing with uncertainties. The characteristics and values are often conflicting, which means that individual stakeholders will not be totally satisfied with the decision. Decision criteria should reflect the types of thinking that is necessary to solve a complex problem. Therefore, within an organization, employees need to be educated about thinking types, so that they can select the most appropriate thinking type to solve each part of the problem.

Leaders need to assess the thinking ability of new employees to ensure that a novice is knowledgeable about the various types of thinking that are beneficial in solving the types of problems that the organization is usually responsible for solving. The leader of the organization should give the novice some general reading on thinking and provide some completed organizational reports that show the breadth in the types of problems that the organization addresses. The material will enable the reader to recognize the connections between the type of thinking and the characteristics of the problem. The exposure to the potential uses of the material will encourage the novice to recognize the importance of the thinking types that are needed to effectively practice in the organization. Reading organizational reports that use a breadth of thinking types will encourage the novice to seek appropriate educational material.

14.6   ORGANIZATIONAL MINDSET AND LEADERSHIP

Being successful in problem-solving requires a combination of abilities, attitudes, and experiences. This requirement is true for organizations as a whole as well as the individuals within the organization. Success depends on both the organization's mindset and the individual mindsets of the employees, especially the leader.

14.6.1   DEFINITION

An organizational mindset reflects the collective mindsets of all individuals in the organization, but the mindsets of those in leadership positions are often principal factors. The organizational mindset is also a reflection of the quality of the practices and policies that have been instituted by the leadership. The collective mindset depends on the mindsets of all individuals and can be inefficient even if one person is not functioning as needed. Therefore, a leader needs to be able to detect dispositions and moods that positively or negatively influence organizational problem-solving effectiveness and efficiency. Negative moods of individuals can be just as detrimental to progress as are an organization's inhibiting policies related to imaginative problem-solving. Minimizing employee negativity is a leader's responsibility.

A person's mindset is a collection of interrelated dispositions and attitudes that govern his or her actions at a specific time. While a person's ability to control his or her mindset is itself important, it is the responsibility of the leader to control the mindsets of each individual employee. While a person's mindset at the time of his or her decision-making needs to be properly controlled, the same is true for an organization's responsibility to control its mindset. This is a responsibility of the leader.

Having a critical attitude can be an important state of mind as it influences problem-solving efficiency; however, the mindset may be subject to change over short periods of time with the length of time dependent on the person's mental capacity. Fluctuations can be detrimental to the progress of the problem-solving. Only when a person recognizes the detrimental effects of mood swings can he or she exert control at times when control is necessary for the group to be collectively efficient and to ensure that the best decisions are made. These simple principles apply to the organization. It must avoid adverse fluctuations of its mindset.

Just as an organization has a value system, an organization also has a mindset. The policies and practices of the organization are derived from the mindset, but the mindset influences the policies and practices. The mindset also greatly influences the decisions made by the organization. Therefore, the word attitude also applies to an organization. With respect to critical problem-solving, the term organizational critical attitude could be defined as *a collective state of mind that values and emphasizes careful, maybe severe, assessment usually based on a reasonable level of knowledge and experience, but influenced by the group's collective mental state*; however, like a person's mindset, a group's mindset is not a permanent pattern of thinking. It actually varies as the policies and practices of the organization change, but is subject to much more temporal variations. Changes made to important personnel can also influence, positively or negatively, the organization's mindset. Since thinking is a process from idea conception to judgment, the general pattern of thinking is likely a relatively stable characteristic over time as long as the mood stability is not fluctuating. If influential group members have a change of professional mood, then the thinking pattern of the organization can change. Critical thinking is especially important in organizational problem-solving and, therefore, it is important for the leadership of an organization to have a stable attitude over time.

The innovative development of novel ideas depends on an organization having a collective critical mindset that becomes ingrained in the organization's decisions and actions; a mindset is more than an array of disconnected attitudes, especially for a group. The organizational mindset involves the integrated connection between values, skills, thinking types, and experiences of all members of the group. The elements of an organization's mindset are embedded in the policies and practices of the organization. A critical thinker generally has good control of his or her own moods, thus minimizing any negativity that could result from changes in mood. If a person is under undue stress to produce a solution, the critical mindset will not materialize to the extent needed. The same is true for the organization as a whole. Stress can serve as an inhibiting block on decision-making, introduce a constraining feeling of nervousness, and prevent critical thinking. Stress is a primary inhibitor to both critical and creative thinking. Having prior multiple experiences with critical idea development generally provides the essential experience that is essential to have the ingrained mindset useful for solving problems when they arise. An organization should not wait until the collective critical mindset is needed; a belief in the need for prior experience with the mindset is necessary and should serve as motivation for all employees within the organization to gain that experience prior to it being needed.

14.6.2 Influencing a Subordinate's Mindset

The mood of a person can be defined as a temporary state of mind or feeling. A person's mood influences his or her decisions and actions. An individual's mood may not be overtly evident; instead, it is the actions that are the outward evidence of the mood. The word temporary is an important part of the definition, as a change in behavior and thinking can easily occur. A pessimistic mood can adversely influence a leader's decision. When an individual recognizes an extreme mood, he or she should be cautious in making decisions.

A person's mindset can influence his or her performance at a task for which he or she is responsible. If a leader detects that a subordinate generally is not in a mindset that promotes efficiency, then action by the leader is needed. Obviously, a pessimistic mindset will adversely influence the individual's ability to apply critical thinking skills and attitudes. The leader should identify any specific attitude or mood that distracts the subordinate in order to decide the best means of altering the subordinate's frame of mind, such that progress can then be made. The following are possible ways that a leader can help a subordinate overcome a frame of mind that is responsible for inefficient productivity on a responsibility:

- If the person is by nature quantitative, ask him or her to rate the effect of his or her mindset. A 1–10 Likert scale could be used, with a 10 indicating very efficient. This activity may make the person aware of the effects of his or her frame of mind and thus motivate the person to improve his or her productivity.
- Develop a reward system for completing responsibilities on time.
- Provide reading materials on the specific issue or general reading on moodiness.
- Discuss the benefits of completing a task to both the employee and the organization.
- Tell the person about the dependency of other team members on the timely completion of his or her work.

It is quite possible that the subordinate does not realize that he or she allows moodiness to detract from the organization's productivity. Subtly making the person aware of his or her inefficiency may be sufficient for the subordinate to overcome his or her moodiness and make better progress on task completion. The person may take the incidence of moodiness as a partial failure and seek a general solution by way of an introspective analysis.

Creative idea generation is most likely to be successful when both the employee's mindset and the organization's culture promote critical thinking. Gaining experience with imaginative problem-solving requires both a receptive mindset of the employee and organizational policies that provide appropriate opportunities. An individual's abilities and attitudes may not be useful if organizational policies do not promote a critical problem-solving environment. A creative individual may believe that it is difficult to function within an organization when the organization's policies and practices suppress employee imaginative thinking. An employee will be more highly motivated if he or she believes that the organization rewards the inclusion of critical thinking into its decision-making; however, the employee's mindset is equally important, as discussed in Chapter 12.

14.7 MINDSET CONTRAST: OPTIMISM VERSUS PESSIMISM

Attitude is a general term that refers *to any feeling with respect to some matter.* For example, it could refer to a person's position on either drug use, protecting endangered species, or social media. Both the attitude of each individual and the collective attitude of a group toward events of the future are important to an organization. Specifically, some individuals always feel that an event will have a positive outcome regardless of the initial indications of the outcome; such individuals are referred to as optimists. Other individuals expect that the worst outcome will occur even when the initial indications suggest that a positive outcome will result; such individuals are referred to as pessimists. These attitudes influence decisions and actions and are central issues to a critical thinker.

14.7.1 AN OPTIMISTIC MINDSET

An individual's attitude can greatly influence the outcome of an effort. Thus, the person's attitude is important to his or her own success and that of an organization. Specifically, studies show that success is highly associated with organizations where the individual employees, as well as the organization itself, have a positive outlook about the future. This starts with the leaders of an organization.

The benefits to an organization where optimism is the dominate mindset within the organization include the following:

- Greater office efficiency.
- Better external reputation.
- Greater company success.
- Lower office employee turnover.
- Employees work harder.
- Employees are happier.

Optimism is an especially important attitude to an organization that is involved in solving complex problems. Compared to pessimists, optimistic employees are generally more creative and willing to apply imaginative idea generation methods. They also are more accepting of new, unconventional ideas, and therefore, they are usually more efficient problem solvers. These benefits suggest that a pessimist should seek help to transition from pessimistic thinking to having thoughts that suggest an optimistic view of the future.

14.7.2 OPTIMISTIC WORK ENVIRONMENTS

A leader who wants a work environment that is characterized by optimism can take action to ensure that the employees act accordingly. The following are a few ways of creating an environment where optimism is the norm:

- Identify the practices that motivate each employee as an individual and encourage and use resources to support those practices.

- To create a harmonious work environment, identify and resolve all internal conflicts within the company; this includes conflicts between individuals and conflicts between divisions of the company.
- Identify the strengths and weaknesses of each employee and make work assignments that reflect these factors.
- Provide advanced training to help employees overcome weaknesses.
- Reward positive work, but try to spread the rewards to all employees so that individuals do not feel that their work is unappreciated.
- Set company goals with bonuses given when the goals are met; these can be bonuses for individuals or for groups.

When hiring new employees, the collective attitudes of a candidate should be assessed. Other factors being equal, the more optimistic candidate will likely be the better long-term option. Since company policies and practices can influence the individual's outlook, the policies should be designed to encourage optimism.

14.7.3 OPTIMISM AND INNOVATION

Innovation can be a significant causal factor to increases in production output. Innovation is often necessary to maintain a change in the work environment of today. A primary responsibility of an organization's leadership, then, is to assure that policies and work practices cultivate critical thinking attitudes, as such thinking encourages innovation. An optimistic attitude leads to a willingness to innovate and improves creative output for both the leaders and the subordinates. Optimistic employees are extremely important in supporting a culture of innovation and critical thinking. Employees who are involved in a creative work environment are usually less likely to seek employment elsewhere.

Attitudes refer to thoughts and feelings of the individual. When a leader appears to be optimistic, subordinates will tend to adopt the same attitude. A person's attitude will direct his or her thoughts and influence subsequent actions. As a result, an organizational climate that is positive toward innovation and critical thinking will contribute to the achievement of organizational goals. The leader's attitude is paramount to this and can greatly influence the productivity of individual subordinates, as well as the group as a whole.

An employee's optimistic attitude and his or her propensity for innovation and critical thinking develop together, as optimism stimulates a creative attitude and creativity stimulates optimism. A leader's support of critical thinking infuses a more positive, more optimistic attitude within the organization. An organization can benefit from this interactive relation by establishing practices and policies that stress optimism and at the same time emphasize innovation in the workplace. The leader must focus on improving both the creative environment and the overall attitude of optimism. Success may not be fully achieved if the leader concentrates on only one of the two. Leaders must acknowledge the synergy between optimism and effective problem-solving.

14.8 LEADERSHIP APPLICATIONS OF CRITICAL THINKING

Leaders can advance the goals of the organization with the proper use of critical problem-solving methods. For example, if a manager recognizes that one of the subordinates has difficulty in initiating his or her writing assignments, such as project reports or work proposals, the manager could introduce the novice to the benefits of using a checklist (see Chapter 6). The manager could stress that a checklist on initiating writing or other assignments would be a useful tool for overcoming the fear of getting the writing assignment started. Table 14.1 provides an initial draft of a checklist that a subordinate could use when he or she has difficulty with writing assignments. Procrastination on writing responsibilities is actually a common problem. Other items could be included in the checklist. Any topic that would help a person overcome the fear of starting any activity could be stated in general terms and included in the checklist; it is not necessary for the topic to be related specifically to writing responsibilities. Additional details, such as overcoming procrastination, details of progressive outlining, or ways of summarizing important points could be directly included in the checklist. Since individuals differ in the reasons why they have difficulty starting assignments, the checklist should cover a broad range of reasons. Also, the characteristics of writing assignments can influence whether or not a person has trouble initiating the responsibility.

TABLE 14.1
Checklist for Initiating Writing Assignments

Getting Started
- Begin with an outline.
- It is not necessary to start writing the introduction.
- Start writing the easiest section or the most interesting section.
- Initially, completed work is not necessary; write to develop the ideas and the organization.
- Initially, only make rough drafts of figures and tables.

Outlining
- Use progressive outlining.
- Use phrases, not complete sentences.
- Initially use the steps of the problem-solving process as headings.

Introduction
- Include background, problem statement, goal, and objectives.
- May include literature summary to establish the state of knowledge.
- Possibly the last part written.

Conclusions
- Include the implications of the results.
- Identify additional work needed.
- Include a summary of the results related to each objective.

Overcoming procrastination
- Procrastination reduces the quality of the end product.
- Becomes more difficult to get started following procrastination.
- Procrastination does not detract from the responsibility to complete the assignment on time.

The Delphi method is a method for generating ideas. Knowledge of the method adds breadth to a person's repertoire of creative thinking methods. Also, the knowledge of brainwriting (see Chapter 3) can be a valuable asset for a novice to problem-solving. Combining brainwriting and iterative questioning (see Chapter 10) is a valuable combination for generating ideas. The greater the variety of knowledge of problem-solving methods, the more likely the person will succeed in solving complex problems.

14.9   LEADERSHIP AND SUBORDINATE CONFIDENCE

Many studies have demonstrated that confident employees are happier with their jobs than those employees who lack self-confidence. Happy employees stay with the job for longer periods of time, which reduces a company's need to perform job searches and train new employees. Happy employees also create a better work environment for the other people in the office. Unhappy colleagues can be inhibitors. Of course, a leader who appears to his or her subordinates as being self-confident serves as a positive role model and inspires the employees to be more confident in themselves.

Confident employees approach complex problems with greater optimism and achieve greater success in solving problems. Therefore, the leader needs to ensure all company policies and practices promote employee confidence. Open lines of communication can be a primary avenue for a leader to assess any problems that may be detracting from employee on-the-job satisfaction. Leaders often conduct annual one-on-one meetings with subordinates and discuss reasons for any job dissatisfaction, which should be a primary point of discussion. A leader should create an individual training program for an employee who lacks self-confidence.

14.10   LEADERSHIP AND STRESS MANAGEMENT

Worker stress is a primary organizational inhibitor. Many workers feel the effects of work-related stress, so one responsibility of a leader is to reduce or eliminate the sources of worker stress. The focus here is not the general stress of the work environment, but instead the stress associated with the innovation and problem-solving responsibilities in the work place and more specifically stress related to critical problem-solving. This responsibility would also apply to a research environment, as most research involves critical thinking for the advancement of knowledge.

Stress, in general, refers to *a mentally disruptive influence or where emphasis is placed on something.* In the case of critical thinking responsibilities, the stress is likely associated with the need to solve complex problems when the person believes that he or she lacks the multi-dimensional knowledge that will be required to solve the problem. A definition for stress may be more easily understood by associating stress with distress. Distress is *the anxiety or worrying about the responsibility to solve a problem.*

To discuss a solution to the problem of stress or distress, it may be instructive to discuss the issue from the standpoint of basic systems theory. This theory identifies three elements: the inputs, the outputs, and the system itself. Stress that is evident in the output can result from the stress-induced inputs or from complications of the system itself, which in this case is the individual. Solving stress-related feelings needs to investigate both possible sources. Is the anxiety felt by the employee due to factors associated with the work environment, i.e., the input, or due to the system, i.e., the self-inhibiting feelings of a problem solver? Potential factors related to the work environment could be that the worker has too heavy a workload and doesn't believe that he or she can devote the time necessary to critical problem-solving. A peer-inhibited environment could also be an "input" cause of the stress. For example, colleagues who tell you that you do not have the necessary experience to solve a problem are considered to be peer inhibitors. Internal feelings, or self-inhibitors, can also be responsible. The internal factor could include a general lack of self-confidence or a lack of knowledge, which could be either technical knowledge or knowledge of problem-solving methods.

One responsibility of a leader is to identify employees who feel distressed because of the work environment, such as stress associated with their responsibilities to assist in solving a complex problem. The cause of the distress must be identified before a solution is possible. Is the stress due to input issues, system issues, or both? The leader needs to develop a solution to the subordinate's problem. If a leader provides optimistic assessments of the employee's efforts, the distress of the subordinate may lessen. Certainly, guidelines on ways to solve problems are a good initial step to relieving distress. Additionally, a more experienced employee can assist on the project; however, this may be interpreted by the employee as a lack of confidence in his or her ability.

It may also be instructive to examine the problem of stress from the standpoint of the steps of the decision process. The need is to identify the reasons for the stress and establish a plan for eliminating the causes of the stress. Using the steps of the decision process could assist in organizing a solution:

1. *Problem statement*: A subordinate can be distressed because of the assigned responsibility to solve a complex problem when he or she believes the solution is beyond his or her capabilities. The stress indicates a lack of courage, which is the first skill in the skill set. Positive feelings about being able to solve complex problems are an indication of positive courage.
2. *Identify resources*: Obtain knowledge of the causes of a subordinate's feeling of distress, which may be for internal or external reasons. Meet with the subordinate to identify his or her feelings as to the state of the work environment, the support provided by the organization, whether or not the colleagues have created an inhibiting environment, or the excessive workload for which the company has made him or her responsible. Only when the reasons for the stress are known can a plan be developed. Performance criteria should be developed and used in analyzing the alternative solutions. These criteria would reflect measures of project-related stress.
3. *Experimental plan*: Identify all possible solutions to each cause of distress and fully analyze the contribution of each one to the overall stress. A decision criterion needs to be developed, with the criterion being used to identify the best alternative for eliminating the stress. The decision criterion will need to properly balance both internal and external sources of stress.

4. *Analysis*: For each alternative solution, identify changes in office polices and office practices that would be needed to alleviate the employee's external stress. If internal stress is a factor, identify changes in attitudes that the employee needs to make. Evaluate each of the alternatives to identify the likelihood that it would successfully alleviate the stress. The decision criterion can be applied to each alternative, from which the best alternative can be identified.

5. *Synthesis*: Transform the potentially best alternative into an actual plan, including some measures aimed at the short term and other elements that will ensure long-term success in stress reduction.

6. *Decision*: Activate the program and periodically assess the successes or failures that result from the program. Make modifications to the stress management plan if short-term stress reduction is not achieved.

Whether worker stress is analyzed by way of the systems model or the decision model, the solution, if effective, will lead to improved production and greater worker satisfaction.

Some employees automatically fear company managers; this fear can create stress and prevent an employee from accepting advice from a company principal. In such cases, a leader should assign a mentor to the young employee. The employee may be more willing to share his or her feelings of stress with a mentor than with a company principal. While the principal is not directly helping to control the employee's stress, the principal is solving the problem through the mentor.

14.11   PROFESSIONAL BENEFITS OF CREATIVE HUMOR

Leaders often have public speaking commitments. Keynote and graduation speeches often begin with a humorous opening. Such beginnings are intended to make the audience believe that the speaker is worth listening to and that those in the audience should stay awake during the presentation. It is preferable for the audience to make a direct connection of the humor to the content of the speech. Humor can also be beneficial when developing material to advertise or promote a new organizational activity. I once saw a poster that was promoting the opening of a new engineering program at a university that did not previously offer engineering. The poster was intended to attract high school students to consider applying to the new engineering program. The poster included a variety of practical reasons for choosing the university over other options. These were interspersed with a series of humorous reasons for choosing that school's program. I imagined that a prospective student who started to read the poster would have continued to read the entire poster and recognize all of the practical benefits of attending that program. The humorous entries promoted the serious entries and acted as encouragement for the person to continue reading; thus, the poster was likely successful in achieving its objective in part because of the humor. Similarly, a good opening to a speech can get the attention of those in the audience.

Humor promotes, maybe requires, glibness. The latter certainly has many professional benefits and often depends partly on humor. When a person has a good sense of humor, most other people recognize it as a positive characteristic. People gravitate to glib people. It certainly is useful in sales, including selling the technical capabilities of a company. During recruitment of new employees, glibness is useful for presenting the merits of being employed by the company. Thus, a humorous personality indirectly enhances a leader's ability to make professional connections.

What is humor? Humor is *anything that is intended to induce amusement*, but the behavior may not have to encourage laughter. Capricious or peculiar behavior can induce amusement, but the behavior may not be something that should encourage laughing. A person should not use humor that is negative in content.

Humor can improve a person's communication responsibilities. Humor is an art that is very dependent on creativity; the creation of humor, such as developing cartoons, requires and enhances the creator's imaginative ability. The creation of humor presents the opportunity to practice divergent-convergent thinking. The development of humor may be spontaneous, but those who create humor for a living believe it is most efficiently and effectively accomplished by following a process. The following series of steps and decision points could be used by someone to develop humorous material for a speaking engagement:

1. Identify the situation/person to be mocked.
2. Decide on an unusual background or setting.
3. Identify an unexpected outcome.
4. Develop the storyline.
5. Select an appropriate deportment for the presentation.
6. Assess whether or not a surprise outcome was created.

If the outcome would be a surprise, then laughter would result and the development of the humor could be declared a success.

14.12   CONCLUDING COMMENTS

Critical problem-solving and innovation are principal constructs of organizational success. A leader is principally responsible for creating a work environment that promotes efficient critical problem-solving. The leader must personally display characteristics and attitudes that enable him of her to serve as a role model. The leader must also institute policies and practices that will clearly demonstrate and promote subordinate adoption of critical problem-solving attitudes. Where the leader believes that the attitudes and practices of an employee are not in the best interests of organizational success, he or she must provide the offending subordinate with the guidance that is necessary to correct the problem.

A principal responsibility of a leader is to improve each subordinate's knowledge of all dimensions of critical thinking; this includes the four dimensions that have been discussed herein as well as the other dimensions such as those for technical, economic, political, and risk dimensions. The leader should establish educational opportunities on the relevant dimensions. All inhibitors should be recognized and controlled to ensure that inhibiting deficiencies are overcome. Employees should be given the time, resources, and opportunity for appropriate educational activities. The leader may need to establish a mentoring program to ensure the appropriate transfer of knowledge and experience from the established employees to the newer employees. Where team work is required, the leader must ensure a harmonious work environment. Inhibiting factors should be corrected, and any causes of conflict should be eliminated. The leader is responsible for stress management and the creation of an optimistic environment to ensure high job satisfaction within the company.

A leader is possibly the most important person on a team or in an organization. The types of roles that a leader plays, of course, depend on the goals of the organization or team. In general, leaders have responsibilities for short-term and long-term planning, the development of the organizational or team structure, and directing the activities of the subordinates. The knowledge development of the organization's personnel may be a primary factor in the success of the organization. Therefore, a leader is responsible for motivating the employees and ensuring that the work is completed on time. In many cases, the most important role of the leader is to act as a mentor, especially to new employees and novices to the technical discipline. To be successful at all of these responsibilities requires a breadth of experience, which may include failures that led to a self-introspective analysis and changes in attitudes and beliefs.

In order for an organization to develop a reputation for being capable of solving complex problems, good leadership is necessary. Leaders need to identify the principal problems within the organization and develop appropriate solutions for each problem. These solutions can be changes to policies within the workplace, educational programs to improve technical knowledge, stress control programs to cope with stress, or reduction of any organizational inhibiting factor. A leader must develop and enhance both positive attitudes and skills of subordinates and help them overcome weaknesses such as procrastination and any lack of self-confidence.

14.13 EXERCISES

14.1. Define leadership. In terms of problem-solving, what roles must effective leaders play?

14.2. Assume that you are a leader of a large company that produces farm machinery. Using the eight ways of promoting creativity identified in Section 14.2, discuss the actions of the leader to promote creativity.

14.3. Make a list of peer-induced inhibitors to creativity. For each one, indicate the actions that a leader can take to prevent the inhibitors from adversely influencing productivity of a company.

14.4. Make a list of self-induced inhibitors to creativity. For each one, indicate the actions that a leader can take to prevent the inhibitors from adversely influencing productivity of a company.

14.5. Why is routine work an inhibitor to creative efforts by employees?

14.6. What are the elements of an organization's mindset? How does a leader set the organization's mindset? What factors would cause change in an organization's mindset?

14.7.   Six benefits of optimism were identified. Discuss the ways that each relates to the efficiency of an organization.

14.8.   Identify and explain the ways that optimism increases the likelihood of successful innovations.

14.9.   How could a person who is applying for a job identify the likelihood that he or she will be assigned work that would represent significant advancements in knowledge?

14.10. Add two or more major items to the checklist of Table 14.1, with two or three subtopics included for each of the major topics.

14.11. What are the organizational inhibitors to the self-confidence of employees?

14.12. Why might work on complex projects cause workplace stress?

14.13. Find the definition of an inhibitor (see Chapter 7). Why could stress be considered as an inhibitor? How can a leader minimize the effects of stress?

14.14. What should a subordinate do if he or she believes that the work assignments are based on biased decision-making by the leader?

14.15. Develop a rating system that could be used to rate the moodiness of an employee.

14.16. How are the steps of the process of humor similar to and different from the processes in Chapter 2?

14.14   ACTIVITIES

14.14.1   ACTIVITY 14A: TEACH ME!

Create a mentoring program for an organization, with the program centered about overcoming primary inhibitors to organizational success.

14.14.2   ACTIVITY 14B: YOU HAVE THE EFFICIENCY OF A LEVER

Create a model that could be used to quantify or rate the efficiency of an organization with respect to some characteristic related to leadership.

14.14.3   ACTIVITY 14C: YOU ARE ALWAYS DOWN!

Create an educational program to enhance the mindset of pessimistic employees.

14.14.4   ACTIVITY 14D: THE MADDEN OF LEADERSHIP

Outline the creation of a videogame on paper in which the objective is to enhance a person's skill at leadership.

14.14.5   ACTIVITY 14E: A SOLDIER HAS COURAGE

Develop a process that a leader could use to improve a subordinate's courage to initiate project responsibilities.

14.14.6   Activity 14F: Wanted Dead or Alive: Reward

Develop a reward system that a leader could use to encourage a procrastinator to initiate reporting of responsibilities.

14.14.7   Activity 14G: The Sheriff and His Posse

Assume that you are the head of a small company, with responsibility for all leadership decisions. Identify and briefly discuss the primary responsibilities that you face.

15 Critical Thinking
Applications to Research

CHAPTER GOAL

To show the ways that critical thinking is important in every step of the research process.

CHAPTER OBJECTIVES

1. To discuss methods of critical analysis and synthesis as they apply to the identification of a research topic.
2. To show ways that weaknesses in existing research can be used to develop new novel research ideas.
3. To present guidelines for research planning.

15.1 INTRODUCTION

Research brings about change. It has been responsible for changes from the seventeenth-century bloodletting to the twentieth-century laser surgery. Research has replaced Galileo's telescope with the Hubble and Webb telescopes. It has changed oar-powered shipping and eighteenth-century naval warfare frigates to nuclear powered aircraft carriers. Research continues to change warfare, as we only need to compare the medieval catapult to the use of drones in the twenty-first century. Would Napoleon have reached Moscow if research had provided him with HUMVEEs? The lack of advancements due to a lack of research has consequences. Many advancements based on research have improved everyday life, and it seems that the rate of change continues to increase. As many have said, "Nothing is permanent but change."

Research can be defined as (1) a *scholarly or scientific investigation*, (2) *an inquiry*, or (3) *a thorough study*. Research is problem-solving. It is the process used to advance a state of knowledge. Research is rewarding. It leads to positive changes, but also some negative changes. Good research is not easy because it requires the completion of challenging experiences and mastering the many dimensions of critical problem-solving. Combining the methods of critical thinking with diverse research experiences provides depth to a person's ability to advance the state of knowledge. Knowing the methods of critical problem-solving and the steps of the research process should lead to higher quality research output and greater happiness with both the effort and the end product.

DOI: 10.1201/9781003380443-15

Critical and creative thinking are tools that can be used to facilitate positive research efforts. Many experiences of successful research involve critical thinking, which goes beyond one-dimensional logical thinking. As a multi-dimensional method of problem-solving, critical thinking applies skills and controls the researcher's mindset to add breadth of knowledge that is needed when confronted by a complex problem. Planning and trained thinking can greatly improve the output quality and the efficiency of any research effort. Being able to apply a variety of methods of thinking represents a breadth of knowledge that can improve the quality of research output. Critical thinking encourages an optimistic mindset to ensure confidence. A multi-dimensional approach to research should produce positive advances to the state of knowledge of any discipline.

People who have had experiences with critical problem-solving readily agree with the premise that the ability to think critically is very worthwhile. Since the central concern in this chapter is research, the following questions about research and critical problem-solving are of special interest:

1. How can knowledge of the dimensions of critical thinking improve research quality and productivity?
2. How does a person identify a weakness in the current state of knowledge?
3. What are the steps of the process that will lead to the most effective solution to a research topic?

Given these questions, it is important to show that critical thinking can be used to generate novel ideas for research and ultimately yield more useful research products through application of the research process. Showing ways that critical thinking can increase research efficiency will support the importance of gaining knowledge about becoming a critical thinker.

15.2 RESEARCH PLANNING

Planned research will yield better results than research conducted haphazardly. Efficiency will also be greater. When research is properly conducted, it has many benefits; however, poorly planned research may not result in positive experiences. What elements of planning will yield the best results?

15.2.1 CAUSES OF POOR PLANNING

Poorly planned research has many causes, including inferior experimental designs, inadequate data, the incorrect selection of research methods, poor decision criteria, and poor supervision of those new to research. Good research cannot result when the problem statement does not truly reflect the weakness in knowledge. Some students who have been involved in research found it to be a bad experience just because they had not been properly introduced to the full research process. Education is part of research planning. Very often, just presenting knowledge of basic research methods could have prevented a feeling of failure. Failures of a novice's initial effort have resulted because the mentor did not provide the novice with adequate education about the process of research; the novice was started at Step 4 of the research process rather than Step 1. The novice was assigned the analysis of data without having knowledge of the problem, the research objective, the origin of the database, the basis for the experimental design, or the benefits in the advanced knowledge that would result from the research. This approach to mentoring can lead to frustration. By not properly introducing the mentee to the overall scope of the work, the mentor is failing to properly advise the novice for a career in research. The start of any research effort should begin with Step 1. Initiating the research at any step other than Step 1 can lead to a poor research experience.

15.2.2 PHASES OF RESEARCH

We can start by viewing research as a two-phase analysis. Phase I is the process of identifying a problem or topic while Phase II is the development of a solution to the problem using the six steps of the research process (see Chapter 2). If a person knows the research topic that needs to be investigated, then the research effort starts with Phase II. If the topic of the research is not known, then a topic must be identified before starting the research process. The material in Section 15.3 or an activity such as introspective idea innovation (see Chapter 17) can be applied to identify a researchable topic.

For someone who needs a researchable topic, a broad area of research interest must first be identified. The objective of Phase I will be to provide a topic that can be studied using the research process. Then literature relative to the topic is collected. The material is reviewed and the state of knowledge is established. Weaknesses in the current state of knowledge are identified; weaknesses could be a poorly defined objective, an inadequate experimental design, or constraints or limitations made by the authors of the research. Limitations to the data that were used to establish the current state of knowledge can also be a prime source of ideas for new research to advance the state of knowledge. The outcome of the Phase I search is a researchable topic. Planning is the key to success in research!

15.3   A MODEL OF THE RESEARCH PROCESS

To what do we associate the word research? Medical research would be one common answer. Some people may use the word research to describe the effort that they put forth to prepare for placing bets on weekend football games. Obviously, research on medical issues and research on football games are quite different examples of research. Yet surprisingly, these two examples of the concept of research share many of the same general activities. The same steps that are used in medical research can be applied to research being conducted on the improvement of the safety of infrastructure and in solving environmental problems, such as climate change or improving the quality of water in small ponds. The diversity of these topics for research suggests that it may be worthwhile to have a general model of the research process and adapt the steps to the specific problem.

The need for researchers has grown considerably as the problems of society have become more complex, but also because research leads to advances that can have considerable economic implications. Research is not limited to the sciences. It is important in a variety of fields from criminal justice to economics, from music to education, and from engineering to political science. Certainly, the benefits of medical research are evident to all of society. Most graduate degrees have a research requirement. In spite of the differences in the outcomes of research from one discipline to another, the same procedure is followed. Those involved in research want it to be completed as efficiently as possible; research is less troublesome when a systematic solution procedure is followed.

The quality of research can vary considerably, with the quality being measured in part by the worth of the outcomes and depend on the excellence of both the ideas and the effort. The quality of the research will be influenced by the thoroughness of the investigation, the conscientiousness of the researcher, and the research leader's strength of commitment. Both the foresight of the investigator and his or her knowledge will influence the quality of the result and the efficiency of the effort. The quality of the results will be enhanced through the practice of good planning, the self-confidence of those involved, and having an optimistic mindset.

15.3.1   THE STEPS OF THE RESEARCH PROCESS

Many versions of the research process have been proposed. A six-step model of the research process is given in Table 15.1. In some ways, Step 1 is the most critical, as a poor problem statement cannot lead to a worthwhile outcome. Step 2 involves the establishment of the current state of knowledge on the subject, including a review of the literature that summarizes all past research on the subject; the existing database is part of the resources and represents important inputs to successful research efforts. In Step 3, an experimental plan is developed for conducting research. For each research objective of Step 1, the research plan should formulate a research hypothesis and identify both the methods of analysis and the decision criteria that will be used in decision-making. In Step 4, the necessary experimental data are collected and analyzed for the purpose of advancing knowledge that is related to the problem. Step 4 efforts lead to a set of results from each of the experimental analyses. Following the analysis phase, Step 5 provides the synthesis of potential conclusions from the results of Step 4. Step 6 involves the application of decision criteria, which are used to select the most feasible outcome. Identifying the broad implications of the conclusions is an important part of Step 6, as the implications identify the value

---

TABLE 15.1

Steps of the Research Process

1. Problem identification
2. Resource collection (knowledge and data)
3. Research plan development (experimental plan)
4. Analysis of results
5. Synthesis of conclusions
6. Make a decision and file a report

---

of the research effort. Any feedback loop will provide the option for returning to a previous step and modifying the analyses. When a decision does not seem feasible or optimum, a feedback loop should be used to consider alternative actions. For most research efforts, feedback can be initiated at any of the steps.

15.3.2   Step 1: Problem Identification

Problem identification is arguably the most important step of the research process. A good solution cannot result from a poor problem statement. Difficulties encountered during the other five steps of the research process will be minimized if the problem statement clearly indicates the new knowledge that is needed. Solving research problems can be central to achieving the goal of the research. It is important to clearly and unambiguously understand the central issue of the problem for which a solution is needed. Having a critical thinking perspective on the effort can be very helpful to the completion of this step. First, courage to confidently investigate the issue is necessary. Other skills such as curiosity and skepticism will enhance the likelihood of searching beyond the obvious elements of a problem, thus increasing the likelihood of uncovering the unique nature of the problem. Feedback in the form of readjustment of a research plan may be needed later in the research program if the full scope of the problem was not clearly understood during this first step of the research.

15.3.3   Step 2: Resource Collection

The success of research often depends on the quantity and quality of resources, which can include both the available knowledge and measured data. If the existing knowledge or data are inadequate, then gathering the missing resources will be a central part of the research effort. The existing knowledge base will be weak if it does not accurately consider the variables and processes of the actual system. All relevant variables need to be identified and accurately measured. A researcher's creative ability can influence the extent to which important knowledge can be extracted from the measured data. Outliers in the measured data and a specific reason for the occurrence of each one need to be documented. Just reproducing the research activities of past work reported on a topic but with new data does not usually lead to novel outcomes or worthwhile results. Innovative analyses of data or restructuring of existing knowledge are the research efforts that reflect novelty and lead to recognition. Where the database seems to be inadequate, which is probably close to the norm, creative ability can be valuable for successfully mining the knowledge inherent to the available data and combining it with relevant knowledge.

15.3.4   Step 3: Experimental Design

Novel research depends on having a comprehensive research plan, which requires good thinking skills, i.e., the ability to critically and creatively analyze problems. A worthwhile approach to developing a plan is to imagine the ideal outcome of the research effort and then work backward through a reasonable pathway to the initial phases of the research. Connecting research planning to the outcomes of the research will generally produce an efficient and effective research plan and increase the likelihood of achieving worthwhile outcomes, which implies advancing the state of knowledge. The practice of planning the next step only after completing a step represents inefficient research and results in outcomes that lack novelty. A good research plan should identify the performance criteria and the decision criterion that will be used to evaluate alternative outcomes in the making of decisions. All of the alternative courses of action that would be taken when the projected best alternative does not prove to be feasible must also be detailed.

In Step 1, a general statement of the research goal was presented, along with a series of specific objectives. If the research succeeds in producing worthwhile results for each objective, then the goal will be met. Therefore, the central focus of Step 3 is to develop a research plan for each of the objectives of Step 1. This involves establishing a research hypothesis, gathering any necessary input data, verifying the hypothesis, and providing the information needed for the analyses of Step 4.

15.3.5   Step 4: Analysis of Results

Step 4 centers on the assembling of knowledge from Step 1 into a useful format and performing analyses of the experimental data. The word *analysis* implies *to break apart*. In terms of the fourth step of the research process, the definition could be applied to identifying the most valuable knowledge that could be extracted from the data. It could refer to the analyses of the existing knowledge base for the development of a model that will be used to represent a system. Given that measured data cannot be perfectly accurate or cover all range of conditions of the system, analyses should include the assessment of the uncertainty of measurements and the ultimate effects of the uncertainty on the research results. Step 4 is an important part of the research process; analyses have a significant influence on the quality of the ultimate decision.

15.3.6   Step 5: Development of Conclusions

Step 5 centers on *synthesis*, with the word meaning *to put together.* With respect to research, the word *synthesis* would mean the establishment of a set of general conclusions using the results of Step 4. Synthesis could also refer to the identification of an additional set of conclusions based on simulations with a model. The knowledge that is obtained from experimentation and data analysis forms the basis for synthesizing conclusions, which are assessed in Step 6 using pre-specified decision criteria.

15.3.7   Step 6: Make a Decision

As suggested, the development of criteria for assessing the quality of research output is difficult. Important assessment criteria that can be quantified are difficult to develop, but they may have a significant influence on the results. The number of citations of the papers generated from the research is a commonly used criterion, but this criterion is regularly criticized as not reflecting research quality, only research quantity. Ideally, a metric should reflect the extent to which research advances the state of knowledge. Unfortunately, it would be difficult to quantify, and it would likely be decades before the true value of the research output could be accurately assessed. The breadth of the implications of research results is important, but again, an assessment criterion of breadth would be nearly impossible to quantify; it would also suffer from the problem of the time frame needed for assessment. In summary, good indicators are difficult to identify and assess in the short term, and easily quantified criteria are likely not accurate indicators. Research assessment criteria are very uncertain.

15.3.8 RESEARCH EFFICIENCY

Research is a problem-solving activity where efficiency is very important; however, it is difficult to measure the intellectual effort and resources that research requires. Thus, it would be difficult to evaluate the two variables in Eq. 2.2. The quantification of the short-term and long-term benefits of research is an impossible task. Thus, objective quantification of research efficiency would be difficult, so its assessment is often quite subjective and estimates are subject to considerable variation. However, when a student is searching for a job, prospective employers certainly feel able to assess the quality and quantity of the applicant's research. Also when research results are submitted to professional journals for possible publication, the reviewers focus on the novelty of the work, which is their way of assessing the creative effort that was involved in the research. Thus, those who conduct research should at least appreciate the concept of research efficiency. A sensitivity to the principle of efficiency will have many benefits to the researcher.

In spite of the lack of objective assessment criteria for the measuring efficiency of Eq. 2.2, students must recognize that any improvement in their problem-solving skills will translate into improvement in research efficiency, even though accurate measurements may be nearly impossible to make. Little doubt exists that research efficiency has a positive influence on the quality of the results. Critical thinking skills are a central factor in all phases of the research process; therefore, developing a critical attitude and learning to apply methods of idea innovation to develop a worthwhile topic will generally lead to novel research outcomes and greater research efficiency.

The mathematical basis of Eq. 2.2 emphasizes the quantity of the research output, but research efficiency should also be judged on the quality of the output. Some may believe that research quality is inversely proportional to quantity, and in some cases that may be true. However, a researcher who is proficient in idea generation and innovation may prove to be more efficient in terms of both the quantity and quality of the outcome. Objective metrics for both the quality and quantity of research could be developed and used in assessing one's efficiency as a researcher; however, the robustness of any metric will likely be uncertain and legitimately debated.

15.4 METHODS OF IDENTIFYING NEW RESEARCH TOPICS

To identify new research topics, critical analysis skills can be applied to each section of a published paper. The introduction and conclusion sections of a journal paper are often potential sources of new innovative research ideas. When reading a journal article, the conclusions section of the paper should be reviewed in detail, as authors often identify needed extensions to their analyses and results. For example, the authors might specify constraints that were imposed on the development of their ideas, or state limitations due to the data that were available to them. Such extensions and constraints could be relaxed to generate new research, as the relaxation of constraints can broaden the applicability of the conclusions.

The introduction of a paper should also be critically analyzed because this section often includes the problem statement that motivated the authors. The problem as stated in the introduction could be viewed as being too narrow, which then suggests that the broadening of the problem statement could lead to new research ideas that would provide significant advancements. While the authors may have adopted an experimental design that focused on one possible path to solve the problem, other experimental designs could lead to productive research ideas with quite different results and conclusions. Curiosity and critiquing are primary skills for identifying new research ideas. The ability to think critically and creatively is central to identifying these paths for generating novel research ideas. While critical analyses focus on the strengths and weaknesses of published research, it is usually the weaknesses that lead to the best advancements of knowledge.

When faced with a requirement to conduct research, it is important to have a plan. It is best to select a general topic of real interest to you; however, you can choose any area of general interest, such as climate change, plastic trash in the oceans, debris in outer space, or homelessness. The ideas presented below attempt to summarize ways that could be used to identify a specific topic and set up an experimental plan to address a very general problem area.

During the time when a perusal of professional journals or commercially available articles is being undertaken to establish the state of knowledge, the time should also be used to generate ideas that can potentially extend the reported research to make future advances in knowledge. When reading journal articles, the following critical assessments based on the following ideas can foster new research paths:

1. To develop a new model for use to solve a different problem by integrating two existing but independent models.
2. To diversify the use of a model by identifying new model components that are not currently included in an existing model.
3. To make an existing model a function of new variables, which would be introduced into the model components.
4. To improve the accuracy of model predictions by recalibrating the empirical coefficients using a more sophisticated method of calibration.
5. To make an existing model applicable to other data conditions by calibrating the model with data for that condition.
6. To assess, recalibrate, and reduce the uncertainty of the empirical coefficients by significantly increasing the quantity and breadth of data.
7. To assess the effects of constraining assumptions of an existing model structure such as by replacing a linear model with a nonlinear form that might yield better accuracy and/or be more theoretically justified than the existing linear relation.

8. To assess the effect of relaxing limitations of an existing model by elimi-nating constraints that were specified by the authors, such as the range of model applicability.

9. To create a dynamic model from a temporally static model, i.e., introduce time as a factor.

10. To address spatial variations of model components by introducing spatial dependency into a model that was not formulated for spatial variations.

11. To transfer statistical methods or modeling methods that are common to one specialty to the specialty of interest to you.

12. To adopt a classification scheme or policy used in one specialty for use in another specialty.

13. To evaluate the relative importance and uncertainty of model predictions through sensitivity analyses.

14. Question the question! Did the investigators of the existing literature really recognize the real problem? If not, clarify the problem and redesign the experimental plan.

15. Read the conclusions section of a paper and ask whether or not the results could be more useful if the statement of the research objectives has been broader. Start with the conclusion and work back to the objectives and mod-ify the objectives, so that the conclusions yield advances in knowledge of the subject.

When reading the research results of others, each of these 15 practices and many others should be an ingrained way of thinking. A critical thinker automatically adopts this critique-and-question mindset when reviewing reported research results. These methods of searching represent the practice of critical analysis, as they break apart existing knowledge in search of a need for advancement. Critical analysis leads to critical synthesis, which is the process of acting on the weak-nesses and developing experimental designs that will advance the state of knowl-edge. Poor critical analysis skills will limit one's potential for success in the search for a topic.

Guidelines, such as the 15 ideas previously presented, can be used to practice identifying weaknesses of existing research. Weaknesses of any research can be identified throughout published reports, including the statement of the research objectives, the statement of an experimental design, the interpretation of the results, and the conclusions drawn from the results. Some experience in the technical area is usually necessary to be proficient at critical analysis, but any attempt at critical analysis should instill the courage to make attempts at discovery and to help develop confidence in the ability to generate new ideas.

The extent of success with idea development depends on the person's innate char-acteristics and experience with methods of idea innovation. A person's efficiency at developing new research ideas can be improved by having working knowledge of the previously discussed weakness detection guidelines; experience with critical think-ing methods for transforming weaknesses into productive research solutions is also important and influences research efficiency.

15.5 CRITICAL ANALYSIS IN RESEARCH

Let's revisit the concept of critical problem-solving, but now with the specific reference to conducting research. After all, research is one example of problem-solving.

15.5.1 STATE OF KNOWLEDGE

A primary element of research is establishing the current state of knowledge of a
topic of interest often for the purpose of conducting research to make significant
advancements in the state of knowledge. Identifying the current state of knowledge
is usually accomplished by conducting a literature review. Obviously, research efficiency is important and can be measured both by the time required to do the literature review and by the amount of useful knowledge gained from the review. Applying
a critical mindset will enhance one's ability to complete both of these aspects of
reviewing the literature and, therefore, maximize efficiency. Having the proper state
of mind when doing such analyses will also increase the likelihood of success. The
literature review enables both the strengths and weaknesses of the literature to be
recognized. Critical thinking can be used to identify the weaknesses that exist, so
that the state of knowledge can be advanced.

In addition to the technical knowledge dimension, all dimensions of critical
problem-solving are important to the establishment of the state of knowledge. The
values most relevant to a critical analysis of the literature are efficiency, fairness,
unbiasedness, and knowledge. When conducting novel research, questioning and
curiosity are essential skills. Confidence, optimism, and diligence are important
elements of a mindset for conducting research. Multiple thinking types will be
necessary to match the thinking type with the nature of the reported knowledge.
For example, upward-downward thinking is very useful for research that involves
trends. Stochastic thinking is very useful for cases that involve a high level of uncertainty. Using all of the dimensions of critical problem-solving should provide the
most efficient plan, especially when compared with a unidimensional plan. If all
dimensions are efficiently applied, the potential knowledge abstracted from the literature will be maximized.

15.5.2 BENEFITS OF THINKING TYPE AND SKILL SET

If research efforts are to be successful, the critical thinker needs to be capable of
applying several types of thinking. He or she will need to recognize that solutions
to different types of problems are best solved using the most appropriate type of
thinking. Every research problem cannot be solved with just one type of thinking.
In the term critical thinking, the word *thinking* is a special type of activity where
potential solutions depend on the strengths and/or weaknesses of an issue and are
based on experiences and beliefs that have proven to be successful. Additionally, the
critical thinker recognizes the potential effects of variations in frame of mind and
has learned to control variations. Developing a critical thinking attitude is a central
element of intellectual growth. Table 15.2 indicates the potential value of the different types of thinking to the research process.

TABLE 15.2

Research Uses and Value of Thinking Types

Thinking Type	Use/Value in Research
Abstract	Researchers want novel work; abstract thinking suggests wonderment, possibly consider characteristics of abstract art as a source of novelty in the research.
Additive-subtractive	Think about possible research outcomes that could not be expected to result and generate ideas on ways that they could be made to occur; similarly, subtract negative concerns to improve flexibility.
Analogical	Think about a related but distinct type of problem or activity, and use logical thinking to generate ideas for your topic; relates to the synectics method.
Cognitive	Generate ideas using logical thinking, but make the assumption that data that does not currently exist but would enhance the analysis if the data were available.
Computational	Think about other topics that are correlated with the topic of interest and focus on them.
Counterfactual	Imagine results that could not be obtained and use these to generate ideas that have potential to provide feasible solutions.
Creative	Identify imaginative or fantasy ideas that can be transformed into novel research ideas.
Integrative	Use both logical and creative ideas to generate novel ideas.
Lateral	Look for similar real-world situations and transfer the ideas to your area of interest.
Logical	Avoid using methods based on imaginative thinking; focus only on ideas that arise directly from facts, experiences, and existing knowledge.
Stochastic	Access words from a dictionary or from your immediate surroundings and transfer them to the topic of interest.
Upward-downward	Think about an end product that could not occur and show how changes could be made in order to make it occur; this is especially useful for analyses that involve increasing or decreasing trends.
Visual	View something in your local environment and transform it into a solution.

The usefulness of critical thinking skills that are central to research should be evident. The skill set of Chapter 13 should be applied. Skills, such as curiosity, questioning, and critiquing, are central to critical problem-solving and important in analyzing the professional literature. When conducting a literature review, a critical mindset will enable the reviewer to identify both the fundamental elements of the relevant literature and the aspects of the existing knowledge base that can be improved.

15.5.3 KNOWLEDGE OF CRITICAL THINKING

Recall that analysis means *to break apart*. With respect to the literature review, critical analysis would refer to the identification of knowledge inadequacies in the reported results. Each of the ideas that reflect an inadequacy can be used to generate a problem statement for novel research. Completing the research will subsequently advance the state of knowledge, which can be published as new findings. For a critical thinker to achieve these results, he or she must have the requisite qualifications, which means having a good understanding of the dimensions of critical thinking. In critical analysis, the knowledge and resources parts must first be analyzed to understand the real need. The value dimension is also relevant to critical analysis. A critical thinker must have a strong, unbiased knowledge of both human values and the subject matter of the research, which can be referred to as the fifth dimension. Note that technical knowledge is not the only requirement. Values are important even when identifying a topic from the existing literature. Without a critical attitude, a person who reviews journal articles might be overwhelmed by the depth of the technical content. Simultaneously, they will likely lack either the confidence or courage to identify ways that the published material could be improved. An attitude of skepticism will promote a belief that the material is not a finished product. Curiosity will enable the development of specific ways to improve the published work. In summary, knowledge of the idea innovation methods of critical thinking is essential for critical analysis and subsequent critical synthesis to complete research that is based on identified weaknesses.

When analyzing completed research work, decisions are made about the quality of the research. Some analyses that have been published are accurate and of good quality, while others are of minimal value. Good research ideas can be obtained from either good or bad research. It takes a number of skills to recognize the deficiencies that are most likely to lead to the development of new knowledge and the advancement of the state of knowledge. Talent for and experience with critical problem-solving can be valuable in identifying the weaknesses of others that are more likely to lead to high-quality ideas for new research. Application of the skills model (see Chapter 10), which extends skills from courage to consequence offers the greatest likelihood of identifying new research ideas. The systematic application of important skills is also the most efficient means of achieving this objective.

15.5.4 QUESTIONING IN THE RESEARCH PROCESS

Questioning, which is a principal skill, often results from curiosity and critiquing. When reviewing published material, the person's mindset must be sensitive to the critiquing-questioning skills. As sentences and paragraphs are read, it is important to be mentally asking questions of the material. To provide some organization to any review and to identify where weaknesses in the state of knowledge exist, it can be helpful to relate questions to the steps of the research process, which was outlined in Chapter 2. Table 15.3, which is separated by the six steps of the research process, provides a few generally relevant questions that could lead to ideas for new research. These are just a few of the questions that could be considered. The important point is that questioning every aspect of a paper can lead to new ideas.

TABLE 15.3

Questions to Assess the Quality of the Research

Step 1: Problem identification

- Does the statement of the goal and objectives directly follow from the stated problem?
- Is it even possible to provide a solution to the problem, at least with some level of assurance?
- Would it be necessary to answer other questions before an answer to the stated problem should be investigated?
- Is the problem statement too narrow in scope?

Step 2: Resource collection

- Will the pre-existing database when combined with any additional data generated for the proposed research be sufficient to provide reasonably accurate results?
- What constraints, restrictions, and assumptions did the authors make?
- Is the current state of knowledge sufficient to approach solving the stated problem?
- Were important variables or processes not included in the model?

Step 3: Research plan development

- Will the stated experimental designs be adequate to assess the research objectives?
- What limitations are placed on the problem-solving because of data restrictions?
- What performance and decision criteria should be used?

Step 4: Analysis

- Are the results of the analyses strong enough to continue?
- Should other statistical methods have been used, such as those that are more statistically powerful?
- What graphical or tabular formats will be helpful in summarizing the results?

Step 5: Synthesis

- Do the conclusions really answer the stated problem?
- Do the conclusions seem conclusive or do they only partially solve the stated problem?
- Did the experimental analyses show sufficient validity to justify conclusions that can be supported?

Step 6. Decision

- Do the implications of the results and conclusions really help to advance the state of knowledge?
- Do uncertainties exist in the analyses that cause the conclusions to be suspect?
- Do the conclusions and implications suggest that additional research is needed?

15.6 CHARACTERISTICS RELEVANT TO CRITICAL THINKING

Several of the characteristics of critical thinkers (see Section 9.6) are relevant to the task of generating new research ideas. First, everyone in research should have a positive attitude. In spite of a pessimist's ingrained critical attitude, optimists will be more successful than will be pessimists. Negative attitudes need to be controlled. This perspective reflects the mindset dimension of critical thinking. Second, procrastination is a primary inhibitor to progress and would obviously be counterproductive to research accomplishment. Third, when compared with unidimensional logical thinking, the use of multiple thinking types (see Section 13.2) provides greater problem-solving flexibility and increases both productivity and research efficiency. The most appropriate thinking types will vary with the steps of the research process and the type of problem. Fourth, skills such as critiquing (see Section 10.7), questioning (see Section 10.8), and curiosity (see Section 10.9) are very important analytical skills that are necessary to advance knowledge. These skills enable complex problems to be solved after properly evaluating alternative decisions. Fifth, external pressures from sources such as the organization or peers can act as an inhibitors and decrease research productivity (see Chapter 7); inhibiting pressures should be recognized and avoided.

15.7   A MINDSET FOR CONDUCTING RESEARCH

Being successful in research requires a combination of skills, attitudes, and experiences; however, knowing these characteristics but not applying them with the proper frame of mind will render the skills less helpful. A person's mindset needs to be properly controlled. A person's mindset is a collection of interrelated moods, dispositions, principles, and attitudes that govern actions at each phase of the research. A person's state of mind can significantly influence the degree of success in research. This mindset would consist of a comprehensive group of characteristics (see Section 12.3) that are important to a critical thinker. However, some elements of a person's mindset are subject to change over short periods of time with the length of time dependent on the person's nature. Such fluctuations can be detrimental to the progress and efficiency of the research. Only when a person recognizes the detrimental effects of swings in the mindset can he or she exert mindset control at those times when control is necessary to ensure the best decisions are made.

With respect to research, the term critical attitude could be defined as *a state of mind that values and emphasizes careful, maybe severe, assessment usually based on a reasonable level of knowledge and experience but influenced by the person's mental state.* However, a person's mindset is not a permanent pattern of thinking and successful research depends on the researcher's ability to adapt his or her mindset to the mental need at a particular time. Conversely, since thinking is a process from idea conception to judgment, a person's general pattern of thinking is likely a relatively stable characteristic over time. Additionally, a person who is involved in research should have a mindset that recognizes the benefit of using various types of thinking to solve problems. The broad spectrum of methods of imaginative thinking is especially important in research and, therefore, it is usually an important element of critical problem-solving, as creative thinking stresses broad-based idea generation. The critical attitude-thinking mindset is a primary basis for the production of novel research.

The following point needs to be emphasized: the innovative development of research ideas depends on the investigator having a critical mindset that becomes ingrained in a person's character and that the person has learned to control. A mindset is more than an array of disconnected attitudes. It involves the integrated connection with other dimensions that are related to values, skills, and thinking types, as well as others. A critical thinker generally has good control of his or her own moods, thus minimizing any negativity that could result from uncontrolled changes in mood. If a person is under undue stress to produce new research, the critical mindset will not materialize to the extent needed. Stress can serve as a creative block or inhibitor, introduce a constraining feeling of nervousness, and prevent critical thinking from taking place. Stress, which can be internally motivated or from external inhibitors, is a primary inhibitor to both critical and creative thinking. Having prior multiple experiences with critical idea development generally provides the essential experience that is necessary to develop the ingrained mindset useful for solving problems. A person should not wait until the critical mindset is needed to be able to apply it; a belief in the need for prior experience with the mindset and its control is necessary and should serve as motivation to gain that experience.

The mood of a person can be defined *as a temporary state of mind or feeling*. A person's mood can influence his or her actions and decisions in either positive or negative ways. Moods can be unproductive or worthwhile. Moods conducive to research may not be overtly evident; instead, it is the resulting actions and the ability to control the actions that are the evidence of the mood. The word temporary is an important part of the definition, as changes in behavior and thinking will vary with time, but can be altered when the individual recognizes the effect of the mood in terms of efficiency and output quality.

15.8 ASSESSMENT USING THE INTELLECTUAL DEVELOPMENT MODEL

The Intellectual Development (ID) model can be used as a metric that indicates a person's level of knowledge on a particular dimension of critical problem-solving. Just as the ID model can be applied to the dimensions of critical thinking, it can be applied to the conduct of research to indicate a person's knowledge and application of the ability to conduct research.

The ability of a researcher to think is important primarily because it will likely be reflected in the quality of the research. That is, research output is a reflection of input, and the thinking level is an important input. Those who function at the higher levels of ID thinking are more likely to recognize the broader connections of their work and its limitations. They are more likely to recognize the potential implications of the results of the research, so the research plan should be formulated so that the research will have broader applications. The research objectives should reflect a greater understanding of the state of knowledge and a better understanding of the best direction for conducting the research.

The solution of complex problems will require excessive effort as addressing the concerns of a wide array of stakeholders is usually necessary when solving complex problems. A person who only functions at a low level of ID thinking about research may not recognize the broad array of stakeholders that can be affected by his or her research. Thus, a low-ID researcher may not recognize the relevance of issues like public safety, environmental quality, and aesthetics, which makes the research appear less valuable. Critical thinkers are more likely to address such concerns as they may have developed a checklist (see Chapter 6) that reminds them of the need to discuss the implications of their research. Such thinking indicates a more mature level of research ability. Critical thinking also influences the efficiency of the research efforts, and the ability to apply creative problem-solving methods occurs only at the higher levels of ID thinking.

A person's skill set, as discussed in Chapter 10, can advance to higher levels of thinking using the principles of the Intellectual Development model (see Chapter 9). According to the ID taxonomy, questioning is a skill that can be used to rise to a higher level of thinking. An improper question or one that does not reflect the potential for greater knowledge would indicate that the asker of the question has not achieved thinking levels at the next higher ID level. Of course, the question posed should address thinking at the next higher level in order to support achievement of the rating that reflects the next higher level.

15.9   VALUES IN RESEARCH

Good research depends, in part, on good values. A value can be defined as *a principle that is desired or is an important quality*. Some values such as honesty and diligence have obvious connections to research, but many others are important, such as care, accountability, and promptness. Volumes have been published on both unethical behavior in research and the effects of the failure to maintain high ethical standards. Ethical failures have been the reason for many failed careers in research.

Values closely related to honesty include care, unbiasedness, respect, integrity, and accountability. Honesty is *the condition of being trustworthy* or *the capacity to act truthfully without deception or fraud*. In many cases of unethical behavior in research, honesty was the value about which the conflict was centered. Cases involving plagiarism, data falsification and fabrication, and conflict of interest violations are not uncommon. Redirection of research is one consequence of unethical research conduct. Reporting research results that are not truthful could cause other researchers to re-direct their research away from the beneficial production of knowledge to pursue the direction based on the unethical behavior. This redirection would represent a waste of resources and inefficiency of research in the broad area of interest. Those who are starting to get involved in research should read a relevant code of ethics to learn the conduct expected of them. Dishonesty in research can cloud the reputations of others who are currently conducting research in the same discipline.

The following are examples of ethical problems:

- Plagiarism is a common ethical concern, as guidelines on the exact nature of plagiarism are not clear. Plagiarism also includes the use of ideas of others without providing a proper citation. It is often a problem when individuals are required to submit their work in a language that is different from their native tongue. Plagiarism also occurs when wording in the reported material is too much like the wording in the source material. Paraphrasing with a citation is the way to avoid a charge of plagiarism. Plagiarism violates values such as honesty, accountability, and respect.
- Data fabrication is another example of an unethical research practice. In such cases, the researcher invents values for the research data instead of actually collecting data. Actual measurements for the numerical values were never made, so the researcher just created numerical values that were approximately the expected values. This practice violates values such as truth and fairness.

- Dual submission refers to the unethical practice of submitting supposedly original material to two publication outlets without notifying either publisher that the two papers are essentially the same. Such acts violate values such as fairness and accountability.
- Ghost authorship occurs when the name of a person is listed as an author of research results when, in fact, the person was not actually active and directly involved in the development of the ideas that underlie the research, the execution of the research, and finally the detailed writing of the results. The practice violates values such as honesty, credibility, accountability, and pride.
- Date falsification is the practice of changing the values of measured data, when the values do not conform to expectations or agree with the values of other measurements collected. Such acts violate values like honesty, truth, and self-respect.

Diligence can be defined as persistent effort. Procrastination contributes to a lack of diligence. Research often involves the use of resources; therefore, the lack of diligence may suggest a waste of resources. A lack of diligence can constrain knowledge from being advanced. Where research is conducted in a group environment, a lack of diligence by one member of the team can adversely influence the production and happiness of other group members. A group leader always expects and is responsible for ensuring the diligence of group members to avoid any negative consequences that can result from a lack of diligence.

15.10   SELF-CONFIDENCE IN THE CONDUCT OF RESEARCH

In Section 10.9, self-confidence was discussed in general terms. Self-confidence is often a primary determinant of success in research. With respect to research, the confident attitude is one of "I can do," not one of "I cannot improve on the works of others." Successful researchers always believe that they can do better research than the research completed by others in the past. This attitude does not imply that the previous work was faulty, but instead that the previous work could be improved in some way. One's confidence to improve upon past research is greatly influenced by past successes in making advancements in research, especially research that is considered by others as novel or unique. For those new to research, a feeling of confidence can also be achieved indirectly through faith in the abilities of superiors who have successfully conducted research in the past. The successes of mentors serve as a resource that will enable the novice to succeed in his or her current research work and then begin to gain experience in developing critical problem-solving ability.

How does having a feeling of self-confidence influence a person's application of the research process? The most obvious benefit is in the selection of the research topic. A self-confident person will pursue a more challenging topic than will a person who lacks self-confidence; a confident attitude will invoke a more focused effort and lead to a more significant advancement in the state of knowledge. The less self-confident person will pursue a safer, less challenging topic, which will lead to a result that more substantiates the current state of knowledge rather than advances the state of knowledge. High-quality, novel research leads to greater self-confidence and has broader implications than research that results from less challenging work.

A person's self-confidence improves with success. However, an optimist's confidence will likely advance more quickly than that of a pessimist after completing comparable research activities. Attitudes and a person's state of mind influence the effects of the success on the development of self-confidence. Having a positive outlook has benefits beyond those solely related to confidence. Happiness is also affected by a feeling of success that results from the completion of a research effort, and the happiness will contribute to further improvement in self-confidence. It is a cyclical process: happiness leads to success which leads to confidence which leads to more happiness.

Self-confidence is an extremely important feeling for a researcher. A lack of self-confidence is usually the result of a lack of experience in a research environment, multiple failures, or a lack of success at related activities. For example, one or two mediocre performances at making an oral presentation will lead to an apprehensive attitude toward future speaking engagements. The lack of confidence will likely initiate an apprehensive attitude in the early stages of preparation for future speaking engagements. Failure breeds a depletion of self-confidence. Even partial failures can encourage pessimistic thinking, which reduces self-confidence. So failure leads to a loss of confidence which leads to unhappiness which leads back to failure. Both of these cases are cyclical, so the trick is to take the action needed to be in the confidence-success-happiness cycle, rather than in the lack of the confidence-failure-unhappiness cycle.

Self-confidence is a primary characteristic of critical thinkers who are successful in research. If self-confidence appears to be a person's weakness, then developing confidence will be a precursor goal to becoming a critical thinker. Initially, being successful at solving minor short-term problems may be the most beneficial action for improving self-confidence, as such achievements can often be completed within a short period of time; however, minor successes generally only lead to minor advancements in self-confidence, so self-improvement plans must involve a willingness to pursue increasing more challenging tasks. Following each problem-solving endeavor, the degree of success should be assessed and the factors that contributed to the success noted. Also, each failure should be evaluated to identify the reasons for the failure, and then a plan should be developed for avoiding such failures in the future. Counterfactual thinking (see Section 13.10) can be a valuable resource in assessing failures.

15.11   CRITICAL THINKING AND RESEARCH EFFICIENCY

Equation 2.2 provided a definition of research efficiency (ε_R) that can be applied to the conduct of research. Specifically, the numerator R_0 is a measure of research output, which can be modeled using one of any number of numerous metrics. The denominator R_i is the input to the research effort, which is measured by the resources expended. Thus,

$$\varepsilon_R = R_0/R_i \qquad (15.1)$$

Research input would include the monetary cost of performing the research, the manpower effort expended, and the existing knowledge and databases, such as that provided by the existing literature. If the research involves the use of data, time and monetary costs would be included as part of the input. Note that many of the inputs are qualitative and cannot easily be transformed to a quantitative metric. Initially, R_i can be assumed constant, which allows the value of the efficiency to vary only due to the quality of the output, which might depend on the effort of critical thinking. If critical thinking is beneficial, then the numerator and the research efficiency should increase. This improvement may be reflected in simple criteria such as the increased number of publications and greater knowledge being available to both the public and the profession. Society could benefit from the increased output by way of better products such as new medical devises or safer infrastructure. It may also benefit through increased employment in research industries and increased economic competitiveness. If the nature of research causes the inputs to change, such as them becoming cheaper, then the denominator will decrease accordingly and the efficiency would increase. The point is: if researchers use the principles of critical thinking, research efficiency will increase.

If critical thinking produces a decrease in the denominator of Eq. 15.1, then the efficiency will increase because of worthwhile changes in both the numerator and the denominator. For example, instruction in knowledge of critical thinking could reduce the cost and effort of doing research and, therefore, an increase in efficiency can be expected. Critical thinking produces better research efficiency because it adds depth to the thinking and breadth to the value of the outcome.

15.12 RESEARCH ASSESSMENT

Research is undertaken to advance the state of knowledge. The outcomes of research can have significant effects on the lives of the public. So, should the people doing the research be assessed or is it the outcomes of the research that should be assessed? How would either of these options be assessed and what decision criteria should be used? Financial backers of the research would likely be interested in the assessment of the researchers who they support. Those affected by the research results would be interested in the quality of the research output. Researchers have a responsibility to disseminate the conclusions so that the users of the new knowledge understand the ways that the research would improve their lives.

It would be impossible to try to develop a single assessment worksheet that would apply to all cases. Medical research is different from engineering research even though both types may use the same research process. Therefore, a reasonable approach to research assessment might be to identify criteria that might be important at each phase of the research, which can be represented by the six steps of the research process. The following identifies some important assessment criteria:

Step 1: Problem Statement
- Originality and clarity of the problem statement.
- Meaningfulness of goal and objectives.
- Rationality of expectations.

Step 2: Resources
- Data adequacy and quantity.
- Validity of the existing knowledge base.
- Appropriateness of the performance criteria.

Step 3: Experimental Design
- Quality of hypotheses.
- Suitability of methods of analysis.
- Appropriateness of decision criterion.

Step 4: Analysis
- Accuracy level of results.
- Significance of results.
- Knowledge of the uncertainty of the outputs.

Step 5: Synthesis
- Novelty of conclusions.
- Fairness of statements.
- Relevance of findings.

Step 6: Implications
- Breadth of implications.
- Significance of the advancement of the state of knowledge.
- Understanding the effects of uncertainties of the outputs.
- Accuracy of applications.

While it may seem impossible to develop one set of objective criteria that could represent each of these criteria for every research project regardless of the discipline, these general criteria are a reasonable place to start regardless of the discipline. Attempts to develop objective metrics for each criterion should be made. Statistical criteria such as the prediction bias or the root mean square error could be used for the assessment in Step 4, while a benefit-cost analysis would be appropriate for use in Steps 5 and 6. Qualitative criteria may be the only option for some steps, especially Steps 1 and 2. Additionally, the assessment rating would depend on the stakeholder making the assessment and the person's personal biases. The above list of criteria can serve as a starting point for assessment.

15.13   CONCLUDING COMMENTS

When developing and publishing research results, it is important to emphasize the novel aspect of the work. Other researchers may have approached the same problem and provided a solution to their interpretation of the problem, but the solutions would not be universal. They may have had a more narrow focus on the issue or less experience in conducting research. Having critical thinking skills can enable a researcher to identify the lack of universality of solutions previously published, and then develop a more universally valid outcome. Understanding the critical analysis-critical synthesis duality is a central requirement for the formation of a novel solution, as well as for developing new research ideas.

Research productivity and efficiency can be greatly increased if good research practices are followed. Planning is a key, but confidence and the proper mindset can greatly contribute to success. Critical thinking skills can be used to identify research topics based on weaknesses of past research and then ensure that the results of the new work serve to advance the state of knowledge. Following the steps of the research process supplemented with good questioning and critiquing practices reflects the activities of a critical thinker.

15.14 EXERCISES

15.1. How will having a positive attitude help research output? How would a negative attitude hurt a person's research effort?

15.2. How does a person's mood influence research efficiency?

15.3. Identify three skills that are important to a critical thinker. Discuss how each influences research productivity.

15.4. How would critiquing a movie or TV show be similar to critiquing an article from a professional journal or magazine article?

15.5. Why do weaknesses of a published research paper lead to more productive research ideas?

15.6. How do the words *analysis* and *synthesis* relate to the conduct of research?

15.7. How would increasing one's self-confidence increase one's ability to conduct research?

15.8. What factors have positive effects on research? Negative effects?

15.9. In a report that documents completed research, what characteristics could be interpreted as weaknesses? Hint: consider the steps of the research process.

15.10. What inhibitors could limit a person's ability to do novel research?

15.11. Select a paper that discusses current research in your area of interest. Use any three of the 15 problem identification methods to propose new research topics.

15.12. Select an abstract from any professional journal. Identify ways that it could be improved.

15.13. For each of the 15 items in Section 15.4 propose a general modification that could produce new research. For example, Point 4 might suggest new research where conditions under climate change occur.

15.14. How does procrastination influence research productivity? Research efficiency?

15.15. What would be the benefits of critiquing a journal article to show its strengths?

15.16. Discuss the connection between research novelty, the production of research papers, and the factors that would inhibit novelty and productivity.

15.17. With respect to a mentee who is involved in research, what should be the objectives of a mentor?

15.18. Find a dictionary definition of each of the following values and briefly state the relevancy of the value to research productivity: care, integrity, and fairness.

15.19. Why is honesty in research important?

15.20. Why is tolerance an important characteristic in developing research ideas?

15.21. Why is perseverance important when conducting research?

15.15   ACTIVITIES

15.5.1   ACTIVITY 15A: RESEARCH PLANNING

Let me propose an activity that can help to ensure that you will meet your goal of experiencing the two-phase planning of research. The proposal is a one-hour activity that I refer to as *research planning*, which is a set of steps that can be used to develop an initial plan that reflects the "big picture" of your task. The intent of this activity is only to be introduced to the steps; the intent is not to actually do the research.

First, gather a couple of magazine-type articles on a topic that interests you. Now use your senses of curiosity and questioning. Briefly critique the articles and make notes of points that you feel do not adequately address the common issue of the articles. These might be aspects of the articles that cause you to be curious, possibly about a point not adequately addressed in the article. Based on your critiques of the articles, write out a brief two- or three-sentence statement of the specific problem that you would want to investigate. This statement should focus on a deficiency in the current state of knowledge of the topic, at least with respect to your knowledge, or it could be the output of an I^3 analysis (see Chapter 17). Weaknesses of past research are useful starting points for new research. The statement should not be a statement of the research goal or the expected end product of the subsequent research. Focus on the issue that seems incomplete, i.e., the problem or deficiency in knowledge.

Second, begin Phase II by setting both a general goal and a set of specific research objectives that would be needed to address the problem stated in Phase I; these would be vital to efficiently completing the research if you were actually performing the research. The goal should be a very general statement that will suggest an advance in knowledge related to the problem. While the goal is a very broad statement, the objectives should be very specific statements, ones that will collectively contribute to fulfilling the research: goal. For this activity, identify two objectives; in practice, research activities often involve three to five objectives. Each objective should be stated specifically concise so that it can be studied with a single experimental design. As research progresses, it may be necessary to change the statements of the objectives or add new objectives. As you progress through the steps of the research process and gain knowledge of the subject, your perspective on the objectives will mature and thus changes may be needed.

Third, the intent of this step is to seek out additional literature needed to continue with the research. This should be a brief review of literature that relates to your goal. In this planning exercise, the intent should not be to actually review literature. Instead, the review could be the creation of a list of keywords that will reflect on the scope of ideas that you believe would help with your research. Typical journals or books that may be relevant could also be listed.

Fourth, for each objective, develop a preliminary experimental plan. Initially, these statements should be brief, maybe even written in outline form. They should identify both the knowledge required to continue with the research and the type of data that would possibly be needed to complete the research related to each of your objectives. These plans may identify theory that you believe will be important to the solution. Make a preliminary assessment of the likelihood that the needed data would be available and of reasonable quantity and quality.

Fifth, based on Steps 3 and 4, the feasibility of the objectives stated in Step 2 should be assessed. If you have questions about their feasibility, appropriate changes should be made; this is a feedback loop from Step 4 to Step 1. It is an opportunity to improve the statements of the research objectives.

Sixth, assume that the research was completed, with data analyzed to identify results. Identify the potential outcomes from each objective that could result if you had performed the research. Assume that the conclusions that you would develop will make an adequate advance of the state of knowledge. Specify these advancements in knowledge and their benefits to relevant stakeholders. Again, the intent of the activity is solely the experience of going through the steps of the research process. It may produce some questions about which you should consult your mentor.

15.5.2   ACTIVITY 15B: BASEBALL MODEL

A brief abstract is shown in the below box. The purpose of this activity is to show ways to find research topics by extending the work of other research. Read the abstract and identify as many weaknesses as you can. Pose each weakness as a question that reflects the problem. Then briefly outline the way that you would provide a solution to the problem.

Abstract: A model that relates the baseball-to-bat coefficient of restitution e to a standard coefficient of restitution (e_0) and the dynamic stiffness (k) of the ball is presented: $e = k\,e_0$. Values of e are related to values of e_0 and k for a "standard ball." The efficiency is demonstrated by comparison with experimental data. The proposed model is vastly superior to the method based on the physically unjustified assumption that the ratio $e/e_0 = 1$ is independent of both e_0 and k.

15.5.3   ACTIVITY 15C: SOIL PARTICLE CHARACTERIZATION

The amount of stream bed erosion depends partly on the shapes of the soil particles. Geomorphologists typically use three parameters to describe soil particle shape: the sphericity, the roundness, and the form. The form can be quantified by the regularity of the three dimensions: length, width, and height. The sphericity is a comparative measure relative to the diameter of a sphere with the same volume. The roundness is a function of the two-dimensional profile of the particles. For each of these general measures, develop three parameters that would reflect the three characteristics. Note that geomorphologists have developed several parameters for each of these.

The purpose of this activity is to develop new parameters that could be experimentally compared with actual streambed erosion rates.

15.5.4   ACTIVITY 15D: THE THREE LITTLE PIGS

Curly Tail, one of the three little pigs, attended two imaginative idea generation sessions, where the problems were as shown in the table. He generated ideas for each

Critical Thinking, Idea Innovation, and Creativity

problem. Porky and Fatty, the other two little pigs, were in the same two sessions. For each of the two problems provide piggybacked ideas for both Porky and Fatty using Curly Tail's idea.

Problem	Generate Ideas on Ways to Increase the Sales of New Bath Tubs	Generate Ideas on Removing a Large Concrete Dam from a River to Improve the Environment
Curly Tail's idea	Build the bathtub into cars so that people can use the tub while traveling to work.	Develop a chemical that will dissolve the concrete.
Porky's piggybacked idea		
Fatty's piggybacked idea		

15.5.5 Activity 15E: Award Activities

A student chapter has in the past won national awards for its activities, including academic excellence, service to the university, service to the community, and service to the profession. In the last two years, the ranking of the chapter has seriously declined. For each hat of de Bono's thinking process, provide an idea for returning the chapter to national prominence. The sequence of hats is: blue, white, red, black, green, yellow, blue.

15.5.6 Activity 15F: Coefficient of Restitution

The coefficient of restitution (e) is used as a measure of the liveliness of a baseball, where e is estimated using the following relation:

$$e = \left(V_{b2} - V_{a2}\right)/\left(V_{b1} - V_{a1}\right) \tag{15.2}$$

where V = velocity, a = object A, b = object B, 1 = just prior to the collision, and 2 = just after the collision. Let A be the ball just thrown by the pitcher just before it reaches the bat and B is the ball as it just leaves the bat, which was swung by the hitter. Assume that you believe the ball from the current baseball season is livelier than the ball that was used last season. Design an experiment to test your hypothesis.

16 Critical Decision-Making

CHAPTER GOAL

To demonstrate that knowledge of critical problem-solving increases the effectiveness of decision-making.

CHAPTER OBJECTIVES

1. To outline and discuss the steps of the decision process.
2. To define failure as it applies to decision-making and stress the importance of introspective analyses to understand failures.
3. To identify inhibitors to the successful use of critical problem-solving in an organizational setting.
4. To discuss the importance of prioritization of responsibilities.
5. To imbed critical time management into the efficient solution of complex problems.

16.1 INTRODUCTION

Individuals and groups make decisions on a daily basis. Some of these decisions have significant, sometimes long-term consequences. Consider the following cases:

- On Columbus's first voyage to the Americas, many of his crew had mutinous thoughts because they feared not finding land; they wanted to turn around and return to Europe. What if the crew had overpowered Columbus and reversed course? How would the course of exploration have changed if the sailors on the Nina, Pinta, and Santa Maria had decided to go through with their mutinous behavior?
- Because of the cost of the Seven Year's War between Britain and France, the British government passed numerous legislative acts that passed a significant portion of the financial burden of the war onto the 13 colonies. Legislation such as the Sugar Act, the Navigational Acts, the Quartering Act, the 1765 Stamp Act, and the Townshend Duties were opposed by many colonialists. These levies led to rebellious events such as the Boston Tea Party. How would history have changed if the British government had made the decision to lessen or eliminate these imposts?

Decisions can lead to actions, and these actions can influence the course of history, which can be one's personal history, the history of an organization, or the history of a global movement. Even the short-term consequences of decisions cannot be predicted with certainty because we live in an uncertain world; however, understanding rules that are expected to lead to good decisions can improve the likelihood that decisions will have favorable outcomes. Long-term consequences of decisions are even less predictable. Having the characteristics of a critical thinker should improve the likelihood of making decisions that lead to positive outcomes. The dimensions of critical thinking promote good decision-making.

Decision-making is at the heart of change, whether it is technologic advancements, the beneficial innovations of commercial products, the outcomes of scientific research, or successful financial actions. The consequences of good decision-making can have positive global impacts; however, poor outcomes often result from decision-making that is not based on good thinking. Some decisions have short-term implications, while other decisions have long-term consequences. The expected life of a decision can influence the uncertainty of the outcome. Decisions can be made by an individual, while others are collective decisions made by large groups, such as those that result from large organizations. Regardless of the type of change or its consequences, most instances of decision-making follow a set of general steps. While these general steps are essentially identical to processes like the research process, the specifics for any one problem can require deviating from the normal decision process; however, knowing the general steps of the decision process is a good place to start. It should be evident that the decision process is complicated by all of these factors: the cause of the change, the potential impact, the number of decision makers, and whether or not the decision is for the short term or the long term. Each of these factors can introduce uncertainty into the outcome.

The goal of this chapter is to introduce a model of the decision process and clearly show the ways that the process can be enhanced by having knowledge of critical problem-solving. Such knowledge can improve the effectiveness and efficiency of personal and organizational decision-making. Knowledge of the dimensions of critical thinking yields personal and professional benefits.

16.2   THE DECISION PROCESS

Every organization uses a general decision process. For any one problem, a general model of the process can be modified to fit the specific needs or any new restrictions. A general set of steps for decision-making is summarized in Table 16.1. When a performance or opportunity gap is recognized, organizations apply their version of the decision process; the gap acts as a business stimulator. The company then sets specific goals and formulates decision criteria that can be used to set targets for reducing or overcoming the gap. A set of alternative solutions are developed, data are collected to be used in evaluating the different alternative solutions, and the expected performance of each alternative is evaluated using the performance criteria that were developed as part of Step 2. The alternative that shows the best value of the decision criterion of Step 3 is chosen as the alternative to be implemented. While the process is presented here as consisting of six steps, other versions of the process could legitimately be presented, some with fewer steps and others with more steps.

TABLE 16.1
The Decision Process

1. Identify the stimulus.
2. Identify the goals and performance criteria.
3. Formulate the decision criteria.
4. Identify alternative solutions.
5. Collect relative data to use in evaluating the alternatives.
6. Make a decision and assess the results after implementation.

The stimulus is the event that motivates an organization or person to use resources. This task corresponds to identifying the problem, which is the first step of the research process. To use the decision process, an organization requires one or more performance criteria, each of which attempts to summarize the extent to which an alternative solution would satisfy a goal or an objective. Performance criteria are metrics that can quantitatively or qualitatively measure the worth of any alternative decision. Quantitative metrics are usually preferred over qualitative metrics, as comparisons of quantitative metrics are more objective. The third step of the decision process is to identify and evaluate the usefulness of one or more decision criterion, which is a statement that describes the way that the performance criteria are weighted and used to select the optimum solution. The decision criterion of Step 3 is more easily established when the performance criteria are expressed quantitatively. If the problem only involves one performance criterion, then the decision criterion would be to select the alternative that shows the best value of the performance criteria. In cases where multiple performance criteria are used to describe the objectives, the decision criterion must include a means of prioritizing and weighting the values of the different performance criteria, i.e., the decision criterion is a weighting scheme. The objective of Step 4 is to develop all possible solutions to the case, i.e., the results. This step can incorporate idea generation methods for each objective in such a way that every dimension is considered. As many alternative solutions as possible should be generated. In Step 5, the data can be used to quantify each of the performance criteria and then used with the decision criterion in Step 6 to select the best of the alternative solutions. Step 6 could incorporate a feedback loop as part of the decision process. If after implementation, the initial decision did not meet the original goals, then changes would be necessary. Control is returned to the step most likely to produce an effective solution.

16.3   CRITICAL TIME MANAGEMENT

Time itself cannot be controlled or managed; however, the use of time can be controlled and managed. The efficient use of time influences the effectiveness of the interval of time. Critical thinking is usually needed to solve complex problems. Complex problems are often subject to a number of constraints. Time is often one of those constraints. Thus, effective time management will likely lead to more efficient use of resources; however, if time is a serious factor in developing a solution to a complex problem, then the principles of critical thinking will need to be applied through proper time management. Procrastination is one example of the inefficient use of time that most often leads to ineffective results. Pessimism and a lack of confidence are other time wasters. Learning to control these inhibitors can increase program efficiency and improve the effectiveness of the decision.

16.3.1 Definitions

Let's start by considering definitions of the words that comprise the term critical time management:

Critical: Characterized by careful judgment; degree of intensity.
Time: A clock interval devoted to an activity; a continuum during which an event occurs.
Management: The manner of controlling an activity.
Putting the essence of these ideas together yields the following definition:
Critical time management: Using careful judgment and efficiently applying one's resources during a clock interval to intensely control actions related to the completion of a task.

Leaders who apply critical time management will likely provide more effective solutions to problems.

16.3.2 Time as a Resource

Time is a resource. Money and equipment can be resources. Misuse of these latter two can lead to inefficiency and ineffectiveness. The same is true of time. It must be managed properly if the project objectives are to be met. The management of time follows many of the same principles as used in the management of any resource. An effective time manager sets priorities, including the ranking of the importance of each task. Managing time involves identifying each task, the time required for completing each task, and the sequence of working on the tasks so no one task becomes a bottleneck to progress on other tasks. Therefore, it is important for the leader to assign personnel in a way that bottlenecks are least likely to occur. In many cases, time management may include the use of facilities, not just personnel. If some facilities will be needed for multiple project tasks, then a separate but coordinated schedule will be needed for these other resources.

16.3.3 The Influence of Dimensions

The value dimension has a direct influence on critical time management. A critical time manager must be responsive to a number of values, including:

Fidelity: Unfailing fulfillment of one's duties.
Accountability: Answerable for duties.
Industriousness: The quality of being diligently active in work.

Promptness: The practice of being punctual.
Reliable: Dependable in meeting duties.

While a reference to time does not appear in these definitions, an applicable duration of time could be inserted into each of these definitions; for example, accountability could be defined as answerable for all duties over the duration of the project. When time is imbedded into each value, the values would be fundamental to critical time management. Other values could be included in the list. Many of these values are interdependent. The point is: time interacts with values, both of which directly relate to the success of decision-making.

The skill dimension is also important to time management. Just being on time is not proof that the person is a critical time manager. Courage, which is a primary skill, takes on a slightly different perspective than it did when it was applied to the general framework of critical problem-solving. Now, courage takes on the added constraint of time; specifically, courage reflects the need to have the confidence to deal in a timely manner with any constraint placed on the solution of a complex problem. Perception is the second skill in the skill dimension. A critical time manager must be able to identify the effects of incomplete knowledge and uncertainties and address the issues under the time constraint. Emergence (see Section 10.10) is another skill that will be sensitive to the effective management of time. It is important to recognize that applying skills within the scheduled time is not the sole basis for assessing whether or not the manager is a critical time manager. The manager must have the foresight to anticipate time distractions that contribute to inefficiency and to effectively make the necessary revisions to the plans to avoid delays. The manager must use careful judgment of time to efficiently complete the requirements of the complex problem.

16.3.4   PRIORITIZATION OF RESPONSIBILITIES

A critical time manager learns to set priorities that reflect the responsibilities of the position. Some responsibilities require immediate attention while actions to other responsibilities can be postponed to a later time. Actions on which other people are dependent may take priority over activities that only require your immediate attention, especially when any delay will cause others to be delayed in meeting their responsibilities. The former is referred to as a *crucial need* while the latter is referred to as a *critical need*. Activities that are needed but not crucial or critical will be labeled as *decisive*; an action of decisive priority may involve settling a dispute between other team members or responsibilities that are not critical. Actions that can be temporarily delayed will be labeled as *serious*, as they do not need immediate attention, but must be completed on time to meet the current responsibilities on another assignment. In summary, the action priorities of a critical thinker are given as follows:

1. *Crucial*: Important for your needs and the needs of others.
2. *Critical*: Addressing your own responsibilities.
3. *Decisive*: Addressing the needs of others for their critical responsibilities.
4. *Serious*: Actions that can be delayed.
5. *Planning*: Action on future projects.

Rarely does a person work on only one project at a time, so having other responsibilities is the norm. Work performed on the other responsibilities is referred to as a planning activity. The lowest priority action is planning for the future. The completion of a planning task may not be necessary to meet responsibilities of the future and action on such responsibilities can be completed at appropriate times.

Many managers use prioritized TO-DO lists. They may formally set priories on actions. The five categories listed above may be useful for making decisions about the order for completing tasks on a TO-DO list. Prioritizing the workload can improve efficiency. Priority numbers can be inserted on the left-hand side of a TO-DO list. These numbers will control the order of the actions.

A priority list can be influenced by one's mindset. If a person is not in the mindset to work on a high priority item, it may be better to work on a less stressful, lower priority activity. This plan of action will make the overall best use of time, as the time during a poor mindset state will not be totally lost; however, it is important to ensure that revising the priorities of the activities does not adversely influence the completion of a co-worker's responsibilities.

16.4 THE ASSESSMENT OF DECISION FAILURES

Failures can be reversed. In August 1776, George Washington lost a major battle in New York to General Howe, the British leader. Washington recognized that his troops were not a battle-ready army, so he corrected the failure through training, and in December 1776 Washington crossed the Delaware River and routed the British forces at Trenton, NJ. A few days later, he repeated the victory at Princeton. Washington did an introspective analysis following the New York defeat and determined the actions that he needed to take to correct the causes of his failure in the New York campaign. The defeat was not reversed, but the failure was reversed.

16.4.1 Definitions

Failure is one possible outcome of any decision situation. Failure can be defined as *not achieving the desired end*, but the word should be viewed as a continuum as the end result is not limited to two outcomes: total failure or total success. The success-failure continuum can extend from total success to total failure, but the separation distance along the continuum offers the opportunity for many outcomes. Even terms like *partial failure* or *partial success* may not adequately portray the range of possible assessments on the success-failure spectrum. As used herein, failure will refer to any level of failure, not necessarily a total failure. When dealing with complex problems, especially those that involve conflicting stakeholders, it is quite possible that outcomes of either a complete success or a total failure are unlikely because some stakeholders will be disappointed with the outcome. Not achieving a complete success would not imply that the effort was a failure. Following any assessment where some level of failure is evident, an introspective analysis should be conducted in order to obtain knowledge that will help ensure that future efforts in solving similar problems will result in better outcomes.

16.4.2   THE EFFECTS OF FAILURES

Those who fail at an activity should subsequently critically review all aspects of the case and determine the causes of the failure. Any failure, even a partial failure, can influence a person's attitudes and maybe even cause a change to his or her value system. Any activity that has a negative outcome should encourage a self-introspection, identify the cause of the failure, and make the necessary changes in beliefs and attitudes to avoid failures in future problem-solving activities. If the cause of the failure is evident, then the needed change in beliefs and attitudes can be identified and corrected. In most cases, a complete understanding of the causes of the failure will not be immediately evident. Self-introspection will enable the person to identify a set of potential causes, specifically the most likely attitudes and actions that contributed to the failure. The failure to perform such an assessment magnifies the significance of the failure; in other words, the failure to learn from a failure is in itself a failure. When even a personal failure does occur, prudency calls for a change in attitude. Mindset issues are especially relevant to failures. A person who generally acts on values like knowledge and truth is generally more willing to change his or her attitude in a way that will help overcome the negative feelings that may follow a failure. Negative feelings can cause stress and, more importantly, a reluctance to make the necessary changes.

Failures that are internal to an organization can have consequences that are external to the organization. For example, jobs completed for public works projects may adversely influence public health or safety. Organizations that avoid critical problem-solving may not provide the best solutions even to the point of requiring follow-up work identified by post-project assessment activities. Such failures would contribute to increased costs and possibly organizational embarrassment. An organization that tries to minimize the time spent on completing a project may find that more time is actually required when the time for the follow-up activity is considered.

16.4.3   CAUSES OF FAILURES

It is important to identify and fully appreciate any cause of failure. Failures can be due to poor or outdated beliefs, poor planning, a pessimistic mood or negative attitudes that dominated actions at the time of the decisions, a lack of knowledge, or external pressures from either the organization or peers. Failure to address the principles of the dimensions of critical problem-solving is likely to lead to failure. Therefore, a problem solver's first task following a failed activity is to identify the causes of failure and then make all necessary changes to prevent future failures that stem from the same causes. External stressors, such as inhibiting peers, are usually the most difficult to correct, as it is difficult to get the individuals to change their thinking. Personal discussions with the external stressors should precede the drafting and distribution of written memos. Memos should be the last resort. Personal-life stresses are often troublesome to overcome, but like external stresses, a failure to correct any stressor may increase the likelihood of future failures for the same reasons.

16.4.4 INTROSPECTIVE ANALYSIS AND SYNTHESIS

Improvements in decision-making are often the result of introspective analyses. Keeping in mind that the word *analysis* means to break apart and the word *introspective* means to assess one's own feelings, the term *introspective analysis* means to contemplate both (1) the reasons that certain actions were taken or decisions were made and (2) the outcomes of the actions and decisions. All connections between the decision/action and the outcomes must be identified. One potential outcome of an introspective analysis is the improvement of one's own long-term decision-making. Whether or not the end result of a decision was a failure, a partial failure, or even a partial success, an introspective analysis can provide guidance for both personal growth with respect to decision-making and reinforcement of the worth of existing decision criteria.

Self-introspection has many benefits. An introspective analysis can be in broad terms or for a very specific issue, such as a specific aspect of a critical problem-solving task. In general, self-introspections should lead to improvements in one's weaknesses, especially related to decision-making, so as to decrease the likelihood of repeated failures in the future. Introspective thinking can generate enthusiasm for self-improvement and the revision of an attitude toward a task. Introspective analyses give a person the opportunity to take a different perspective on his or her future as the time that is needed to integrate past events into a more coherent picture of recent events can be taken. Events that did not end as well as they were expected to end can be analyzed in a perspective that provides a more realistic picture of events of the future.

16.5 DECISION-MAKING INHIBITORS

Organizational inhibitors can occur at any step of the decision process. Inhibitors of the environment, which in this case is the organization, were discussed in Chapter 7. The following inhibitors can prevent organizations from applying critical problem-solving in their decision-making:

- *An unwillingness of the organization to perform introspective analyses*: While we usually view introspective analysis as an activity for an individual, the leaders of an organization should apply the principles of introspection to regularly conduct in-house assessments of policies and practices. Introspective analysis should also be used to identify weaknesses in past critical problem-solving activities.
- *A reluctance to provide the resources that are necessary to apply the critical problem-solving approach*: Resources includes financial backing of critical problem-solving efforts, funds to support staff assistance, free time for creative learning and thought, and support for educational activities for enhancing critical thinking abilities. Failures to provide resources can reduce the overall efficiency of organizational activities.
- *Failure to ensure that discouraging attitudes toward imaginative thinking are suppressed*: The organization should discourage peer pressure that is

aimed at stifling creative problem-solving. All employees should be made aware of the organization's commitment to creativity and prevent any criticism of effort.

- *A hesitancy to discourage an a priori pessimistic attitude*: Optimism toward critical problem-solving should be widely known and strongly encouraged.
- *Policies that cause excessive pressure to minimize job-completion times*: Since novel solutions generally require more time to complete than do non-imaginative solutions, organizational policies must prevent policies and practices that emphasize minimizing job-completion times. While diligence is encouraged, novelty should not be sacrificed.
- *An inability to recognize that better solutions can result from more modern thinking*: Organizations that have solved problems and made decisions using the same methods for a long period of time should evaluate their practices and revise them to include critical problem-solving. Special efforts, including the development of a reward system, could be used to overcome this problem.
- *Organizational practices that do not try to reduce stressful environments*: Individuals who work in a stressful environment believe that finishing a job in the minimum amount of time will make the stress disappear, which perpetuates an environment where critical problem-solving is discouraged. A creative workplace is often less stressful because of greater employee job satisfaction.

Additional organizational inhibitors include a distrust that creative activities can provide better results, not considering creative ability when hiring new employees, unenlightened management that fails to recognize the benefits of critical problem-solving, a resistance to change, excessive routine work, not providing recognition to those who successfully use creative thinking, and the failure to discourage peer pressure aimed at avoiding novel problem-solving. Unless an organization makes an effort to eliminate inhibitors to critical problem-solving practices, they can expect low employee job satisfaction, a higher than expected rate of employee turnover, and clients who are not overly satisfied with the organization's work.

16.6   MINDSET AND DECISION-MAKING

A person's mindset can certainly influence the ways that he or she approaches each of the steps of the decision process. Of special interest would be the extent to which the counter-skills of pessimism and procrastination could hinder progress. A decision maker who is in a pessimistic mood might provide more negative assessments of the performance criteria. Pessimism could also influence a decision maker into undervaluing the implications of the decision. Obviously, a procrastinator could limit productivity and efficiency throughout all steps of the decision process. Specifically, any unnecessary delay in collecting data could result in data of poor quality, which would negatively influence the potential benefits of the effort. Counter-skills generally have a negative influence on any effect and the resulting decisions.

If a person makes an introspective analysis at a time when he or she is under stress, then the outcome may reflect the stress more than the failure that led to the stress. Actually, the mood of a person can be the dominant factor that determines the outcome of a post-project introspective analysis. Stress may even cause distortions of the knowledge about one's decision-making, i.e., the wrong lesson may be learned. Thus, the mindset at the time of an introspective analysis can be as important as the mindset at the time of the action or decision. The mindset dimension of critical thinking is an important factor in the assessment of failures, even partial failures. A person's mindset is relevant both when making a decision and at the time of the introspective analysis.

The benefits of an introspective analysis will be greatly influenced by the person's mindset at the time of the analysis. If an experience of failure is viewed when the person is in a pessimistic mood, the pessimism may bias the assessment toward negativity and thus distort the views of the future. Thus, a person should begin a self-introspective analysis by acknowledging his or her own mood. Whether the person is in an optimistic or a pessimistic mood should not be allowed to distort the self-introspective assessment. At the end of the analysis, the person should summarize the results and create an action plan to guide future decisions and actions. If a person believes that the introspective analysis was adversely influenced by his or her mood at the time when the decision was made, then a re-analysis should be undertaken at a later date to refute or substantiate the findings of the original introspective analysis.

16.7   VALUES IN DECISION-MAKING

The value dimension plays an important role in decision-making. It may be best to identify the role that individual values play as they can be influential in each of the six steps of the decision process. While many values are relevant to each of the six steps, only two of the more important values will be used to demonstrate the importance of values in each step of the decision process:

1. Identify the stimulus.
   - *Unbiasedness*: The performance gap that reflects the motive for the effort should be accurate and unbiased. Any bias will introduce error and/or uncertainty into any decision.
   - *Respect*: Stakeholders can influence the performance gap. Since stakeholders can have very conflicting goals, the statement of the problem should attempt to accurately incorporate the goals of each stakeholder.
2. Identify the goal and performance criteria.
   - *Truth/Honesty*: The goal must be based on an honest assessment of the performance gap and the implications of the potential decision that will be made in Step 6.
   - *Accountability*: The performance criteria should be developed so that the final decision is sensitive to the goals of all stakeholders.
3. Formulate the decision criteria.
   - *Fairness*: Since the decision criterion will be used to make the final decision, it should show fairness to all stakeholders, including society.

- *Competence*: The decision criterion is formulated to competently reflect the problem identified in Step 1 and the implications of Step 6.

4. Identify alternative solutions.
   - *Wisdom*: Wisdom is the ability to use good judgment in ensuring understanding of what is true or right. Wisdom is important in generating alternative solutions and assuring that all ideas adequately reflect solutions to the goal of Step 2 and the factors that were responsible for the performance gap.
   - *Fairness*: For complex problems, multiple decision criteria may be necessary to match each of the dimensions and the breadth of the problem statement. The decision makers will need to be fair to all stakeholders when assigning weights that measure the importance to each of the decision criteria.

5. Collect relative data to use in evaluating the alternatives.
   - *Accountability*: It is important to select data that reflect the functioning of the real system as it currently exists and will reflect future states of the system. Those who collect and process the data must be accountable, i.e., answerable for the duties that their position invests in their role as a decision maker.
   - *Prudency*: The decision maker must exercise good judgment in all analyses with the data and in selecting the outcome that is considered best.

6. Make a decision and assess the results after implementation.
   - *Accountability*: The performance and decision criteria must be accountable to all stakeholders, including society, for both the short term and the long term.
   - *Responsibility*: The decision makers have a responsibility to make a post-implementation analysis to ensure that the final decision fairly respected the expectations of the stakeholders.

Other values that are relevant across all steps of the decision process include confidentiality (i.e., important information will not be improperly disclosed), integrity (i.e., adherence to a code of behavior), and excellence (i.e., the condition of superiority in conducting the process). A critical thinker must apply all of these values when making decisions in the solution of complex problems.

16.8   A SKILL SET FOR DECISION-MAKING

Given the similarity of the steps of the decision process and the steps of the research and design processes, many of the skills used for these processes would also apply to the steps of the decision process; however, the role of each skill for the decision process would likely differ from the role of the skills for the research process. Table 16.2 summarizes the relevance of the skills to each step of the decision process. Some of the skills shown in Column 3 also apply to more than just one step, not just the steps shown; a certain amount of overlap is expected.

TABLE 16.2

The Skills and Uncertainty of Each Step in the Decision-Making Process

Step	Objective	Skills	Use of Skill	Sources of Uncertainty
1	Identify the stimulus	Courage and perception	Essentially define the problem; have the courage to approach the problem and establish the goals and objectives; provide clarity to the nature of the performance gap through a perceptive understanding of the problem.	Inaccuracies of projected change; reluctance to change; difficulty in forecasting market conditions.
2	Identify the goals and performance criteria	Skepticism	Collect and critically analyze the existing knowledge related to the performance gap; for each objective, develop a performance criterion that is related to the gap.	Deficiencies in knowledge; sparse data; ambiguity of data.
3	Formulate the decision criterion	Critiquing and questioning	Given both the complexity of the problem and the uncertainties, critiquing is needed to develop decision criterion that reflects the lack of knowledge; match the decision criterion to the goals.	Conflicting goals of the stakeholders; the simplicity of the decision criterion.
4	Identify alternative solutions	Curiosity	The search for truthful knowledge of all possible outcomes and accurate estimates of the uncertainties for all possible solutions.	All inputs to the alternatives are uncertain; the functional forms of model components.
5	Collect data; evaluate alternatives	Emergence	The results emerge from the analyses; the decision criterion should be sufficiently reliable to assess each alternative.	The selected performance criteria may not fully address the performance; difficulty in using qualitative criteria.
6	Decision, implementation, and assessment	Consequence	Balance value issues and incorporate them into the implementation plan; use decision criterion to identify the best solution. Assess and verify the goal and objectives.	Difficulty in weights of decision criteria; accounting for stakeholder conflicts.

16.9 CONCLUDING COMMENTS

One aspect of Step 6 of the decision process is to conduct an assessment or provide a verification of the decision. The best alternative would be selected, and the decision and actions would be based on the decision undertaken. For example, the decision could be to institute the best alternative, or, if the decision criterion suggested that none of the alternatives were accurate, the action might be to collect more data and return to Step 2, i.e., feedback. Assessment should also include an evaluation of the extent to which the goal and objectives of Step 2 were achieved. This task may be somewhat subjective, but the assessment should be unbiased. Verification is another recommended Step 6 activity. Verification requires additional data, with the new data being independent of the data initially used. These new data can be analyzed to quantify the performance criteria originally established in Step 2. If the performance criteria fail to verify that the decision solved the problem, then follow-up activities should be used to ensure that the goal and objectives were accurate indicators of the problem.

The Panama Canal construction project was considered to be a complex project, especially given the knowledge and technology that existed in the late nineteenth and early twentieth centuries. deLesseps' failures on each of the four dimensions of critical problem-solving point to the importance of including each of the four dimensions as part of the decision process. George Goethals (1858–1928) was successful at the monumental task of constructing the Canal, and the evidence shows that his actions met all of the restrictions of the four dimensions. Evidence did not show any financial mismanagement by Goethals. Goethals was very concerned with the welfare and safety of the workers, which indicates that values were important to him. He was able to use broad-based critical thinking to solve all of the unique technical and health problems that surfaced during construction. His vision of future demands on the Canal was evident in the design of the Canal. Major changes to the Canal were not needed for more than 100 years following the completion of the Canal in 1914. Goethals used critical problem-solving practices and successful leadership principles to successfully complete a project that had not been finished by others who had not been sensitive to the dimensions of critical problem-solving.

16.10 EXERCISES

16.1. Of the ten characteristics of a critical thinker (see Chapter 9) identify the three that are most relevant to good decision-making. Provide a brief justification for your selections.

16.2. Compare/contrast the six steps of the decision process (Section 16.2) with the Scientific Method (Section 2.3).

16.3. Identify and define one additional value that is relevant for each of the six steps of the decision process (see Section 16.3).

16.4. How could a pessimistic mood cause a failure, such as failing a test in college or failing a drunk driving test when pulled over by a trooper?

16.5. Define uncertainty and discuss the ways that it could be applied to taking an examination at the college level.

16.6. For each of the seven inhibitors of Section 16.6, discuss ways that it could be applied to the issue of preventing failures in an organization.

16.7. How can knowledge of the assessment of a brainstorming session help with the assessment of organizational failures at critical thinking?

16.8. Provide definitions for each of the five levels of prioritization of responsibilities and give examples to distinguish between the different levels.

16.11   ACTIVITIES

16.11.1   ACTIVITY 16A: YOU ARE A FAILURE AT LEVEL 5

Create an ordinal scale metric to define intermediate points on the success-failure spectrum. Label each point and indicate the performance criterion for assessing values of the metric. This activity is an exercise in the creation of a new metric that could be used in a critical problem-solving exercise.

16.11.2   ACTIVITY 16B: BAD BOY!

Identify a major company that failed, such as ENRON. Identify the reasons for the failure and relate each cause to the dimensions of critical problem-solving.

16.11.3   ACTIVITY 16C: PULLING OUT OF A NOSE DIVE

Create a process for conducting introspective analyses that could be used following a failure of a critical- thinking activity. Identify the goal, objectives, and the steps of the process.

16.11.4   ACTIVITY 16D: MIND OVER FAILURE

Create a metric that can be used to evaluate the effects of a person's mindset on organizational failures.

16.11.5   ACTIVITY 16E: A SCALE OF INHIBITION

Seven inhibitors of organizational critical decision-making were listed in Section 16.6. Create a quantitative performance criterion for each one of the seven inhibitors and then create a weighted decision criterion that can be used to measure the overall level of organizational inhibition.

16.11.6   ACTIVITY 16F: EFFICIENCY IN MANAGING TIME

Develop a model of problem-solving efficiency that incorporates time into the calculation.

17 Introspective Idea Innovation

CHAPTER GOAL

To introduce a method of critiquing existing research results for the purpose of extending the general idea to new novel research.

CHAPTER OBJECTIVES

1. To introduce a method of developing new research topics.
2. To emphasize the importance of the ability to critique the works of others.
3. To briefly discuss the innovation of existing models.
4. To discuss values that are important to idea innovation.

17.1 INTRODUCTION

The steam engine was a primary source of power to propel ships in the eighteenth and nineteenth centuries. The steam engine was the result of continual innovation of the power of steam generated pressure. In 1707, Denis Papin (1647–1712) showed that the power of steam was able to lift a lid off of a pot. But to control the steam required the development of a pressure valve and other innovations. The automobile is another example of innovations in the transportation industry. Some of the early developments of automobiles came from innovations in the knowledge of the bicycle and the basic engine. Late nineteenth and early twentieth centuries saw innovations of the steering wheel, ignition systems, shock absorbers, and bumpers. The innovation of weather forecast models, which have been based on knowledge of the physical processes and variables based on the availability of sensor measurements, is now better able to forecast the characteristics and tracking of hurricanes, which enables the public to be better prepared for limiting the damages from wind events and flooding. The innovation of design flood models to include watershed characteristics, such as the watershed slope, can influence estimated values of flood magnitudes and thus improve engineering designs. These innovations have been the result of problems being recognized and actions taken to make the necessary innovations. The innovations led to continual improvements and advancement of the products and designs.

Innovation of electronic equipment is very common. We rely on such innovations each day and we expect innovations to continue in the future, maybe even at a faster rate. We know that the innovative changes that are continually made cost us additional

DOI: 10.1201/9781003380443-17

money and they produce profits for many industries. Therefore, the result of innovation is a trade-off, with both the public and the industry benefitting. The public gets more useful products and the manufacturers make profits. Innovation is also profitable in the research community, sometimes financial benefits accrue, but in many cases the true profit is in terms of professional recognition. Therefore, knowing ways to innovate can be useful and profitable to both the researcher and the business person.

The need to know of problems that need solutions is common in research environments. Researchers regularly recognize weaknesses in the existing state of knowledge of a discipline, and such weaknesses can be a source of ideas for important advancements of knowledge. Researchers recognize that their own research does not completely solve the intended problem and that additional investigations will lead to new research results and hopefully be closer to the ideal of truth. If this is true of their own research, the researcher must be aware that it is true for the results of the research of others. A person may need to complete a research requirement, but he or she does not have a specific topic to use to fulfill the requirement. Such a situation is common to students who are in need of a topic to fulfill an academic research requirement. Both graduate and undergraduate students who have a research responsibility may be searching for a topic that would acceptably fulfill their academic research requirement. In such cases, they cannot initially use creative thinking methods, such as brainstorming or synectics, as the specific topic is not known; then the first task is to find a general topic. Finding a topic that is novel is an important skill. The Introspective Idea Innovation (I^3) method is a useful method for finding a topic based on a critical analysis of existing research.

The focus of this chapter is on the introduction of a procedure that can be used to find extensions of someone else's research result to make a significant advancement in knowledge of the topic without plagiarizing the works of others. The extension is made through the application of skills associated with the researcher's mental processes. For this reason, the method is referred to as introspective idea innovation, or I^3. The outcome of an application of the I^3 method is a new idea that can lead to a problem statement that is a worthy extension of the works of another researcher and will lead to a novel outcome that advances the state of knowledge, i.e., a move toward the truth.

17.2   DEFINITIONS

The word introspective implies *contemplative* or *turning one's thoughts inward*. An idea can be defined as *the gist of the situation*; an idea is a product of mental activity and exists in the mind. The term innovation means *to renew, to change, to make new, to alter,* or *to make changes to anything established*. If these three terms are combined, a definition for introspective idea innovation, or I^3 would be:

> To use one's own mental processes to get the gist of someone else's ideas and make significant changes to develop a novel idea.

The I^3 method is a useful way to advance knowledge through the modification of an existing concept to create new knowledge and hopefully significantly advance the state of knowledge. As such, it is useful method for developing new research ideas based on the existing state of knowledge.

17.3   PHASES OF THE I³ METHOD

Innovation was presented in Chapter 2 as a four-step process. Only four steps were needed because in the case of innovation the problem statement is usually obvious since it is relatively simple. Step 6 of the decision process is also not needed because the outcome of Step 5 is used as the decision of Step 6. Thus, a four-step process for innovation can be summarized by:

1. Resource collection.
2. Weakness identification.
3. Innovation analysis.
4. Innovation synthesis.

Again, cases may arise where it is useful to include Steps 1 and 6 of the problem-solving process, but innovations are generally not of sufficient complexity that a complete critical problem-solving assessment is needed.

Introspection idea innovation is a three-phase procedure that employs both critical and creative thinking. The objectives of the phases are given as follows:

- *Phase I*: Identify a problem that reflects a weakness in the current state of knowledge and one that suggests a need for new research.
- *Phase II*: Generate alternative ideas for overcoming the inadequacy identified in the stated problem.
- *Phase III*: Use the ideas of Phase II to develop a novel solution to the originally stated problem such that the state of knowledge is advanced.

The first two phases of I³ reflect specific aspects of analysis, while Phase III is an exercise in synthesis. A variety of thinking methods can be used in all phases. The principles of critical thinking are most important in Phase I. Creative thinking methods dominate in Phase II.

To properly complete Phase I requires obtaining the reported works of others. The I³ method requires a researcher to have various attitudes and skills, which would be used in the first part of Phase I. Application of the skill-set model in Chapter 10 is a reasonable approach, as the skill-set model involves the systematic sequential application of useful skills, especially critiquing, curiosity, and questioning. The basis for Phase II is the adoption of a method of creative thinking to use in generating unique solutions. While logical thinking could be used, solving new problems usually requires an imaginative thinking method. Phase III emphasizes the use of various thinking types to transform the potential solutions identified in Phase II into a solution to the original problem of Phase I. The best alternative of the Phase II options forms the basis for the solution of Phase III.

17.4   SKILLS ESSENTIAL FOR I³ APPLICATIONS

The I³ method differs from the more recognizable methods of creative thinking such as brainstorming. In creative thinking activities, the specific problem is initially known and the purpose is to generate possible solutions. With the I³ method, a problem needs to be identified before possible solutions are developed. The objective of Phase I is to define a problem that needs to be investigated. Critical thinking is a key to meeting the I³ objective. The skill and mindset dimensions of the critical problem-solving model are focal parts of Phase I.

To be successful at Phase I, the investigator should apply the skill set abilities in the following way:

- *Courage*: Courage is the skill of having the confidence to approach a problem about which the researcher has little knowledge. In this case, the person needs to select a theme of interest and have the confidence that he or she can advance the state of knowledge.
- *Skepticism*: Skepticism is an attitude that requires approaching the existing works of established researchers with the belief that their reported work does not fully establish the state of knowledge and that advancements can be made to advance the state of knowledge.
- *Critiquing*: With respect to the I³ method, critiquing involves having the knowledge of ways of analyzing the works of others with a critical mindset and identifying weaknesses in the current state of knowledge that can lead to advancements in knowledge. Some methods were presented in Section 15.4.
- *Curiosity*: Curiosity is the skill of transforming weaknesses into an experimental plan in a way that can potentially yield a substantial advancement in knowledge.
- *Questioning*: A skill based on an inquisitive attitude where questions about the weakness in the knowledge base are asked, with the question being used to initiate the steps of the research process.

To apply these skills, the following abilities are applicable:

- *Conscientiousness*: In general, the practice of being thorough and careful when undertaking a specific responsibility; in the I³ case, the responsibility is to develop a research topic that will fulfill a need and lead to an advance in knowledge. A conscientious attitude should ensure that the most fruitful idea will be identified.
- *Diligence*: The general practice of providing persistent effort when undertaking a responsibility with a responsibility to conduct the I³ procedure in an efficient manner.
- *Self-discipline:* The control of one's attitudes, moods, and conduct, especially having the discipline to avoid succumbing to counter-skills such as procrastination and pessimism that reflect inefficient effort.
- *Fairness:* To act without showing favoritism or bias; to be impartial. With respect to the use of I³, an important aspect of fairness is to avoid taking credit for the words and ideas of others; credit needs to be given for all source materials used to directly obtain the underlying idea for the problem.

It is important to approach any research effort with the mindset of adhering to these attitudes and values.

The solutions of very simple problems may involve minimal skill requirements and only a few value conflicts, such that the technical and economic dimensions dominate the search. In such cases, only a minimum skill set is needed. Skills such as curiosity and critiquing are largely sufficient to meet the demands for innovation in such cases; thus, use of the entire skill set of Chapter 10 may not be necessary. The problem statement will propose an idea that is of minimal complexity such that the knowledge requirement for the necessary innovation would be minimal. Similarly, the skepticism element of the skills set would also be an inconsequential factor because the problem is immediately evident. Imaginative idea generation may be needed to identify possible alternative solutions, but the commonly used generation methods would usually be adequate. Simple problems do not require all elements of the critical solving process; thus, adequate solutions are easily developed even by those who have not developed all of the critical problem-solving dimensions.

17.5 CONDUCTING PHASE I OF I³

Keep in mind that the problem to be researched is not initially known, so the objective of Phase I of the I³ method is to identify a researchable problem. It is assumed that the investigator has some interest in a very broad area and the task is to narrow the investigation to one specific aspect of the general topic, where knowledge of the topic is deficient in some way. For example, the person might be very interested in the general topic of plastic in the oceans. Obviously, such a problem is likely much too broad to form the basis for a single research effort, so a specific aspect of the problem would need to be identified in order to realistically advance the state of knowledge relative to the general topic. This is the basis for Phase I.

The first step of Phase I is to obtain important literature that is relevant to the general topic of interest; this literature could include technical material such as trade journals, as well as information readily available to the non-technical public, such as commercial magazines. The second step of Phase I is to peruse the literature and critically assess parts of the documents that identify any of the following:

- *Weaknesses*: Aspects of the problem that appear to be deficient, i.e., points that fail to provide knowledge needed to draw accurate conclusions. This could include ideas such as the following: (1) the reported conclusions are not fully justified by the experimental analyses; (2) the database was inadequate in quality or quantity to justify the results stated; or (3) the stated research objectives were too limited.
- *Constraints*: Assumptions that were made by the authors seemed restrictive and would limit both the scope of the investigation and the breadth of applicability of the outcomes. This would include restrictions such as the following: (1) the functional forms used for equations limited the variation; (2) the calibration method that was used to fit the model was inadequate; or (3) the breadth of the data was limited (e.g., missing data, outliers not addressed, seasonality not accounted for).

- *Omitted factors*: A critical review of the literature indicated that important variables, conditions, physical processes, or modeling practices were not addressed in the study. The following are specific examples of omitted factors: (1) the failure to include other principles into the analysis; (2) the analyses lacked proper verification; (3) relevant physical processes were not accounted for; (4) some relevant theory was missing from the experimental analyses; or (5) potentially valuable variables were not included.

In the review of the published material, all relevant deficiencies should be identified. In Step 3 of this phase, one issue of those identified in Step 2 will appear to be the most fruitful for investigation; this issue would be the basis for the problem to be addressed in Phase II of the I^3 method. The other generated ideas can be retained for later review and investigation.

As an example, assume that a person wanted to conduct research that was related to the environmental impact of plastic waste materials that are being discharged into the oceans. A perusal of the literature might yield the following deficiencies:

- *Weakness*: The study only considered the specific effect of the plastics on the tuna fish population; thus, the overall effects could not be approached.
- *Constraint*: The effects of seasonal variations in ocean currents were not discussed; this constraint could limit any model from properly modeling the dispersion of the plastic materials.
- *Omitted factor*: The potential for capture of plastic in tidal areas was not addressed; the omission of this factor could bias the magnitude of the problem in open-water areas due to over-estimates of the materials discharged into the ocean.

Issues such as these could be used as the basis for developing a new problem statement, which would be the outcome of Phase I and the basis for starting Phase II.

How does an investigator critically analyze the general topic selected for study? Questioning was an important skill that was discussed in Chapter 10. Therefore, one easy way to identify weaknesses is to ask questions that arise during the reading of the literature. For example, with the plastic-into-the-ocean topic, a question might be: what size and shape of plastic is most damaging to each size of fish? To identify a variable that was not referenced in the literature but could be important, the question might be: What is the effect of tides on the entrainment of plastics and the dispersal of plastics into the ocean? It may be that the influence of tides has not been investigated. Note that by nature the questions are critical as they suggest that an important issue is currently not known.

Another method of identifying topics is to transfer solutions recommended for a similar problem to an analogous solution in the topic of current interest. For example, the plastic-into-the-ocean problem is in some ways similar to the movement of space debris (i.e., small parts of old satellites floating in space). Both are forms of uncontrolled trash. This example illustrates the potential value of knowing the synectics method (see Chapter 4). It may be possible to use methods of collecting the space debris to develop ideas for ridding the oceans of plastic debris. Information on the space debris would likely be found in totally different sources of literature, so the investigation of plastic trash could involve the transfer of ideas from aerospace engineering sources to a topic on oceanography. This pathway to meeting the need for a research topic illustrates the value of believing in breadth.

17.6   USE OF I³ FOR MODEL DEVELOPMENT

Models are used in many of the "hot" research topics, such as climate change modeling or the modeling of environmental degradation. Models for such topics already exist, but novel research can be based on modifications of the existing models for the purpose of improving their performance or broadening their applicability. In applying the I³ procedure to the modification of an existing conceptual model, the process can be summarized by the following steps:

1. Conduct a Phase I I³ analysis to identify the research objective or problem statement.
2. Assemble all past work relevant to the objective of the new research, i.e., conduct a literature review.
3. Identify the specific models that are potentially most applicable to the research problem.
4. Identify the variables or components of the existing model selected for adaption and decide the ones that should be adapted to the new model; eliminate components that are not relevant to the new research objective.
5. Create a basic structure that would incorporate the new variables or be appropriate for the new model component(s).
6. Incorporate the new variables or components with the components of Step 4 into a new model structure, which yields the innovated model.

The new model structure could involve a series of linear or nonlinear components assembled using the systems process. The new model could include decision nodes and feedback loops such that intermediate decisions could be modified based on the results of intermediate steps.

17.7   CONCLUDING COMMENTS

Assume that a researcher's objective is to identify a problem that could lead to developing a statement for a novel research effort. Research papers published in professional journals are summary reports of the application of the research process and can be valuable sources of material for developing research ideas that extend beyond the reported results. The practice of generating new research ideas can be learned without formal instruction. First, a person who is interested in learning the skill must believe in the premise behind the I³ method; specifically, that existing research results are good starting places for new research. Second, the person must have the attitude that new research ideas can be developed and that skills such as critiquing and questioning are important in developing novel outcomes. Third, the person needs to develop experiences in the application of critical and creative thinking methods to analyze the works of others. In order to gain confidence in using the procedure, several research ideas should be developed by executing the phases of the I^3 method.

The success of a research effort can hinge on having a quality topic to research. Recall that the first step of the problem-solving process is to identify the problem. If a topical area is not known, then none of the processes of Chapter 2 can be started. The primary intent of research is the advancement of the state of knowledge. Thus, the problem statement should clearly indicate that a unique problem exists, and the statement of the research goal should be worded to show that fulfilling the goal will advance the state of knowledge. The failure to produce a significant advancement with the research will likely result in the rejection of the ideas and any reports based on the research.

While the purpose of the I^3 method is the development of something new, we cannot consider it as an approach to discovery, at least within a scientific framework. Discovery is based on mental analyses of unexplainable observations. Innovation is based on the mental analysis of existing knowledge. I^3 is more of an idea development approach, but not in the same way as the brainstorming method. In scientific investigations, the word discovery suggests a new idea with its use not immediately recognizable. Discovery is something that the Einstein's of the world perform, not the local researcher who gets involved in idea innovation. The discoverer's idea is often derived from observation. The discovery of penicillin by Alexander Fleming in 1928 resulted because of his interest in molds and his curiosity when he observed an event that he could not explain. The actual discovery was the result of an accidental observation of a bacterial culture. His general knowledge and intense curiosity enabled him to make the observation; however, it was another decade before the marketing of penicillin occurred.

In summary, introspective idea innovation can be a useful approach for conducting an original investigation, such as an academic research study when the investigator is not given a specific topic. It is introspective because it requires the person to use his or her intellectual abilities and interests to select and ultimately solve a new problem. The ideas of others are innovated to produce this work. Thus, introspective idea innovation!

17.8 EXERCISES

17.1. Discuss the way that each of the following values is relevant to idea innovation; also, define each of the values: truth, prudency, fairness, and accountability.

17.2. Find information on the history of cell phone innovation that have occurred over time and the ways that the changes influenced economics, convenience to the user, and other related factors. Summarize the importance of innovation in product development.

17.3. What innovations have been made to the automobile to make it more energy efficient?

17.4. Propose changes to the TV show Law and Order that would make it more entertaining. More thrilling.

17.5.   Why is it not plagiaristic to innovate someone else's research?
17.6.   Where does creative thinking enter into idea innovation?
17.7.   Propose an innovation of the Delphi method.
17.8.   What aspects of the I³ method represent analysis? Synthesis?
17.9.   Create a new model for idea innovation that is based on five phases rather than the three phases used here.
17.10. Provide a comprehensive discussion of the role of courage in identifying a research topic when using the I³ method.
17.11. Explain the use of the skill set in using the I³ method.
17.12. Based on assessment (5) of Table 8.1, identify three other assumptions that could lead to a new functional form.
17.13. Based on assessment (7) of Table 8.1, what is meant by data condition?
17.14. Based on assessment (13) of Table 8.1, what is meant by the term relative importance of model predictions?
17.15. Is the I³ method a creative thinking method? Discuss.
17.16. Is the use of the checklist approach (see Chapter 6) an act of innovation? Is the development of a new checklist an act of innovation? Is modification of an existing checklist an act of innovation? Explain.

17.9 ACTIVITIES

17.9.1 ACTIVITY 17A: INNOVATION FOR CREATIVITY

Create a series of steps for Phase II of the I³ method to specifically include creative thinking methods. This active involves a form of innovation.

17.9.2 ACTIVITY 17B: MODEL INNOVATION

The innovation of an existing model is a common academic exercise. Assume that the following equation is used for estimating the streambed erosion rate (q_b):

$$q_b = 2\left(V/V_c\right)^3 (d\,/\,h)^{0.5} \left(V - V_c\right)^{1.3}$$

where V is the water velocity (m/s) in the stream, V_c is the critical velocity (m/s) for detachment, d is the mean sediment diameter (m), and h is the mean flow depth (m). The coefficient and exponents are values that were fitted using measured data. Propose experiments and changes to the model that could possibly improve the accuracy of estimates of q_b. Discuss your reasoning.

17.9.3 ACTIVITY 17C: APPLIED I³

In a subject area that is of interest to you, find an article from a magazine that covers the topic. Use the I³ method to develop a realistic research topic. Discuss your application of the I³ method.

17.9.4 ACTIVITY 17D: I³ AND THE SCIENTIFIC METHOD

Induction and deduction are two principles that are important to the application of the Scientific Method. Induction is a principle of reasoning to a conclusion about all of the members of a class from examination of only a few members of the class (reasoning from the particular to the general).

Deduction is the process of reasoning in which a conclusion follows necessarily from the stated premises; inferences by reasoning from the general to the specific. Various versions of the Scientific Method have been proposed. A four-step version is

1. Inductive data analysis and hypothesis.
2. Deduction of possible solutions.
3. Test alternative solutions.
4. Implement best solution.

While the above four-step version may be acceptable for those versed in Scientific Methods, it may not be well understood by middle school students who lack a formal introduction to the fundamentals of scientific practices. The task is to use the I³ method to develop a model of the Scientific Method that would have greater meaning and interpretation by middle-school students.

17.9.5 ACTIVITY 17E: THE RULE REFORMER

The purpose of this activity is to innovate an existing game to create a new game. Hopefully, the end product will be a better game even though the relation to the existing game is evident. The procedure to follow is given below:

1. Select a game, such as Old Maid, hangman, solitaire, and blackjack.
2. Identify the rules of the game as it is commonly played.
3. Play the game a few times to ensure that you know all of the rules.
4. Now, brainstorm or brainwrite on ways of changing each rule listed in instruction 2, so that the rule changes would make the game better.
5. Create a new list of rules based on your recommended change(s).
6. Play the game a few times to ensure that the new rules are complete.
7. Now create a creative name for the new game.

17.9.6 ACTIVITY 17F: PUTTING ON WEIGHT

One proposal for a model of the innovation process included just the following three steps:

1. Recognition of a problem.
2. Implement change.
3. Market the new product.

Using this set of steps as a starting point, expand this three-step model to a greater number of steps while including all of the ideas that are imbedded in the three-step model. The purpose of this activity is to demonstrate the process of expanding an existing model to show more specific understanding of a process.

17.9.7   ACTIVITY 17G: KNOWLEDGE FOR SALE

Assume a model of consumer product development and sales can be represented by the following six steps:

1. Product idea.
2. Identify consumer behavior about the product.
3. Identify the channels of distribution of the product.
4. Establish guidelines for pricing the product.
5. Identify ways of advertising the product.
6. Assess the competition.

Using the ideas that underlie this model, innovate the steps to create a knowledge-sales model.

17.9.8   ACTIVITY 17H: RECONCILING RIGHT VERSUS WRONG

At one point in time, the American Management Association proposed the following five-step model on conflict reconciliation:

1. Define the source of the conflict.
2. Look beyond the incident.
3. Request solutions.
4. Identify solutions that both disputants can support.
5. Obtain agreement on a solution.

Identify values that are relevant to conflict management. Modify the five-step model to specifically incorporate the values into the steps. Other steps can be added, as needed. The intent of this activity is to apply the principles of innovation for the improvement of an existing model.

18 Conclusions

CHAPTER GOAL

To provide a broad overview of essential abilities needed for success in critical problem-solving.

CHAPTER OBJECTIVES

1. To provide summary statements of the important elements of critical thinking.
2. To emphasize the value of critical thinking to problem-solving efficiency.
3. To identify the learning outcomes from the coverage of this book.

18.1 INTRODUCTION

Several times, the following question has been posed: what is critical thinking? Critical thinking is the art of analyzing the existing state of knowledge to identify known weaknesses that can be overcome by using imaginative mental processes that involve multiple types of thinking, an array of important values, and a sequence of skills to identify potential solutions from which a decision criterion will identify the best solution to a complex problem. This question is often asked when a person confronts a technically complex problem. Truly complex problems are not limited to unidimensional issues such as technical difficulty. Complex problems are usually subject to multi-dimensional constraints with the constraints reflecting factors such as economics, values, and risk. To solve a complex problem, a decision maker must address value issues, especially those where stakeholders have very contentious biases. Skills such as curiosity and questioning are usually very relevant to the solutions of complex problems. If a decision maker has trouble controlling mindset states, such as moodiness or pessimistic thinking, then the thinking dimension can suppress the development and analysis of potential solutions. These considerations suggest that critical problem-solving is a more descriptive concept than critical thinking.

The second word, thinking, refers to the mental processes that are used during thought about some event or issue that could be a problem. Problem-solving involves more than thinking, as action in the form of decisions is necessary following the actual thinking. Actions usually follow decisions. Problem-solving implies the assembling of decisions based on thinking; therefore, as opposed to critical thinking, critical problem-solving is a more descriptive statement of the thoughts, decisions, and actions made in the solution of a complex problem.

DOI: 10.1201/9781003380443-18

A central theme of this book has been that the most novel solutions to complex problems will most likely result from broad-based thinking that involves more than competency in the underlying technical knowledge of the field of study. Knowledge of the four universal dimensions stressed in this book encourages both breadth and depth of problem-solving, planning, and execution. Overcoming inhibitors is important to become a critical thinker and to improve problem-solving efficiency. The necessity to have quality experiences at problem-solving is also stressed. The quality of the experiences is probably more important than the quantity. Knowing the way to conduct an introspective analysis following experiences that led to failures and even partial failures is necessary to become a critical thinker. Following failures of any magnitude, it is necessary to identify the reasons for the failure and make changes to the ways of thinking that were used to solve the problems that led to the poor results. Updating one's knowledge of decision-making following each experience is essentially an unstated element of Step 6 of the problem-solving process.

18.2   THE UNIVERSAL DIMENSIONS

The foundation of each complex problem will be different and based on the fundamental concepts of the primary technical issue. Technical herein refers to the fundamental concepts of any discipline, not just STEM subjects; however, it is likely that a complex problem in any technical discipline will also involve the four dimensions discussed in this book, which can be referred to as the universal dimensions. The four dimensions that provide breadth and depth beyond the dimension that is appropriate for any technical issue have the following objectives:

1. *The mindset dimension*: A decision maker needs to have control of his or her mindset during all phases of the decision-making process; this control requires a depth of experience that enables a person to create a state of mind that is appropriate for solving a specific problem and avoiding inhibitors.
2. *The thinking type dimension*: Broad-based thinking is necessary to ensure that the most effective types of thinking are applied; in order to adopt the thinking type that is most appropriate for the characteristics of the problem, in-depth experiences in the use of each type of thinking are needed.
3. *The skill set dimension*: A systematic application of skills should replace the arbitrary application of skills in order to ensure that the investigation of a problem will be the most efficient and effective. The sequencing of skills provides organization and a breadth of thinking that encourages studying a problem from different perspectives.
4. *The value dimension*: Since value issues are often a primary source of conflict, knowing the proper way to assess value issues and properly balance conflicting goals of multiple stakeholders is necessary to effectively solve complex problems; a value-based perspective is often necessary to balance societal and organizational goals.

These four dimensions supplement other dimensions such as the technical dimension and likely a dimension based on economics as well as dimensions that represent other issues that are relevant to the problem statement. In any one case, dimensions related to politics, risk, and the environment, as well as others, may also be decision factors that are relevant to the complex problem under consideration. Every dimension acts as a restriction, and the restrictions ensure that decisions are not concluded haphazardly; additionally, any dimension involved in solving a problem should add confidence to decisions, as uncertainties due to variations of the dimension would be under greater control. Critical problem-solving is more than the critical thinking component. Focusing on just the thinking element will undervalue these other dimensions. The best solutions to critical problems occur when a person has sufficient knowledge of and experiences with all dimensions that are relevant to the problem.

18.3 ATTITUDES AND ABILITIES OF CRITICAL THINKERS

A summary was provided for the question: what is critical thinking? A follow-up question could be: what abilities and attitudes are important for a critical thinker to possess? A person who wishes to develop critical thinking skills needs an answer to this question. Knowing the abilities and attitudes will enable the person to make a self-assessment, which can provide the basis for developing plans to overcome weaknesses. The following summarizes the needed attitudes and abilities:

1. A critical thinker must understand the steps of the problem-solving process and make decisions to maximize the efficiency of each step.
2. A critical thinker must be able to apply the principles that form the basis for all important dimensions.
3. A critical thinker recognizes the importance of values and knows the way to balance values, even those that are difficult to quantify.
4. A critical thinker knows the benefits of controlling his or her mindset, which ensures that an improper state of mind does not distort decision-making.
5. A critical thinker also knows many types of thinking and knows to match the most appropriate type of thinking with the characteristics of the problems.
6. A critical thinker uses the problem-solving skills in the proper sequence.
7. A critical thinker does not allow problem-solving inhibitors to interfere with the development of the most novel solutions to the problem. The counter-skills of pessimism and procrastination are often the inhibitors that contribute most to inefficiency.
8. A critical thinker recognizes that an attitude of optimism is more beneficial that a pessimistic attitude.
9. A critical thinker searches for opportunities that will enhance his or her self-confidence.
10. A critical thinker practices assessment, especially the use of self-introspection following activities that are less than totally successful.
11. A critical thinker seeks mentoring, as he or she recognizes that mentoring provides knowledge based on other people's experiences.
12. A critical thinker seeks quality experiences, as it is the quality of the experiences that contributes to learning the most effective ways of solving problems.

13. A critical thinker develops the attitude that weaknesses in the existing state of knowledge of any topic can be identified by properly analyzing the knowledge.

Improvements in these attitudes and abilities will enable a person to become a better critical thinker.

18.4   LEARNING OUTCOMES

Problem complexity is much more than the complexities associated with technical issues. While the details of some aspects of problem-solving are specific to the technical nature of the problem, many aspects of complexity are associated with the dimensions discussed herein. Knowledge of critical thinking is important when solving these complex problems. The following learning outcomes attempt to summarize the most important aspects of critical problem-solving that have been stressed in this book:

1. To understand the commonalities and differences among problem-solving processes. Knowing the different characteristics emphasized in the steps of the different processes enables a critical thinker to develop a problem-solving process that is relevant to a specific problem. The process should include all of the important elements that will contribute to greater flexibility. A process should be designed specifically to characterize the important elements of the problem.

2. To recognize and believe that imaginative idea generation is a valuable skill that can be vital in the solution of complex problems. Complex problems likely contain a rare combination of unique issues. Imaginative thinking will encourage flexible thinking, which will increase the likelihood of recognizing a factor that will provide a novel solution to the problem.

3. To understand that the efficiency (output/input) of problem-solving is important but difficult to quantify. In spite of this difficulty, attempts must be made to identify the efficiency of problem-solving. Complex problems likely involve performance criteria that are reported on different statistical scales, which makes it difficult to develop a decision criterion that is not characterized by uncertainty. The ability to think critically will enable more effective management of the uncertainty.

4. To provide accurate and unbiased weights when developing a decision criterion, with the decision criterion weights indicating the importance of an individual performance criterion. Unbiased weights for the decision criterion should minimize stakeholder conflict about the final decision.

5. To recognize that creative thinking methods are capable of providing decisions to non-complex problems and are often a central element of critical thinking activities for solving complex problems. Using imaginative problem-solving processes will encourage decision makers to be cognizant of their mindsets when making decisions. This awareness should lead to more effective and efficient decisions.

6. To understand that creativity and innovation can be learned, in spite of the myth that they are innate abilities. While innate traits play a role in the ability to be creative, education and experience can advance a person's knowledge of critical problem-solving. Anti-creative thinking myths must be suppressed if those involved in the development of novel solutions are going to encourage others to learn the principles of critical thinking.

7. To recognize and overcome creativity inhibitors, those by the organization, by associates, and most importantly by oneself; each inhibitor should be replaced by a creativity stimulator. Inhibitors direct the critical thinker's focus away from the proper use of thinking types and important skills. Therefore, inhibitors distort the mindset and reduce problem-solving efficiency and effectiveness.

8. To recognize the many methods of assessment that are available, including ones applicable to facilitators, groups, individuals, research efforts, and personal development; assessment has many benefits and will lead to improvements in problem-solving.

9. To recognize that critical thinking is a basic framework for problem-solving with a basis in the diverse dimensions, such as those of values, thinking skills, mindset control, and types of thinking. While these four dimensions may seem to be constraints, failure to include them in the problem-solving may lead to less than optimum decisions.

10. To recognize the importance of mindset control to the quality of decisions. Critical thinkers must be capable of controlling the mental processes that engender the state of mind that is most appropriate for a specific problem. A critical thinker continually monitors his or her own mindset to ensure that decisions are made with the best mindset.

11. To recognize that solving complex problems requires having the proper sensitivity to value issues and to be able to demonstrate the responsible balancing of all value issues that conflict. Value issues are difficult to quantify, which greatly complicates decision-making. In order to properly balance the competing values, decision makers must know both the legitimate responsibilities and the selfish biases of stakeholders.

12. To recognize that successful problem-solving depends partly on applying the thinking type that is most appropriate for the characteristics of the complex problem that needs a solution. A critical thinker should have experience with a diverse array of thinking types.

13. To recognize that a sequence of mental skills is important in the solution of complex problems and to show that applying the sequence of skills can improve problem-solving effectiveness.

14. To have the courage to accept the responsibility for finding solutions to complex problems and to have the attitudes needed to confidently apply all steps of the problem-solving process.

15. To recognize the value of critiquing as a skill with its special purpose of identifying strengths and weaknesses of a current state of knowledge. This skill is very relevant to identifying topics that can serve as the basis for novel research.

16. To recognize the skill of questioning as being part of the brainwriting method and as a vital element of the skill set. Questions serve both to challenge problem solvers and to provide a focal point in the search of a solution to a complex problem.

17. To recognize that extensive experience is necessary to effectively use the critical problem-solving approach; the quality of the experiences is just as important as the quantity of the experiences.

18. To recognize that failures can be good learning experiences. Conducting introspective analyses will ensure that faulty aspects of past decisions are corrected so that the failures will not be repeated in the future.

19. To recognize that pessimism is a counter-skill and an attitude that suppresses imaginative thinking and creativity, thus limiting success. Pessimism can be overcome with the adoption of more optimistic thinking.

20. To recognize that the myths surrounding procrastination are not true and to appreciate the detrimental effects of procrastination. Knowledge of ways to overcome this counter-skill should be sought.

21. To recognize that organizations have value systems that are imbedded in their policies and practices; the value system is changed by making changes in the policies. Policies and practices that suppress imaginative thinking are especially detrimental and need to be changed.

22. To understand that focusing solely on short-term consequences of the outcomes of critical thinking can lead to under-valuing the long-term implications of decisions.

23. To recognize the sources as potential implications of all uncertainties on the accuracy of each alternative solution, especially on the final decision, and account for uncertainties in calculating values of performance and decision criteria.

24. To recognize the many ways that knowledge of critical problem-solving can enable a leader to advance the goals of an organization. Leaders are influential in all phases of the problem-solving process. A leader's experiences can ensure the accuracy of the problem statement (Step 1) and ensure that all important implications of the work (Step 6) are identified.

25. To recognize the many benefits of developing strong relationships with mentors, as these will enable the mentee to benefit from the experiences of a more experienced person. These second-handed experiences can provide useful knowledge of problem-solving, as the mentee will recognize that the mentor's experiences are ones that he or she may confront in the future.

18.5 POSTSCRIPT

Throughout this book, I have tried to show that the greatest thinkers of all times worked under constraints much as the top thinkers of today must contend. During past times, the great thinkers were under religious, economic, and knowledge constraints, which were essentially dimensions with which they needed to recognize and contend. They also dealt with the dimensions emphasized in this book, including value conflicts, problems that required different thinking types, mindset constraints, and the requirement to be adept at skills such as courage, curiosity, and questioning. A review of the personal histories of the great thinkers shows a wide range of backgrounds, from differences in religious upbringings to the constraint of poverty, from different levels of formal education to the availability of sources of knowledge, and from different political systems to differences in parental guidance. The bottom line is that each of the greatest thinkers had to overcome constraints in order to achieve the monumental success that each one did. The same is true today: success requires a person to overcome constraints.

What contributed to the success of those who are known as the greatest of all time, e.g., Aristotle, Copernicus, Bacon? It cannot be that they were born into wealthy families, as many of them were born into poverty. It was not because they were enrolled in stellar educational systems, as many lacked formal education. Much of their successes were partially the result of opportunities in their adult years, but it is likely that the most important reasons were partially innate and partially due to attitudes that they developed from their experiences. While inquisitiveness seems to be a common element, some were fortunate enough to receive external funding. For example, King Frederick II of Denmark provided Tyco Brahe (1564–1601) with a well-equipped observatory on a private island. A few of the greats received mentoring that helped them achieve success. For example, Johannes Kepler (1571–1630) was an assistant to Tyco Brahe. A few had the opportunity to receive a broad array of opportunities for training and experience. For example, James Maxwell (1831–1879) was educated in mathematics, but used his uncommon sense for understanding physical phenomena to make significant advances in the field of electromagnetism. Michael Faraday (1831–1879), who was born into poverty and had little formal education, achieved success in electromagnetic induction; as a youngster, he worked in a bookstore and had access to a wide array of books and he associated with people who gave him the opportunity to work as an experimentalist. Sir Isaac Newton (1642–1727) had an innate inquisitiveness and developed his own experiments even as a youngster; for example, at the age of 16, Newton estimated the power of wind by jumping from a tree limb into the wind and then with the wind at his back. Andreas Vesalius (1514–1564), a Flemish anatomist and the father of modern anatomy, dissected small animals during his youth, which provided him with knowledge of experimental methods and an interest in his chosen field. It appears that experiences and an innate inquisitive attitude were common to those who made monumental advances to the collective body of knowledge.

Leonardo da Vinci (1452–1519) is recognized as one of the all-time greats, a genius who is recognized for his creative ability in many disciplines, including the arts (e.g., the Mona Lisa), engineering (e.g., his 730 unique finds about water and rivers), fantasy (e.g., his sketches of drones and other flying machines), medicine (e.g., explanations about light striking the retina), and military hardware (e.g., his tank designs). His creativity laid the groundwork for the Scientific Method. Francis Bacon's (1561–1626) study of da Vinci's research methods led Bacon to formalize the Scientific Method. Leonardo da Vinci used many of the methods that are now recognized as methods of critical thinking. The sketches in his widely documented notebooks indicate his ability with pictostorming, which is the graphical equivalent of brainstorming. Many of his famous paintings involved work with subordinates, much like the creative decision method known as the Delphi method. All of his notebooks indicate that brainwriting was a day-to-day activity for Leonardo. The breadth of da Vinci's use of methods of creativity suggests the potential value of creativity for the researchers of today. Was his informal knowledge of creativity the factor that elevated him to be known as one of the all-time greats? Will monumental advancements by futurists depend on their knowledge of critical thinking or will major advances in thinking be needed to replace critical thinking in order to solve the complex, challenging problems of the future?

Index